T0253715

Differential Forms

Differential Forms

Victor Guillemin
Peter Haine
Massachusetts Institute of Technology, USA

World Scientific
NEW JERSEY • LONDON • SINGAPORE • BEIJING • SHANGHAI • HONG KONG • TAIPEI • CHENNAI • TOKYO

Published by

World Scientific Publishing Co. Pte. Ltd.
5 Toh Tuck Link, Singapore 596224
USA office: 27 Warren Street, Suite 401-402, Hackensack, NJ 07601
UK office: 57 Shelton Street, Covent Garden, London WC2H 9HE

Library of Congress Cataloging-in-Publication Data
Names: Guillemin, Victor, 1937– author. | Haine, Peter (Mathematician), author.
Title: Differential forms / by Victor Guillemin (Massachusetts Institute of Technology, USA) and Peter Haine (Massachusetts Institute of Technology, USA).
Description: New Jersey : World Scientific, 2019. | Includes bibliographical references and index.
Identifiers: LCCN 2018061406 | ISBN 9789813272774 (hc : alk. paper)
Subjects: LCSH: Differential forms. | Geometry, Differential.
Classification: LCC QA381 .G85 2019 | DDC 515/.37--dc23
LC record available at https://lccn.loc.gov/2018061406

British Library Cataloguing-in-Publication Data
A catalogue record for this book is available from the British Library.

First published 2019 (Hardcover)
Reprinted 2020 (in paperback edition with corrections)
ISBN 978-981-121-377-9 (pbk)

Copyright © 2019 by World Scientific Publishing Co. Pte. Ltd.

All rights reserved. This book, or parts thereof, may not be reproduced in any form or by any means, electronic or mechanical, including photocopying, recording or any information storage and retrieval system now known or to be invented, without written permission from the publisher.

For photocopying of material in this volume, please pay a copying fee through the Copyright Clearance Center, Inc., 222 Rosewood Drive, Danvers, MA 01923, USA. In this case permission to photocopy is not required from the publisher.

For any available supplementary material, please visit
https://www.worldscientific.com/worldscibooks/10.1142/11058#t=suppl

Desk Editors: V. Vishnu Mohan/Kwong Lai Fun

Typeset by Stallion Press
Email: enquiries@stallionpress.com

Printed in Singapore

Preface

Introduction

For most math undergraduates one's first encounter with differential forms is the change of variables formula in multivariable calculus, i.e., the formula

$$(1) \qquad \int_U f^\star \phi |\det J_f| dx = \int_V \phi\, dy.$$

In this formula, U and V are bounded open subsets of $\mathbf{R}^n$, $\phi\colon V \to \mathbf{R}$ is a bounded continuous function, $f\colon U \to V$ is a bijective differentiable map, $f^\star\phi\colon U \to \mathbf{R}$ is the function $\phi \circ f$, and $\det J_f(x)$ is the determinant of the Jacobian matrix:

$$J_f(x) := \left(\frac{\partial f_i}{\partial x_j}(x) \right).$$

As for the "dx" and "dy", their presence in (1) can be accounted for by the fact that in single-variable calculus, with $U = (a,b)$, $V = (c,d)$, $f\colon (a,b) \to (c,d)$, and $y = f(x)$ a C^1 function with positive first derivative, the tautological equation $\frac{dy}{dx} = \frac{df}{dx}$ can be rewritten in the form

$$d(f^\star y) = f^\star dy$$

and (1) can be written more suggestively as

$$(2) \qquad \int_U f^\star(\phi\, dy) = \int_V \phi\, dy.$$

One of the goals of this text on *differential forms* is to legitimize this interpretation of equation (1) in n dimensions and in fact, more generally, show that an analogue of this formula is true when U and V are n-dimensional manifolds.

Another related goal is to prove an important topological generalization of the change of variables formula (1). This formula asserts that if we drop the assumption that f be a bijection and just require f to be proper (i.e., that preimages of compact subsets of V to be compact subsets of U) then the formula (1) can be replaced by

$$(3) \qquad \int_U f^\star(\phi\, dy) = \deg(f) \int_V \phi\, dy,$$

where deg(f) is a topological invariant of f that roughly speaking counts, with plus and minus signs, the number of preimage points of a generically chosen point of V.[1]

This degree formula is just one of a host of results which connect the theory of differential forms with topology, and one of the main goals of this book will explore some of the other examples. For instance, for U an open subset of $\mathbf{R}^2$, we define $\Omega^0(U)$ to be the vector space of C^∞ functions on U. We define the vector space $\Omega^1(U)$ to be the space of formal sums

$$f_1\, dx_1 + f_2\, dx_2, \tag{4}$$

where $f_1, f_2 \in C^\infty(U)$. We define the vector space $\Omega^2(U)$ to be the space of expressions of the form

$$f\, dx_1 \wedge dx_2, \tag{5}$$

where $f \in C^\infty(U)$, and for $k > 2$ define $\Omega^k(U)$ to the zero vector space.

On these vector spaces one can define operators

$$d : \Omega^i(U) \to \Omega^{i+1}(U) \tag{6}$$

by the recipes

$$df := \frac{\partial f}{\partial x_1}\, dx_1 + \frac{\partial f}{\partial x_2}\, dx_2 \tag{7}$$

for $i = 0$,

$$d(f_1\, dx_1 + f_2\, dx_2) = \left(\frac{\partial f_2}{\partial x_1} - \frac{\partial f_1}{\partial x_2}\right) dx_1 \wedge dx \tag{8}$$

for $i = 1$, and $d = 0$ for $i > 1$. It is easy to see that the operator

$$d^2 : \Omega^i(U) \to \Omega^{i+2}(U) \tag{9}$$

is zero. Hence,

$$\operatorname{im}(d : \Omega^{i-1}(U) \to \Omega^i(U)) \subset \ker(d : \Omega^i(U) \to \Omega^{i+1}(U)),$$

and this enables one to define the *de Rham cohomology groups of U* as the quotient vector space

$$H^i(U) := \frac{\ker(d : \Omega^i(U) \to \Omega^{i+1}(U))}{\operatorname{im}(d : \Omega^{i-1}(U) \to \Omega^i(U))}. \tag{10}$$

It turns out that these cohomology groups are topological invariants of U and are, in fact, isomorphic to the cohomology groups of U defined by the algebraic topologists. Moreover, by slightly generalizing the definitions in equations (4), (5) and (7)–(10) one can define these groups for open subsets of $\mathbf{R}^n$ and, with a bit

---

[1] It is our feeling that this formula should, like formula (1), be part of the standard calculus curriculum, particularly in view of the fact that there now exists a beautiful elementary proof of it by Peter Lax (see [6, 8, 9]).

more effort, for arbitrary C^∞ manifolds (as we will do in Chapter 5); and their existence will enable us to describe interesting connections between problems in multivariable calculus and differential geometry on the one hand and problems in topology on the other.

To make the context of this book easier for our readers to access we will devote the rest of this introduction to the following annotated table of contents, chapter-by-chapter descriptions of the topics that we will be covering.

Organization

Chapter 1: Multilinear algebra

As we mentioned above one of our objectives is to legitimatize the presence of the dx and dy in formula (1), and translate this formula into a theorem about differential forms. However a rigorous exposition of the theory of differential forms requires a lot of algebraic preliminaries, and these will be the focus of Chapter 1. We'll begin, in §§1.1 and 1.2, by reviewing material that we hope most of our readers are already familiar with: the definition of vector space, the notions of *basis*, of *dimension*, of *linear mapping*, of *bilinear form*, and of *dual space* and *quotient space*. Then in §1.3 we will turn to the main topics of this chapter, the concept of k-tensor and (the future key ingredient in our exposition of the theory of differential forms in Chapter 2) the concept of alternating k-tensor. Those k-tensors come up in fact in two contexts: as *alternating k-tensors*, and as *exterior forms*, i.e., in the first context as a subspace of the space of k-tensors and in the second as a quotient space of the space of k-tensors. Both descriptions of k-tensors will be needed in our later applications. For this reason the second half of Chapter 1 is mostly concerned with exploring the relationships between these two descriptions and making use of these relationships to define a number of basic operations on exterior forms such as the wedge product operation (see §1.6), the interior product operation (see §1.7) and the pullback operation (see §1.8). We will also make use of these results in §1.9 to define the notion of an *orientation* for an n-dimensional vector space, a notion that will, among other things, enable us to simplify the change of variables formula (1) by getting rid of the absolute value sign in the term $|\det J_f|$.

Chapter 2: The concept of a differential form

The expressions in equations (4), (5), (7) and (8) are typical examples of differential forms, and if this were intended to be a text for undergraduate physics majors we would define differential forms by simply commenting that they're expressions of this type. We'll begin this chapter, however, with the following more precise definition: Let U be an open subset of $\mathbf{R}^n$. Then a *k-form* ω on U is a "function" which to each $p \in U$ assigns an element of $\Lambda^k(T_p^\star U)$, T_pU being the tangent space to U at p, $T_p^\star U$ its vector space dual, and $\Lambda^k(T_p^\star)$ the kth-order exterior power of $T_p^\star U$. (It turns out, fortunately, not to be too hard to reconcile this definition with the physics definition above.) Differential 1-forms are perhaps best understood as the dual objects to vector fields, and in §§2.1 and 2.2 we elaborate on this observation, and recall for future use some standard facts about vector fields and their integral curves. Then in § 2.3 we will turn to the topic of k-forms and in the exercises at the end of § 2.3 discuss a lot of explicit examples (that we strongly urge readers of this text to stare at). Then in §§ 2.4–2.7 we will discuss in detail three fundamental operations on differential forms of which we've already gotten preliminary glimpses in equation (3) and equations (5)–(9), namely the wedge product operation, the exterior differential operation, and the pullback operation. Also, to add to this list in § 2.5 we will discuss the interior product operation of vector fields on differential forms, a generalization of the duality pairing of vector fields with 1-forms that we alluded to earlier. (In order to get a better feeling for this material we strongly recommend that one take a look at § 2.7 where these operations are related to the div, curl, and grad operations in freshman calculus.) In addition this section contains some interesting applications of the material above to physics, in §2.8 to electrodynamics and Maxwell's equation, as well as to classical mechanisms and the Hamilton–Jacobi equations.

Chapter 3: Integration of forms

As we mentioned above, the change of variables formula in integral calculus is a special case of a more general result: the degree formula; and we also cited a paper of Peter Lax which contains an elementary proof of this formula which will hopefully induce future authors of elementary text books in multivariate calculus to include it in their treatment of the Riemann integral. In this chapter we will also give a proof of this result but not, regrettably, Lax's proof. The reason why not is that we want to relate this result to another result which will have some important de Rham theoretic applications when we get to Chapter 5. To describe this result let U be a connected open set in $\mathbf{R}^n$ and $\omega = f dx_1 \wedge \cdots \wedge dx_n$ a compactly supported n-form on U. We will prove the following theorem.

Theorem 11. *The integral* $\int_U \omega = \int_U f dx_1 \cdots dx_n$ *of* ω *over* U *is zero if and only if* $\omega = d\nu$ *where* ν *is a compactly supported* $(n-1)$*-form on* U.

An easy corollary of this result is the following weak version of the degree theorem.

Corollary 12. *Let* U *and* V *be connected open subsets of* $\mathbf{R}^n$ *and* $f : U \to V$ *a proper mapping. Then there exists a constant* $\delta(f)$ *with the property that for every compactly supported* n*-form* ω *on* V

$$\int_U f^\star \omega = \delta(f) \int_V \omega.$$

Thus to prove the degree theorem it suffices to show that this $\delta(f)$ is the $\deg(f)$ in formula (3) and this turns out not to be terribly hard.

The degree formula has a lot of interesting applications and at the end of Chapter 3 we will describe two of them. The first, the Brouwer fixed point theorem, asserts that if B^n is the closed unit ball in $\mathbf{R}^n$ and $f : B^n \to B^n$ is a continuous map, then f has a fixed point, i.e., there is a point $x_0 \in B^n$ that gets mapped onto itself by f. (We will show that if this were not true the degree theorem would lead to an egregious contradiction.)

The second is the fundamental theorem of algebra. If

$$p(z) = a_0 + a_1 z + \cdots + a_{n-1} z^{n-1} + z^n$$

is a polynomial function of the complex variable z, then it has to have a complex root z_0 satisfying $p(z_0) = 0$. (We'll show that both these results are more-or-less one line corollaries of the degree theorem.)

Chapter 4: Forms on manifolds

In the previous two chapters the differential forms that we have been considering have been forms defined on open subsets of $\mathbf{R}^n$. In this chapter we'll make the transition to *forms defined on manifolds*; and to prepare our readers for this transition, include at the beginning of this chapter a brief introduction to the theory of manifolds. (It is a tradition in undergraduate courses to define n-dimensional manifolds as n-dimensional submanifolds of $\mathbf{R}^N$, i.e., as n-dimensional generalizations of curves and surfaces in $\mathbf{R}^3$, whereas the tradition in graduate level courses is to define them as abstract entities. Since this is intended to be a text book for undergraduates, we'll adopt the first of these approaches, our reluctance to doing so being somewhat tempered by the fact that, thanks to the Whitney embedding theorem, these two approaches are the same.) In §4.1 we will review a few general facts about manifolds, in §4.2, define the crucial notion of the *tangent space* $T_p X$ *to a manifold* X *at a point* p and in §§4.3 and 4.4 show how to define differential forms on X by defining a *k-form* to be, as in §2.4, a function ω which assigns to each $p \in X$ an element ω_p of $\Lambda^k(T_p^* X)$. Then in §4.4 we will define what it means for a manifold to be *oriented* and in §4.5 show that if X is an oriented n-dimensional manifold and ω a compactly supported n-form, the integral of ω is well-defined. Moreover, we'll prove in this section a manifold version of the change of variables formula and in §4.6 prove manifold versions of two standard results in integral calculus: the divergence theorem and Stokes theorem, setting the stage for the main results of Chapter 4: the manifold version of the degree theorem (see §4.7) and a number of applications of this theorem (among them the Jordan–Brouwer separation theorem, the Gauss–Bonnet theorem and the Index theorem for vector fields (see §§4.8 and 4.9).

Chapter 5: Cohomology via forms

Given an n-dimensional manifold let $\Omega^k(X)$ be the space of differential forms on X of degree k, let $d\colon \Omega^k(X) \to \Omega^{k+1}(X)$ be exterior differentiation and let $H^k(X)$ be the *kth de Rham cohomology group of* X: the quotient of the vector space

$$Z^k(X) := \ker(d\colon \Omega^k(X) \to \Omega^{k+1}(X))$$

by the vector space

$$B^k(X) := \operatorname{im}(d\colon \Omega^{k-1}(X) \to \Omega^k(X)).$$

In Chapter 5 we will make a systematic study of these groups and also show that some of the results about differential forms that we proved in earlier chapters can be interpreted as cohomological properties of these groups. For instance, in § 5.1 we will show that the wedge product and pullback operations on forms that we discussed in Chapter 2 gives rise to analogous operations in cohomology and that for a compact oriented n-dimensional manifold the integration operator on forms that we discussed in Chapter 4 gives rise to a pairing

$$H^k(X) \times H^{n-k}(X) \to \mathbf{R},$$

and in fact more generally if X is non-compact, a pairing

$$H_c^k(X) \times H^{n-k}(X) \to \mathbf{R}, \tag{13}$$

where the groups $H_c^i(X)$ are the de Rham cohomology groups that one gets by replacing the spaces of differential forms $\Omega^i(X)$ by the corresponding spaces $\Omega_c^i(X)$ of compactly supported differential forms.

One problem one runs into in the definition of these cohomology groups is that since the spaces $\Omega^k(X)$ and $\Omega_c^k(X)$ are infinite dimensional in general, there's no guarantee that the same won't be the case for the spaces $H^k(X)$ and $H_c^k(X)$, and we'll address this problem in § 5.3. More explicitly, we will say that X has *finite topology* if it admits a finite covering by open sets $U_1, \dots, U_N$ such that for every multi-index $I = (i_1, \dots, i_k)$, where $1 \leq i_r \leq N$, the intersection

$$U_I \coloneqq U_{i_1} \cap \cdots \cap U_{i_k} \tag{14}$$

is either empty or is diffeomorphic to a convex open subset of $\mathbf{R}^n$. We will show that if X has finite topology, then its cohomology groups are finite dimensional and, secondly, we will show that many manifolds have finite topology. (For instance, if X is compact then X has finite topology; for details see §§ 5.2 and 5.3.)

We mentioned above that if X is not compact it has two types of cohomology groups: the groups $H^k(X)$ and the groups $H_c^{n-k}(X)$. In § 5.5 we will show that if X is oriented then for $\nu \in \Omega^k(X)$ and $\omega \in \Omega_c^{n-k}(X)$ the integration operation

$$(\nu, \omega) \mapsto \int_X \nu \wedge \omega$$

gives rise to a pairing

$$H^k(X) \times H_c^{n-k}(X) \to \mathbf{R},$$

and that if X is connected this pairing is non-degenerate, i.e., defines a bijective linear transformation

$$H_c^{n-k}(X) \xrightarrow{\sim} H^k(X)^*.$$

This results in the *Poincaré duality theorem* and it has some interesting implications which we will explore in §§ 5.5–5.7. For instance in § 5.5 we will show that if Y and Z are closed oriented submanifolds of X and Z is compact and of codimension equal to the dimension of Y, then the *intersection number* $I(Y, Z)$ of Y and Z in X is well-defined (no matter how badly Y and Z intersect). In § 5.6 we will show that if X is a compact oriented manifold and $f: X \to X$ a C^∞ map one can define the *Lefschetz number* of f as the intersection number $I(\Delta_X, \mathrm{graph}(f))$ where Δ_X is the diagonal in $X \times X$, and view this as a "topological count" of the number of fixed points of the map f.

Finally in § 5.8 we will show that if X has finite topology (in the sense we described above), then one can define an alternative set of cohomology groups for X, the *Čech cohomology groups* of a cover $\mathcal{U}$, and we will sketch a proof of the assertion that this Čech cohomology and de Rham cohomology coincide, leaving a lot of details as exercises (however, with some hints supplied).

Appendices

In Appendix A we review some techniques in multivariable calculus that enable one to reduce global problems to local problems and illustrate these techniques by describing some typical applications. In particular we show how these techniques can be used to make sense of "improper integrals" and to extend differentiable maps $f: X \to \mathbf{R}$ on arbitrary subsets $X \subset \mathbf{R}^m$ to open neighborhoods of these sets.

In Appendix B we discuss another calculus issue: how to solve systems of equations

$$\begin{cases} f_1(x) = y_1 \\ \quad\vdots \\ f_N(x) = y_N \end{cases}$$

for x in terms of y (here $f_1, \ldots, f_N$ are differentiable functions on an open subset of $\mathbf{R}^n$). To answer this question we prove a somewhat refined version of what in elementary calculus texts is referred to as the *implicit function theorem* and derive as corollaries of this theorem two results that we make extensive use of in Chapter 4: the canonical submersion theorem and the canonical immersion theorem.

Finally, in Appendix C we discuss a concept which plays an important role in the formulation of the "finite topology" results that we discussed in Chapter 5, namely the concept of a "good cover". In more detail let X be an n-dimensional manifold an let $\mathcal{U} = \{U_\alpha\}_{\alpha \in I}$ be a collection of open subsets of X with the property that $\bigcup_{\alpha \in I} U_\alpha = X$.

Then $\mathcal{U}$ is a *good cover* of X if for every finite subset

$$\{\alpha_1, \ldots, \alpha_r\} \subset I$$

the intersection $U_{\alpha_1} \cap \cdots \cap U_{\alpha_r}$ is either empty or is diffeomorphic to a convex open subset of $\mathbf{R}^n$. In Appendix C we will prove that good covers always exist.

Notational Conventions

Below we provide a list of a few of our common notational conventions.

- $:=$ used to define a term; the term being defined appears to the left of the colon
- $\subset$ subset
- $\to$ denotes an map (of sets, vector spaces, etc.)

- $\hookrightarrow$ inclusion map
- $\twoheadrightarrow$ surjective map
- $\xrightarrow{\sim}$ a map that is an isomorphism
- $\smallsetminus$ difference of sets: if $A \subset X$, then $X \smallsetminus A$ is the complement of A in X

Acknowledgments

To conclude this introduction we would like to thank Cole Graham and Farzan Vafa for helping us correct earlier versions of this text, a task which involved compiling long list of typos and errata. (In addition a lot of pertinent comments of theirs helped us improve the readability of the text itself.)

We would also like to express our gratitude to Doug Stryker who helped us comb through the next-to-final version of the book prior to publication and correct a host of typos that had eluded us earlier.

About the Authors

Victor Guillemin received the B.A. in mathematics from Harvard in 1959, the M.A. from the University of Chicago in 1960, and the Ph.D. from Harvard in 1962, under the direction of Shlomo Sternberg. Following an instructorship at Columbia University, Professor Guillemin joined the mathematics faculty at the Massachusetts Institute of Technology in 1966 (professor in 1973). He was appointed the Norbert Wiener Professor of Mathematics, 1994–1999. Professor Guillemin's research interests are in differential geometry and symplectic geometry. He was elected Fellow of the American Academy of Arts & Sciences (1984) and Member of the National Academy of Sciences (1985). He subsequently received the Guggenheim and Humboldt fellowships. In 2003, Professor Guillemin received the Leroy P. Steele Prize for Lifetime Achievement of the American Mathematical Society.

Peter Haine received the S.B. in mathematics from the Massachusetts Institute of Technology (MIT) in 2016. Since 2016 he has been a Ph.D. candidate in pure mathematics at MIT and expects to complete his Ph.D. in 2021. His research interests center around homotopy theory and its applications to other branches of mathematics. In his Ph.D. thesis research, he has been working with his advisor Clark Barwick to understand stratified homotopy theory and its deep relationship to algebraic geometry.

Contents

Chapter 1

Multilinear Algebra

1.1. Background

We will list below some definitions and theorems that are part of the curriculum of a standard theory-based course in linear algebra.[1]

Definition 1.1.1. A (*real*) ***vector space*** is a set V the elements of which we refer to as ***vectors***. The set V is equipped with two vector space operations:

(1) *Vector space addition*: Given vectors $v_1, v_2 \in V$ one can add them to get a third vector, $v_1 + v_2$.
(2) *Scalar multiplication*: Given a vector $v \in V$ and a real number λ, one can multiply v by λ to get a vector λv.

These operations satisfy a number of standard rules: associativity, commutativity, distributive laws, etc. which we assume you're familiar with. (See Exercise 1.1.i.) In addition we assume that the reader is familiar with the following definitions and theorems.

(1) The ***zero vector*** in V is the unique vector $0 \in V$ with the property that for every vector $v \in V$, we have $v + 0 = 0 + v = v$ and $\lambda v = 0$ if $\lambda \in \mathbf{R}$ is zero.
(2) *Linear independence*: A (finite) set of vectors, $v_1, \dots, v_k \in V$ is ***linearly independent*** if the map

$$\mathbf{R}^k \to V, \quad (c_1, \dots, c_k) \mapsto c_1 v_1 + \dots + c_k v_k \tag{1.1.2}$$

is injective.
(3) A (finite) set of vectors $v_1, \dots, v_k \in V$ ***spans*** V if the map (1.1.2) is surjective.
(4) Vectors $v_1, \dots, v_k \in V$ are a ***basis*** of V if they span V and are linearly independent; in other words, if the map (1.1.2) is bijective. This means that every vector v can be written *uniquely* as a sum

$$v = \sum_{i=1}^{n} c_i v_i. \tag{1.1.3}$$

---

[1] Such a course is a prerequisite for reading this text.

(5) If V is a vector space with a basis $v_1, \dots, v_k$, then V is said to be ***finite dimensional***, and k is the ***dimension*** of V. (It is a theorem that this definition is legitimate: every basis has to have the same number of vectors.) In this chapter all the vector spaces we'll encounter will be finite dimensional. We write $\dim(V)$ for the dimension of V.

(6) A subset $U \subset V$ is a ***subspace*** if it is vector space in its own right, i.e., for all $v, v_1, v_2 \in U$ and $\lambda \in \mathbf{R}$, both λv and $v_1 + v_2$ are in U (where the addition and scalar multiplication is given as vectors of V).

(7) Let V and W be vector spaces. A map $A\colon V \to W$ is ***linear*** if for $v, v_1, v_2 \in U$ and $\lambda \in \mathbf{R}$ we have

$$A(\lambda v) = \lambda A v \tag{1.1.4}$$

and

$$A(v_1 + v_2) = Av_1 + Av_2.$$

(8) Let $A\colon V \to W$ be a linear map of vector spaces. The ***kernel*** of A is the subset of V defined by

$$\ker(A) := \{ v \in V \mid A(v) = 0 \},$$

i.e., is the set of vectors in V which A sends to the zero vector in W. By equation (1.1.4) and item (7), the subset $\ker(A)$ is a subspace of V.

(9) By (1.1.4) and (7) the set-theoretic image $\operatorname{im}(A)$ of A is a subspace of W. We call $\operatorname{im}(A)$ the ***image*** A of A. The following is an important rule for keeping track of the dimensions of $\ker A$ and $\operatorname{im} A$:

$$\dim(V) = \dim(\ker(A)) + \dim(\operatorname{im}(A)). \tag{1.1.5}$$

(10) *Linear mappings and matrices*: Let $v_1, \dots, v_n$ be a basis of V and $w_1, \dots, w_m$ a basis of W. Then by (1.1.3) Av_j can be written uniquely as a sum,

$$Av_j = \sum_{i=1}^{m} c_{i,j} w_i, \quad c_{i,j} \in \mathbf{R}. \tag{1.1.6}$$

The $m \times n$ matrix of real numbers $(c_{i,j})$ is the ***matrix*** associated with A. Conversely, given such an $m \times n$ matrix, there is a unique linear map A with the property (1.1.6).

(11) An ***inner product*** on a vector space is a map

$$B\colon V \times V \to \mathbf{R}$$

with the following three properties:

(a) *Bilinearity*: For vectors $v, v_1, v_2, w \in V$ and $\lambda \in \mathbf{R}$ we have

$$B(v_1 + v_2, w) = B(v_1, w) + B(v_2, w)$$

and

$$B(\lambda v, w) = \lambda B(v, w).$$

(b) *Symmetry*: For vectors $v, w \in V$ we have $B(v, w) = B(w, v)$.

(c) *Positivity*: For every vector $v \in V$ we have $B(v, v) \geq 0$. Moreover, if $v \neq 0$ then $B(v, v) > 0$.

Remark 1.1.7. Notice that by property (1.1b), property (1.1a) is equivalent to

$$B(w, \lambda v) = \lambda B(w, v)$$

and

$$B(w, v_1 + v_2) = B(w, v_1) + B(w, v_2).$$

Example 1.1.8. The map (1.1.2) is a linear map. The vectors $v_1, \dots, v_k$ span V if its image of this map is V, and the $v_1, \dots, v_k$ are linearly independent if the kernel of this map is the zero vector in $\mathbf{R}^k$.

The items on the list above are just a few of the topics in linear algebra that we're assuming our readers are familiar with. We have highlighted them because they're easy to state. However, understanding them requires a heavy dollop of that indefinable quality "mathematical sophistication", a quality which will be in heavy demand in the next few sections of this chapter. We will also assume that our readers are familiar with a number of more low-brow linear algebra notions: matrix multiplication, row and column operations on matrices, transposes of matrices, determinants of $n \times n$ matrices, inverses of matrices, Cramer's rule, recipes for solving systems of linear equations, etc. (See [10, §§ 1.1 and 1.2] for a quick review of this material.)

Exercises for § 1.1

Exercise 1.1.i. Our basic example of a vector space in this course is $\mathbf{R}^n$ equipped with the vector addition operation

$$(a_1, \dots, a_n) + (b_1, \dots, b_n) = (a_1 + b_1, \dots, a_n + b_n)$$

and the scalar multiplication operation

$$\lambda(a_1, \dots, a_n) = (\lambda a_1, \dots, \lambda a_n).$$

Check that these operations satisfy the axioms below.

(1) *Commutativity:* $v + w = w + v$.
(2) *Associativity:* $u + (v + w) = (u + v) + w$.
(3) For the zero vector $0 := (0, \dots, 0)$ we have $v + 0 = 0 + v$.
(4) $v + (-1)v = 0$.
(5) $1v = v$.
(6) *Associative law for scalar multiplication:* $(ab)v = a(bv)$.
(7) *Distributive law for scalar addition:* $(a + b)v = av + bv$.
(8) *Distributive law for vector addition:* $a(v + w) = av + aw$.

Exercise 1.1.ii. Check that the standard basis vectors of $\mathbf{R}^n$: $e_1 = (1, 0, \dots, 0)$, $e_2 = (0, 1, 0, \dots, 0)$, etc. *are* a basis.

Exercise 1.1.iii. Check that the standard inner product on $\mathbf{R}^n$

$$B((a_1, \dots, a_n), (b_1, \dots, b_n)) = \sum_{i=1}^{n} a_i b_i$$

is an inner product.

1.2. Quotient and dual spaces

In this section, we will discuss a couple of items which are frequently, but not always, covered in linear algebra courses, but which we'll need for our treatment of multilinear algebra in §§1.3–1.8.

The quotient spaces of a vector space

Definition 1.2.1. Let V be a vector space and W a vector subspace of V. A W-***coset*** is a set of the form

$$v + W := \{v + w \mid w \in W\}.$$

It is easy to check that if $v_1 - v_2 \in W$, the cosets $v_1 + W$ and $v_2 + W$ coincide while if $v_1 - v_2 \notin W$, the cosets $v_1 + W$ and $v_2 + W$ are disjoint. Thus the distinct W-cosets decompose V into a *disjoint* collection of subsets of V.

Notation 1.2.2. Let V be a vector space and $W \subset V$ a subspace. We write V/W for the set of *distinct* W-cosets in V.

Definition 1.2.3. Let V be a vector space and $W \subset V$ a subspace. Define a vector addition operation on V/W by setting

$$(v_1 + W) + (v_2 + W) := (v_1 + v_2) + W \tag{1.2.4}$$

and define a scalar multiplication operation on V/W by setting

$$\lambda(v + W) := (\lambda v) + W. \tag{1.2.5}$$

These operations make V/W into a vector space, called the ***quotient space*** of V by W.

It is easy to see that the operations on V/W from Definition 1.2.3 are well-defined. For instance, suppose $v_1 + W = v_1' + W$ and $v_2 + W = v_2' + W$. Then $v_1 - v_1'$ and $v_2 - v_2'$ are in W, so

$$(v_1 + v_2) - (v_1' + v_2') \in W,$$

and hence $(v_1 + v_2) + W = (v_1' + v_2') + W$.

Definition 1.2.6. Let V be a vector space and $W \subset V$ a subspace. Define a ***quotient map***

$$\pi \colon V \to V/W \tag{1.2.7}$$

by setting $\pi(v) := v + W$. It is clear from equations (1.2.4) and (1.2.5) that π is linear, and that it maps V surjectively onto V/W.

Observation 1.2.8. Note that the zero vector in the vector space V/W is the zero coset $0 + W = W$. Hence $v \in \ker(\pi)$ if and only if $v + W = W$, i.e., $v \in W$. In other words, $\ker(\pi) = W$.

In Definitions 1.2.3 and 1.2.6, V and W do not have to be finite dimensional, but if they are then by equation (1.1.5) we have

$$\dim(V/W) = \dim(V) - \dim(W).$$

We leave the following easy proposition as an exercise.

Proposition 1.2.9. *Let $A: V \to U$ be a linear map of vector spaces. If $W \subset \ker(A)$ there exists a unique linear map $A^\sharp: V/W \to U$ with the property that $A = A^\sharp \circ \pi$, where $\pi: V \to V/W$ is the quotient map.*

The dual space of a vector space

Definition 1.2.10. Let V be a vector space. Write $V^\star$ for the set of all linear functions $\ell: V \to \mathbf{R}$. Define a vector space structure on $V^\star$ as follows: if $\ell_1, \ell_2 \in V^\star$, then define the sum $\ell_1 + \ell_2$ to be the map $V \to \mathbf{R}$ given by

$$(\ell_1 + \ell_2)(v) := \ell_1(v) + \ell_2(v),$$

which is clearly linear. If $\ell \in V^\star$ is a linear function and $\lambda \in \mathbf{R}$, define $\lambda\ell$ by

$$(\lambda\ell)(v) := \lambda \cdot \ell(v),$$

then $\lambda\ell$ is clearly linear.

The vector space $V^\star$ is called the ***dual space*** of V.

Suppose V is n-dimensional, and let $e_1, \dots, e_n$ be a basis of V. Then every vector $v \in V$ can be written uniquely as a sum

$$v = c_1 e_1 + \cdots + c_n e_n, \quad c_i \in \mathbf{R}.$$

Let

$$e_i^\star(v) = c_i. \tag{1.2.11}$$

If $v = c_1 e_1 + \cdots + c_n e_n$ and $v' = c_1' e_1 + \cdots + c_n' e_n$ then $v + v' = (c_1 + c_1')e_1 + \cdots + (c_n + c_n')e_n$, so

$$e_i^\star(v + v') = c_i + c_i' = e_i^\star(v) + e_i^\star(v').$$

This shows that $e_i^\star(v)$ is a linear function of V and hence $e_i^\star \in V^\star$.

Claim 1.2.12. *If V is an n-dimensional vector space with basis $e_1, \dots, e_n$, then $e_1^\star, \dots, e_n^\star$ is a basis of $V^\star$.*

Proof. First of all note that by (1.2.11)

$$e_i^\star(e_j) = \begin{cases} 1, & i = j, \\ 0, & i \neq j. \end{cases} \tag{1.2.13}$$

If $\ell \in V^\star$ let $\lambda_i = \ell(e_i)$ and let $\ell' = \sum_{i=1}^n \lambda_i e_i^\star$. Then by (1.2.13)

$$\ell'(e_j) = \sum_{i=1}^n \lambda_i e_i^\star(e_j) = \lambda_j = \ell(e_j),$$

i.e., ℓ and ℓ' take identical values on the basis vectors, e_j. Hence $\ell = \ell'$.

Suppose next that $\sum_{i=1}^n \lambda_i e_i^\star = 0$. Then by (1.2.13), $\lambda_j = (\sum_{i=1}^n \lambda_i e_i^\star)(e_j) = 0$ for $j = 1, \dots, n$. Hence the vectors $e_1^\star, \dots, e_n^\star$ are linearly independent. □

Let V and W be vector spaces and $A: V \to W$ a linear map. Given $\ell \in W^\star$ the composition, $\ell \circ A$, of A with the linear map, $\ell: W \to \mathbf{R}$, is linear, and hence is an element of $V^\star$. We will denote this element by $A^\star \ell$, and we will denote by

$$A^\star : W^\star \to V^\star$$

the map $\ell \mapsto A^\star \ell$. It is clear from the definition that

$$A^\star(\ell_1 + \ell_2) = A^\star \ell_1 + A^\star \ell_2$$

and that

$$A^\star(\lambda \ell) = \lambda A^\star \ell,$$

i.e., $A^\star$ is linear.

Definition 1.2.14. Let V and W be vector spaces and $A: V \to W$ a linear map. We call the map $A^\star : W^\star \to V^\star$ defined above the ***transpose*** of the map A.

We conclude this section by giving a matrix description of $A^\star$. Let $e_1, \dots, e_n$ be a basis of V and $f_1, \dots, f_m$ a basis of W; let $e_1^\star, \dots, e_n^\star$ and $f_1^\star, \dots, f_m^\star$ be the dual bases of $V^\star$ and $W^\star$. Suppose A is defined in terms of $e_1, \dots, e_n$ and $f_1, \dots, f_m$ by the $m \times n$ matrix, $(a_{i,j})$, i.e., suppose

$$Ae_j = \sum_{i=1}^{n} a_{i,j} f_i.$$

Claim 1.2.15. *The linear map $A^\star$ is defined, in terms of $f_1^\star, \dots, f_m^\star$ and $e_1^\star, \dots, e_n^\star$, by the transpose matrix $(a_{j,i})$.*

Proof. Let

$$A^\star f_i^\star = \sum_{j=1}^{m} c_{j,i} e_j^\star .$$

Then

$$A^\star f_i^\star(e_j) = \sum_{k=1}^{n} c_{k,i} e_k^\star(e_j) = c_{j,i}$$

by (1.2.13). On the other hand

$$A^\star f_i^\star(e_j) = f_i^\star(Ae_j) = f_i^\star\left(\sum_{k=1}^{m} a_{k,j} f_k\right) = \sum_{k=1}^{m} a_{k,j} f_i^\star(f_k) = a_{i,j}$$

so $a_{i,j} = c_{j,i}$. □

Exercises for §1.2

Exercise 1.2.i. Let V be an n-dimensional vector space and W a k-dimensional subspace. Show that there exists a basis $e_1, \dots, e_n$ of V with the property that $e_1, \dots, e_k$ is a basis of W.

Hint: Induction on $n - k$. To start the induction suppose that $n - k = 1$. Let $e_1, \dots, e_{n-1}$ be a basis of W and e_n any vector in $V \smallsetminus W$.

Exercise 1.2.ii. In Exercise 1.2.i show that the vectors $f_i := \pi(e_{k+i})$, $i = 1, \ldots, n-k$ are a basis of V/W, where $\pi\colon V \to V/W$ is the quotient map.

Exercise 1.2.iii. In Exercise 1.2.i let U be the linear span of the vectors e_{k+i} for $i = 1, \ldots, n-k$. Show that the map

$$U \to V/W, \quad u \mapsto \pi(u),$$

is a vector space isomorphism, i.e., show that it maps U bijectively onto V/W.

Exercise 1.2.iv. Let U, V and W be vector spaces and let $A\colon V \to W$ and $B\colon U \to V$ be linear mappings. Show that $(AB)^\star = B^\star A^\star$.

Exercise 1.2.v. Let $V = \mathbf{R}^2$ and let W be the x_1-axis, i.e., the one-dimensional subspace

$$\{(x_1, 0) \mid x_1 \in \mathbf{R}\}$$

of $\mathbf{R}^2$.

(1) Show that the W-cosets are the lines, $x_2 = a$, parallel to the x_1-axis.
(2) Show that the sum of the cosets "$x_2 = a$" and "$x_2 = b$" is the coset "$x_2 = a + b$".
(3) Show that the scalar multiple of the coset, "$x_2 = c$" by the number, λ, is the coset, "$x_2 = \lambda c$".

Exercise 1.2.vi.

(1) Let $(V^\star)^\star$ be the dual of the vector space, $V^\star$. For every $v \in V$, let $\mathrm{ev}_v\colon V^\star \to \mathbf{R}$ be the ***evaluation function*** $\mathrm{ev}_v(\ell) = \ell(v)$. Show that the ev_v is a linear function on $V^\star$, i.e., an element of $(V^\star)^\star$, and show that the map

(1.2.16) $$\mathrm{ev} = \mathrm{ev}_{(-)}\colon V \to (V^\star)^\star, \quad v \mapsto \mathrm{ev}_v$$

is a linear map of V into $(V^\star)^\star$.

(2) If V is *finite dimensional*, show that the map (1.2.16) is bijective. Conclude that there is a *natural* identification of V with $(V^\star)^\star$, i.e., that V and $(V^\star)^\star$ are two descriptions of the same object.

Hint: $\dim(V^\star)^\star = \dim V^\star = \dim V$, so by equation (1.1.5) it suffices to show that (1.2.16) is injective.

Exercise 1:2.vii. Let W be a vector subspace of a finite-dimensional vector space V and let

$$W^\perp = \{\ell \in V^\star \mid \ell(w) = 0 \text{ for all } w \in W\}.$$

$W^\perp$ is called the ***annihilator*** of W in $V^\star$. Show that $W^\perp$ is a subspace of $V^\star$ of dimension $\dim V - \dim W$.

Hint: By Exercise 1.2.i we can choose a basis, $e_1, \ldots, e_n$ of V such that $e_1, \ldots, e_k$ is a basis of W. Show that $e_{k+1}^\star, \ldots, e_n^\star$ is a basis of $W^\perp$.

Exercise 1.2.viii. Let V and V' be vector spaces and $A\colon V \to V'$ a linear map. Show that if $W \subset \ker(A)$, then there exists a linear map $B\colon V/W \to V'$ with the property that $A = B \circ \pi$ (where π is the quotient map (1.2.7)). In addition show that this linear map is injective if and only if $\ker(A) = W$.

Exercise 1.2.ix. Let W be a subspace of a finite-dimensional vector space V. From the inclusion map, $\iota\colon W^\perp \to V^\star$, one gets a transpose map,

$$\iota^\star\colon (V^\star)^\star \to (W^\perp)^\star$$

and, by composing this with (1.2.16), a map

$$\iota^\star \circ \mathrm{ev}\colon V \to (W^\perp)^\star.$$

Show that this map is onto and that its kernel is W. Conclude from Exercise 1.2.viii that there is a *natural* bijective linear map

$$\nu\colon V/W \to (W^\perp)^\star$$

with the property $\nu \circ \pi = \iota^\star \circ \mathrm{ev}$. In other words, V/W and $(W^\perp)^\star$ are two descriptions of the same object. (This shows that the "quotient space" operation and the "dual space" operation are closely related.)

Exercise 1.2.x. Let V_1 and V_2 be vector spaces and $A\colon V_1 \to V_2$ a linear map. Verify that for the transpose map $A^\star\colon V_2^\star \to V_1^\star$ we have:

$$\ker(A^\star) = \mathrm{im}(A)^\perp$$

and

$$\mathrm{im}(A^\star) = \ker(A)^\perp.$$

Exercise 1.2.xi. Let V be a vector space.

(1) Let $B\colon V \times V \to \mathbf{R}$ be an inner product on V. For $v \in V$ let

$$\ell_v\colon V \to \mathbf{R}$$

be the function: $\ell_v(w) = B(v, w)$. Show that ℓ_v is linear and show that the map

$$L\colon V \to V^\star, \quad v \mapsto \ell_v \tag{1.2.17}$$

is a linear mapping.

(2) If V is finite dimensional, prove that L bijective. Conclude that if V has an inner product one gets from it a *natural* identification of V with $V^\star$.

Hint: Since $\dim V = \dim V^\star$ it suffices by equation (1.1.5) to show that $\ker(L) = 0$. Now note that if $v \neq 0$, $\ell_v(v) = B(v, v)$ is a positive number.

Exercise 1.2.xii. Let V be an n-dimensional vector space and $B\colon V \times V \to \mathbf{R}$ an inner product on V. A basis, $e_1, \dots, e_n$ of V is ***orthonormal*** if

$$B(e_i, e_j) = \begin{cases} 1, & i = j, \\ 0, & i \neq j. \end{cases} \tag{1.2.18}$$

(1) Show that an orthonormal basis exists.

Hint: By induction let $e_1, \dots, e_k$ be vectors with the property (1.2.18) and let v be a vector which is not a linear combination of these vectors. Show that the vector

$$w = v - \sum_{i=1}^{k} B(e_i, v) e_i$$

is non-zero and is orthogonal to the e_i's. Now let $e_{k+1} = \lambda w$, where $\lambda = B(w,w)^{-\frac{1}{2}}$.

(2) Let $e_1, \dots e_n$ and $e'_1, \dots e'_n$ be two orthonormal bases of V and let

$$e'_j = \sum_{i=1}^{n} a_{i,j} e_i.$$

Show that

$$\sum_{i=1}^{n} a_{i,j} a_{i,k} = \begin{cases} 1, & j = k, \\ 0, & j \neq k. \end{cases} \tag{1.2.19}$$

(3) Let A be the matrix $(a_{i,j})$. Show that equation (1.2.19) can be written more compactly as the matrix identity

$$AA^{\intercal} = \mathrm{id}_n, \tag{1.2.20}$$

where id_n is the $n \times n$ identity matrix and $A^{\intercal}$ is the transpose of the matrix A.

(4) Let $e_1, \dots, e_n$ be an orthonormal basis of V and $e_1^\star, \dots, e_n^\star$ the dual basis of $V^\star$. Show that the mapping (1.2.17) is the mapping, $Le_i = e_i^\star$, $i = 1, \dots n$.

1.3. Tensors

Definition 1.3.1. Let V be an n-dimensional vector space and let V^k be the set of all k-tuples $(v_1, \dots, v_k)$, where $v_1, \dots, v_k \in V$, that is, the k-fold direct sum of V with itself. A function

$$T \colon V^k \to \mathbf{R}$$

is said to be ***linear in its*** ith ***variable*** if, when we fix vectors, $v_1, \dots, v_{i-1}, v_{i+1}, \dots, v_k$, the map $V \to \mathbf{R}$ defined by

$$v \mapsto T(v_1, \dots, v_{i-1}, v, v_{i+1}, \dots, v_k) \tag{1.3.2}$$

is linear.

If T is linear in its ith variable for $i = 1, \dots, k$ it is said to be ***k-linear***, or alternatively is said to be a ***k-tensor***. We write $\mathcal{L}^k(V)$ for the set of all k-tensors in V. We will agree that 0-tensors are just the real numbers, that is $\mathcal{L}^0(V) \coloneqq \mathbf{R}$.

Let $T_1, T_2 \colon V^k \to \mathbf{R}$ be linear functions. It is clear from (1.3.2) that if T_1 and T_2 are k-linear, so is $T_1 + T_2$. Similarly if T is k-linear and λ is a real number, λT is k-linear. Hence $\mathcal{L}^k(V)$ is a vector space. Note that for $k = 1$, "k-linear" just means 'linear', so $\mathcal{L}^1(V) = V^\star$.

Definition 1.3.3. Let n and k be positive integers. A ***multi-index of n of length k*** is a k-tuple $I = (i_1, \dots, i_k)$ of integers with $1 \le i_r \le n$ for $r = 1, \dots, k$.

Example 1.3.4. Let n be a positive integer. The multi-indices of n of length 2 are in bijection the square of pairs of integers

$$(i, j), \quad 1 \le i, j \le n,$$

and there are exactly n^2 of them.

We leave the following generalization as an easy exercise.

Lemma 1.3.5. *Let n and k be positive integers. There are exactly n^k multi-indices of n of length k.*

Now fix a basis $e_1, \dots, e_n$ of V. For $T \in \mathcal{L}^k(V)$ write

$$T_I := T(e_{i_1}, \dots, e_{i_k}) \tag{1.3.6}$$

for every multi-index I of length k.

Proposition 1.3.7. *The real numbers T_I determine T, i.e., if T and T' are k-tensors and $T_I = T'_I$ for all I, then $T = T'$.*

Proof. By induction on n. For $n = 1$ we proved this result in §1.1. Let's prove that if this assertion is true for $n-1$, it is true for n. For each e_i let T_i be the $(k-1)$-tensor

$$(v_1, \dots, v_{n-1}) \mapsto T(v_1, \dots, v_{n-1}, e_i).$$

Then for $v = c_1e_1 + \cdots + c_ne_n$

$$T(v_1, \dots, v_{n-1}, v) = \sum_{i=1}^{n} c_iT_i(v_1, \dots, v_{n-1}),$$

so the T_i's determine T. Now apply the inductive hypothesis. □

The tensor product operation

Definition 1.3.8. If T_1 is a k-tensor and T_2 is an ℓ-tensor, one can define a $k+\ell$-tensor, $T_1 \otimes T_2$, by setting

$$(T_1 \otimes T_2)(v_1, \dots, v_{k+\ell}) = T_1(v_1, \dots, v_k)T_2(v_{k+1}, \dots, v_{k+\ell}).$$

This tensor is called ***the tensor product*** of T_1 and T_2.

We note that if T_1 is a 0-tensor, i.e., scalar, then tensor product with T_1 is just scalar multiplication by T_1, and similarly if T_2 is a 0-tensor. That is $a \otimes T = T \otimes a = aT$ for $a \in \mathbf{R}$ and $T \in \mathcal{L}^k(V)$.

Properties 1.3.9. Suppose that we are given a k_1-tensor T_1, a k_2-tensor T_2, and a k_3-tensor T_3 on a vector space V.

(1) *Associativity*: One can define a $(k_1 + k_2 + k_3)$-tensor $T_1 \otimes T_2 \otimes T_3$ by setting

$$\begin{aligned}&(T_1 \otimes T_2 \otimes T_3)(v_1, \dots, v_{k+\ell+m})\\ &\quad := T_1(v_1, \dots, v_k)T_2(v_{k+1}, \dots, v_{k+\ell})T_3(v_{k+\ell+1}, \dots, v_{k+\ell+m}).\end{aligned}$$

Alternatively, one can define $T_1 \otimes T_2 \otimes T_3$ by defining it to be the tensor product of $(T_1 \otimes T_2) \otimes T_3$ or the tensor product of $T_1 \otimes (T_2 \otimes T_3)$. It is easy to see that both these tensor products are identical with $T_1 \otimes T_2 \otimes T_3$:

$$(T_1 \otimes T_2) \otimes T_3 = T_1 \otimes T_2 \otimes T_3 = T_1 \otimes (T_2 \otimes T_3).$$

(2) *Distributivity of scalar multiplication:* We leave it as an easy exercise to check that if λ is a real number then

$$\lambda(T_1 \otimes T_2) = (\lambda T_1) \otimes T_2 = T_1 \otimes (\lambda T_2).$$

(3) *Left and right distributive laws:* If $k_1 = k_2$, then

$$(T_1 + T_2) \otimes T_3 = T_1 \otimes T_3 + T_2 \otimes T_3$$

and if $k_2 = k_3$, then

$$T_1 \otimes (T_2 + T_3) = T_1 \otimes T_2 + T_1 \otimes T_3.$$

A particularly interesting tensor product is the following. For $i = 1, \dots, k$ let $\ell_i \in V^\star$ and let

$$T = \ell_1 \otimes \cdots \otimes \ell_k. \tag{1.3.10}$$

Thus, by definition,

$$T(v_1, \dots, v_k) = \ell_1(v_1) \cdots \ell_k(v_k). \tag{1.3.11}$$

A tensor of the form (1.3.11) is called a ***decomposable*** k-tensor. These tensors, as we will see, play an important role in what follows. In particular, let $e_1, \dots, e_n$ be a basis of V and $e_1^\star, \dots, e_n^\star$ the dual basis of $V^\star$. For every multi-index I of length k let

$$e_I^\star = e_{i_1}^\star \otimes \cdots \otimes e_{i_k}^\star.$$

Then if J is another multi-index of length k,

$$e_I^\star(e_{j_1}, \dots, e_{j_k}) = \begin{cases} 1, & I = J, \\ 0, & I \neq J \end{cases} \tag{1.3.12}$$

by (1.2.13), (1.3.10) and (1.3.11). From (1.3.12) it is easy to conclude the following.

Theorem 1.3.13. *Let V be a vector space with basis $e_1, \dots, e_n$ and let $0 \le k \le n$ be an integer. The k-tensors $e_I^\star$ of (1.3.12) are a basis of $\mathcal{L}^k(V)$.*

Proof. Given $T \in \mathcal{L}^k(V)$, let

$$T' = \sum_I T_I e_I^\star,$$

where the T_I's are defined by (1.3.6). Then

$$T'(e_{j_1}, \dots, e_{j_k}) = \sum_I T_I e_I^\star(e_{j_1}, \dots, e_{j_k}) = T_J \tag{1.3.14}$$

by (1.3.12); however, by Proposition 1.3.7 the T_J's determine T, so $T' = T$. This proves that the $e_I^\star$'s are a spanning set of vectors for $\mathcal{L}^k(V)$. To prove they're a basis, suppose

$$\sum_I C_I e_I^\star = 0$$

for constants, $C_I \in \mathbf{R}$. Then by (1.3.14) with $T' = 0$, $C_J = 0$, so the $e_I^\star$'s are linearly independent. ☐

As we noted in Lemma 1.3.5, there are exactly n^k multi-indices of length k and hence n^k basis vectors in the set $\{e_I^\star\}_I$, so we have proved.

Corollary 1.3.15. *Let V be an n-dimensional vector space. Then* $\dim(\mathcal{L}^k(V)) = n^k$.

The pullback operation

Definition 1.3.16. Let V and W be finite-dimensional vector spaces and let $A\colon V \to W$ be a linear mapping. If $T \in \mathcal{L}^k(W)$, we define

$$A^\star T\colon V^k \to \mathbf{R}$$

to be the function

$$(A^\star T)(v_1, \dots, v_k) := T(Av_1, \dots, Av_k). \tag{1.3.17}$$

It is clear from the linearity of A that this function is linear in its ith variable for all i, and hence is a k-tensor. We call $A^\star T$ the ***pullback*** of T by the map A.

Proposition 1.3.18. *The map*

$$A^\star\colon \mathcal{L}^k(W) \to \mathcal{L}^k(V), \quad T \mapsto A^\star T,$$

is a linear mapping.

We leave this as an exercise. We also leave as an exercise the identity

$$A^\star(T_1 \otimes T_2) = A^\star(T_1) \otimes A^\star(T_2) \tag{1.3.19}$$

for $T_1 \in \mathcal{L}^k(W)$ and $T_2 \in \mathcal{L}^m(W)$. Also, if U is a vector space and $B\colon U \to V$ a linear mapping, we leave for you to check that

$$(AB)^\star T = B^\star(A^\star T) \tag{1.3.20}$$

for all $T \in \mathcal{L}^k(W)$.

Exercises for §1.3

Exercise 1.3.i. Verify that there are exactly n^k multi-indices of length k.

Exercise 1.3.ii. Prove Proposition 1.3.18.

Exercise 1.3.iii. Verify equation (1.3.19).

Exercise 1.3.iv. Verify equation (1.3.20).

Exercise 1.3.v. Let $A\colon V \to W$ be a linear map. Show that if ℓ_i, $i = 1, \dots, k$ are elements of $W^\star$

$$A^\star(\ell_1 \otimes \cdots \otimes \ell_k) = A^\star(\ell_1) \otimes \cdots \otimes A^\star(\ell_k).$$

Conclude that $A^\star$ maps decomposable k-tensors to decomposable k-tensors.

Exercise 1.3.vi. Let V be an n-dimensional vector space and ℓ_i, $i = 1, 2$, elements of $V^\star$. Show that $\ell_1 \otimes \ell_2 = \ell_2 \otimes \ell_1$ if and only if ℓ_1 and ℓ_2 are linearly dependent.

Hint: Show that if ℓ_1 and ℓ_2 are linearly independent there exist vectors, v_i, $i =, 1, 2$ in V with property

$$\ell_i(v_j) = \begin{cases} 1, & i = j, \\ 0, & i \neq j. \end{cases}$$

Now compare $(\ell_1 \otimes \ell_2)(v_1, v_2)$ and $(\ell_2 \otimes \ell_1)(v_1, v_2)$. Conclude that if $\dim V \geq 2$ the tensor product operation is not commutative, i.e., it is usually not true that $\ell_1 \otimes \ell_2 = \ell_2 \otimes \ell_1$.

Exercise 1.3.vii. Let T be a k-tensor and v a vector. Define $T_v \colon V^{k-1} \to \mathbf{R}$ to be the map

$$T_v(v_1, \dots, v_{k-1}) := T(v, v_1, \dots, v_{k-1}). \tag{1.3.21}$$

Show that T_v is a $(k-1)$-tensor.

Exercise 1.3.viii. Show that if T_1 is an r-tensor and T_2 is an s-tensor, then if $r > 0$,

$$(T_1 \otimes T_2)_v = (T_1)_v \otimes T_2.$$

Exercise 1.3.ix. Let $A \colon V \to W$ be a linear map mapping, and $v \in V$. Write $w := Av$. Show that for $T \in \mathcal{L}^k(W)$, $A^\star(T_w) = (A^\star T)_v$.

1.4. Alternating k-tensors

We will discuss in this section a class of k-tensors which play an important role in multivariable calculus. In this discussion we will need some standard facts about the "permutation group". For those of you who are already familiar with this object (and I suspect most of you are) you can regard the paragraph below as a chance to re-familiarize yourselves with these facts.

Permutations

Definition 1.4.1. Let Σ_k be the k-element set $\Sigma_k := \{1, 2, \dots, k\}$. A ***permutation of order*** k is a bijecton $\sigma \colon \Sigma_k \xrightarrow{\sim} \Sigma_k$. Given two permutations σ_1 and σ_2, their ***product*** $\sigma_1\sigma_2$ is the composition of $\sigma_1 \circ \sigma_2$, i.e., the map,

$$i \mapsto \sigma_1(\sigma_2(i)).$$

For every permutation σ, one denotes by σ^{-1} the inverse permutation given by the inverse bijection of σ, i.e., defined by

$$\sigma(i) := j \iff \sigma^{-1}(j) = i.$$

Let S_k be the set of all permutations of order k. One calls S_k the ***permutation group*** of Σ_k or, alternatively, the ***symmetric group on k letters.***

It is easy to check the following.

Lemma 1.4.2. *The group S_k has $k!$ elements.*

Definition 1.4.3. Let k be a positive integer. For every $1 \leq i < j \leq k$, let $\tau_{i,j}$ be the permutation

$$\tau_{i,j}(\ell) := \begin{cases} j, & \ell = i, \\ i, & \ell = j, \\ \ell, & \ell \neq i, j. \end{cases}$$

The permutation $\tau_{i,j}$ is called a ***transposition***, and if $j = i + 1$, then $\tau_{i,j}$ is called an ***elementary transposition.***

Theorem 1.4.4. *Every permutation in S_k can be written as a product of (a finite number of) transpositions.*

Proof. We prove this by induction on k. The base case when $k = 2$ is obvious.

For the induction step, suppose that we know the claim for S_{k-1}. Given $\sigma \in S_k$, we have $\sigma(k) = i$ if and only if $\tau_{i,k}\sigma(k) = k$. Thus $\tau_{i,k}\sigma$ is, in effect, a permutation of Σ_{k-1}. By induction, $\tau_{i,k}\sigma$ can be written as a product of transpositions, so

$$\sigma = \tau_{i,k}(\tau_{i,k}\sigma)$$

can be written as a product of transpositions. □

Theorem 1.4.5. *Every transposition can be written as a product of elementary transpositions.*

Proof. Let $\tau = \tau_{ij}$, $i < j$. With i fixed, argue by induction on j. Note that for $j > i + 1$

$$\tau_{ij} = \tau_{j-1,j}\tau_{i,j-1}\tau_{j-1,j}.$$

Now apply the inductive hypothesis to $\tau_{i,j-1}$. □

Corollary 1.4.6. *Every permutation can be written as a product of elementary transpositions.*

The sign of a permutation

Definition 1.4.7. Let $x_1, \ldots, x_k$ be the coordinate functions on $\mathbf{R}^k$. For $\sigma \in S_k$ we define

$$(-1)^\sigma := \prod_{i<j} \frac{x_{\sigma(i)} - x_{\sigma(j)}}{x_i - x_j}. \tag{1.4.8}$$

Notice that the numerator and denominator in (1.4.8) are identical up to sign. Indeed, if $p = \sigma(i) < \sigma(j) = q$, the term, $x_p - x_q$, occurs once and just once in the numerator and once and just once in the denominator; and if $q = \sigma(i) > \sigma(j) = p$, the term, $x_p - x_q$, occurs once and just once in the numerator and its negative, $x_q - x_p$, once and just once in the numerator. Thus

$$(-1)^\sigma = \pm 1.$$

In light of this, we call $(-1)^\sigma$ the ***sign*** of σ.

Claim 1.4.9. *For $\sigma, \tau \in S_k$ we have*

$$(-1)^{\sigma\tau} = (-1)^\sigma(-1)^\tau.$$

That is, the sign defines a group homomorphism $S_k \to \{\pm 1\}$.

Proof. By definition,

$$(-1)^{\sigma\tau} = \prod_{i<j} \frac{x_{\sigma\tau(i)} - x_{\sigma\tau(j)}}{x_i - x_j}.$$

We write the right-hand side as a product of

$$\prod_{i<j} \frac{x_{\tau(i)} - x_{\tau(j)}}{x_i - x_j} = (-1)^\tau$$

and

$$\prod_{i<j} \frac{x_{\sigma\tau(i)} - x_{\sigma\tau(j)}}{x_{\tau(i)} - x_{\tau(j)}}. \tag{1.4.10}$$

For $i < j$, let $p = \tau(i)$ and $q = \tau(j)$ when $\tau(i) < \tau(j)$ and let $p = \tau(j)$ and $q = \tau(i)$ when $\tau(j) < \tau(i)$. Then

$$\frac{x_{\sigma\tau(i)} - x_{\sigma\tau(j)}}{x_{\tau(i)} - x_{\tau(j)}} = \frac{x_{\sigma(p)} - x_{\sigma(q)}}{x_p - x_q}$$

(i.e., if $\tau(i) < \tau(j)$, the numerator and denominator on the right *equal* the numerator and denominator on the left and, if $\tau(j) < \tau(i)$ are *negatives* of the numerator and denominator on the left). Thus equation (1.4.10) becomes

$$\prod_{p<q} \frac{x_{\sigma(p)} - x_{\sigma(q)}}{x_p - x_q} = (-1)^\sigma. \qquad \square$$

We'll leave for you to check that if τ is a transposition, $(-1)^\tau = -1$ and to deduce the following proposition.

Proposition 1.4.11. *If σ is the product of an odd number of transpositions, $(-1)^\sigma = -1$ and if σ is the product of an even number of transpositions, $(-1)^\sigma = +1$.*

Alternation

Definition 1.4.12. Let V be an n-dimensional vector space and $T \in \mathcal{L}^k(v)$ a k-tensor. For $\sigma \in S_k$, define $T^\sigma \in \mathcal{L}^k(V)$ to be the k-tensor

$$T^\sigma(v_1, \ldots, v_k) := T(v_{\sigma^{-1}(1)}, \ldots, v_{\sigma^{-1}(k)}). \tag{1.4.13}$$

Proposition 1.4.14.

(1) *If $T = \ell_1 \otimes \cdots \otimes \ell_k$, $\ell_i \in V^\star$, then $T^\sigma = \ell_{\sigma(1)} \otimes \cdots \otimes \ell_{\sigma(k)}$.*
(2) *The assignment $T \mapsto T^\sigma$ is a linear map $\mathcal{L}^k(V) \to \mathcal{L}^k(V)$.*
(3) *If $\sigma, \tau \in S_k$, we have $T^{\sigma\tau} = (T^\sigma)^\tau$.*

Proof. To (1), we note that by equation (1.4.13)

$$(\ell_1 \otimes \cdots \otimes \ell_k)^\sigma(v_1, \ldots, v_k) = \ell_1(v_{\sigma^{-1}(1)}) \cdots \ell_k(v_{\sigma^{-1}(k)}).$$

Setting $\sigma^{-1}(i) = q$, the ith term in this product is $\ell_{\sigma(q)}(v_q)$; so the product can be rewritten as

$$\ell_{\sigma(1)}(v_1) \ldots \ell_{\sigma(k)}(v_k)$$

or

$$(\ell_{\sigma(1)} \otimes \cdots \otimes \ell_{\sigma(k)})(v_1, \ldots, v_k).$$

We leave the proof of (2) as an exercise.

Now we prove (3). Let $T = \ell_1 \otimes \cdots \otimes \ell_k$. Then

$$T^\sigma = \ell_{\sigma(1)} \otimes \cdots \otimes \ell_{\sigma(k)} = \ell'_1 \otimes \cdots \otimes \ell'_k,$$

where $\ell'_j = \ell_{\sigma(j)}$. Thus

$$(T^\sigma)^\tau = \ell'_{\tau(1)} \otimes \cdots \otimes \ell'_{\tau(k)}.$$

But if $\tau(i) = j$, $\ell'_{\tau(j)} = \ell_{\sigma(\tau(j))}$. Hence

$$(T^\sigma)^\tau \ell_{\sigma\tau(1)} \otimes \cdots \otimes \ell_{\sigma\tau(k)} = T^{\sigma\tau}.$$

□

Definition 1.4.15. Let V be a vector space and $k \geq 0$ an integer. A k-tensor $T \in \mathcal{L}^k(V)$ is ***alternating*** if $T^\sigma = (-1)^\sigma T$ for all $\sigma \in S_k$.

We denote by $\mathcal{A}^k(V)$ the set of all alternating k-tensors in $\mathcal{L}^k(V)$. By (2) this set is a vector subspace of $\mathcal{L}^k(V)$.

It is not easy to write down simple examples of alternating k-tensors. However, there is a method, called the *alternation operation* Alt for constructing such tensors.

Definition 1.4.16. Let V be a vector space and k a non-negative integer. The ***alternation operation*** on $\mathcal{L}^k(V)$ is defined as follows: given $T \in \mathcal{L}^k(V)$ let

$$\mathrm{Alt}(T) := \sum_{\tau \in S_k} (-1)^\tau T^\tau.$$

The alternation operation enjoys the following properties.

Proposition 1.4.17. *For $T \in \mathcal{L}^k(V)$ and $\sigma \in S_k$,*

(1) $\mathrm{Alt}(T)^\sigma = (-1)^\sigma \,\mathrm{Alt}\, T$;
(2) *if $T \in \mathcal{A}^k(V)$, then* $\mathrm{Alt}\, T = k!\, T$;
(3) $\mathrm{Alt}(T^\sigma) = \mathrm{Alt}(T)^\sigma$;
(4) *the map*

$$\mathrm{Alt}\colon \mathcal{L}^k(V) \to \mathcal{L}^k(V), \quad T \mapsto \mathrm{Alt}(T)$$

is linear.

Proof. To prove (1) we note that by Proposition 1.4.14:

$$\mathrm{Alt}(T)^\sigma = \sum_{\tau \in S_k} (-1)^\tau T^{\tau\sigma} = (-1)^\sigma \sum_{\tau \in S_k} (-1)^{\tau\sigma} T^{\tau\sigma}.$$

But as τ runs over S_k, $\tau\sigma$ runs over S_k, and hence the right-hand side is $(-1)^\sigma \,\mathrm{Alt}(T)$.

To prove (2), note that if $T \in \mathcal{A}^k(V)$

$$\mathrm{Alt}\, T = \sum_{\tau \in S_k} (-1)^\tau T^\tau = \sum_{\tau \in S_k} (-1)^\tau (-1)^\tau T = k!\, T.$$

To prove (3), we compute:

$$\mathrm{Alt}(T^\sigma) = \sum_{\tau \in S_k} (-1)^\tau T^{\tau\sigma} = (-1)^\sigma \sum_{\tau \in S_k} (-1)^{\tau\sigma} T^{\tau\sigma}$$

$$= (-1)^\sigma \,\mathrm{Alt}(T) = \mathrm{Alt}(T)^\sigma.$$

Finally, (4) is an easy corollary of (2) of Proposition 1.4.14. □

We will use this alternation operation to construct a basis for $\mathcal{A}^k(V)$. First, however, we require some notation.

Definition 1.4.18. Let $I = (i_1, \dots, i_k)$ be a multi-index of length k.

(1) I is ***repeating*** if $i_r = i_s$ for some $r \neq s$.

(2) I is ***strictly increasing*** if $i_1 < i_2 < \cdots < i_r$.

(3) For $\sigma \in S_k$, write $I^\sigma := (i_{\sigma(1)}, \dots, i_{\sigma(k)})$.

Remark 1.4.19. If I is non-repeating there is a unique $\sigma \in S_k$ so that I^σ is strictly increasing.

Let $e_1, \dots, e_n$ be a basis of V and let

$$e_I^\star = e_{i_1}^\star \otimes \cdots \otimes e_{i_k}^\star$$

and

$$\psi_I = \operatorname{Alt}(e_I^\star).$$

Proposition 1.4.20.

(1) $\psi_{I^\sigma} = (-1)^\sigma \psi_I$.

(2) *If I is repeating, $\psi_I = 0$.*

(3) *If I and J are strictly increasing,*

$$\psi_I(e_{j_1}, \dots, e_{j_k}) = \begin{cases} 1, & I = J, \\ 0, & I \neq J. \end{cases}$$

Proof. To prove (1) we note that $(e_I^\star)^\sigma = e_{I^\sigma}^\star$; so

$$\operatorname{Alt}(e_{I^\sigma}^\star) = \operatorname{Alt}(e_I^\star)^\sigma = (-1)^\sigma \operatorname{Alt}(e_I^\star).$$

To prove (2), suppose $I = (i_1, \dots, i_k)$ with $i_r = i_s$ for $r \neq s$. Then if $\tau = \tau_{i_r, i_s}$, $e_I^\star = e_{I^\tau}^\star$ so

$$\psi_I = \psi_{I^\tau} = (-1)^\tau \psi_I = -\psi_I.$$

To prove (3), note that by definition

$$\psi_I(e_{j_1}, \dots, e_{j_k}) = \sum_\tau (-1)^\tau e_{I^\tau}^\star(e_{j_1}, \dots, e_{j_k}).$$

But by (1.3.12)

$$e_{I^\tau}^\star(e_{j_1}, \dots, e_{j_k}) = \begin{cases} 1, & I^\tau = J, \\ 0, & I^\tau \neq J. \end{cases} \tag{1.4.21}$$

Thus if I and J are strictly increasing, I^τ is strictly increasing if and only if $I^\tau = I$, and (1.4.21) is non-zero if and only if $I = J$. □

Now let T be in $\mathcal{A}^k(V)$. By Theorem 1.3.13,

$$T = \sum_J a_J e_J^\star, \quad a_J \in \mathbf{R}.$$

Since $k!\,T = \operatorname{Alt}(T)$,

$$T = \frac{1}{k!} \sum a_J \operatorname{Alt}(e_J^\star) = \sum b_J \psi_J.$$

We can discard all repeating terms in this sum since they are zero; and for every non-repeating term, J, we can write $J = I^\sigma$, where I is strictly increasing, and hence $\psi_J = (-1)^\sigma \psi_I$.

Conclusion 1.4.22. *We can write T as a sum*

$$T = \sum_I c_I \psi_I \tag{1.4.23}$$

with I's strictly increasing.

Claim 1.4.24. *The c_I's are unique.*

Proof. For J strictly increasing

$$T(e_{j_1}, \dots, e_{j_k}) = \sum_I c_I \psi_I(e_{j_1}, \dots, e_{j_k}) = c_J. \tag{1.4.25}$$

By (1.4.23) the ψ_I's, I strictly increasing, are a spanning set of vectors for $\mathcal{A}^k(V)$, and by (1.4.25) they are linearly independent, so we have proved.

Proposition 1.4.26. *The alternating tensors, ψ_I, I strictly increasing, are a basis for $\mathcal{A}^k(V)$.*

Thus $\dim \mathcal{A}^k(V)$ is equal to the number of strictly increasing multi-indices I of length k. We leave for you as an exercise to show that this number is equal to the binomial coefficient

$$\binom{n}{k} := \frac{n!}{(n-k)!\,k!} \tag{1.4.27}$$

if $1 \le k \le n$. □

Hint: Show that every strictly increasing multi-index of length k determines a k element subset of $\{1, \dots, n\}$ and vice versa.

Note also that if $k > n$ every multi-index

$$I = (i_1, \dots, i_k)$$

of length k has to be repeating: $i_r = i_s$ for some $r \neq s$ since the i_p's lie on the interval $1 \le i \le n$. Thus by Proposition 1.4.17

$$\psi_I = 0$$

for all multi-indices of length $k > 0$ and

$$\mathcal{A}^k(V) = 0.$$

Exercises for §1.4

Exercise 1.4.i. Show that there are exactly $k!$ permutations of order k.

Hint: Induction on k: Let $\sigma \in S_k$, and let $\sigma(k) = i$, $1 \le i \le k$. Show that $\tau_{i,k}\sigma$ leaves k fixed and hence is, in effect, a permutation of Σ_{k-1}.

Exercise 1.4.ii. Prove that if $\tau \in S_k$ is a transposition, $(-1)^\tau = -1$ and deduce from this Proposition 1.4.11.

Exercise 1.4.iii. Prove assertion (2) in Proposition 1.4.14.

Exercise 1.4.iv. Prove that $\dim \mathcal{A}^k(V)$ is given by (1.4.27).

Exercise 1.4.v. Verify that for $i < j - 1$

$$\tau_{i,j} = \tau_{j-1,j}\tau_{i,j-1}, \tau_{j-1,j}.$$

Exercise 1.4.vi. For $k = 3$ show that every one of the six elements of S_3 is either a transposition or can be written as a product of two transpositions.

Exercise 1.4.vii. Let $\sigma \in S_k$ be the "cyclic" permutation

$$\sigma(i) := i + 1, \quad i = 1, \dots, k - 1$$

and $\sigma(k) := 1$. Show explicitly how to write σ as a product of transpositions and compute $(-1)^\sigma$.

Hint: Same hint as in Exercise 1.4.i.

Exercise 1.4.viii. In Exercise 1.3.vii show that if T is in $\mathcal{A}^k(V)$, T_v is in $\mathcal{A}^{k-1}(V)$. Show in addition that for $v, w \in V$ and $T \in \mathcal{A}^k(V)$ we have $(T_v)_w = -(T_w)_v$.

Exercise 1.4.ix. Let $A\colon V \to W$ be a linear mapping. Show that if T is in $\mathcal{A}^k(W)$, $A^\star T$ is in $\mathcal{A}^k(V)$.

Exercise 1.4.x. In Exercise 1.4.ix show that if T is in $\mathcal{L}^k(W)$ then $\mathrm{Alt}(A^\star T) = A^\star(\mathrm{Alt}(T))$, i.e., show that the Alt operation commutes with the pullback operation.

1.5. The space $\Lambda^k(V^\star)$

In § 1.4 we showed that the image of the alternation operation, $\mathrm{Alt}\colon \mathcal{L}^k(V) \to \mathcal{L}^k(V)$, is $\mathcal{A}^k(V)$. In this section we will compute the kernel of Alt.

Definition 1.5.1. A decomposable k-tensor $\ell_1 \otimes \cdots \otimes \ell_k$, with $\ell_1, \dots, \ell_k \in V^\star$, is ***redundant*** if for some index i we have $\ell_i = \ell_{i+1}$.

Let $\mathcal{I}^k(V) \subset \mathcal{L}^k(V)$ be the linear span of the set of redundant k-tensors.

Note that for $k = 1$ the notion of redundant does not really make sense; a single vector $\ell \in \mathcal{L}^1(V^\star)$ cannot be "redundant" so we decree

$$\mathcal{I}^1(V) := 0.$$

Proposition 1.5.2. *If* $T \in \mathcal{I}^k(V)$ *then* $\mathrm{Alt}(T) = 0$.

Proof. Let $T = \ell_1 \otimes \cdots \otimes \ell_k$ with $\ell_i = \ell_{i+1}$. Then if $\tau = \tau_{i,i+1}$, $T^\tau = T$ and $(-1)^\tau = -1$. Hence $\mathrm{Alt}(T) = \mathrm{Alt}(T^\tau) = \mathrm{Alt}(T)^\tau = -\mathrm{Alt}(T)$; so $\mathrm{Alt}(T) = 0$. □

Proposition 1.5.3. *If* $T \in \mathcal{I}^r(V)$ *and* $T' \in \mathcal{L}^s(V)$, *then* $T \otimes T'$ *and* $T' \otimes T$ *are in* $\mathcal{I}^{r+s}(V)$.

Proof. We can assume that T and T' are decomposable, i.e., $T = \ell_1 \otimes \cdots \otimes \ell_r$ and $T' = \ell'_1 \otimes \cdots \otimes \ell'_s$ and that T is redundant: $\ell_i = \ell_{i+1}$. Then

$$T \otimes T' = \ell_1 \otimes \cdots \ell_{i-1} \otimes \ell_i \otimes \ell_i \otimes \cdots \ell_r \otimes \ell'_1 \otimes \cdots \otimes \ell'_s$$

is redundant and hence in $\mathcal{I}^{r+s}$. The argument for $T' \otimes T$ is similar. □

Proposition 1.5.4. *If $T \in \mathcal{L}^k(V)$ and $\sigma \in S_k$, then*

$$T^\sigma = (-1)^\sigma T + S, \tag{1.5.5}$$

where S is in $\mathcal{I}^k(V)$.

Proof. We can assume T is decomposable, i.e., $T = \ell_1 \otimes \cdots \otimes \ell_k$. Let's first look at the simplest possible case: $k = 2$ and $\sigma = \tau_{1,2}$. Then

$$\begin{aligned} T^\sigma - (-1)^\sigma T &= \ell_1 \otimes \ell_2 + \ell_2 \otimes \ell_1 \\ &= \frac{1}{2}((\ell_1 + \ell_2) \otimes (\ell_1 + \ell_2) - \ell_1 \otimes \ell_1 - \ell_2 \otimes \ell_2), \end{aligned}$$

and the terms on the right are redundant, and hence in $\mathcal{I}^2(V)$. Next let k be arbitrary and $\sigma = \tau_{i,i+1}$. If $T_1 = \ell_1 \otimes \cdots \otimes \ell_{i-2}$ and $T_2 = \ell_{i+2} \otimes \cdots \otimes \ell_k$, then

$$T - (-1)^\sigma T = T_1 \otimes (\ell_i \otimes \ell_{i+1} + \ell_{i+1} \otimes \ell_i) \otimes T_2$$

is in $\mathcal{I}^k(V)$ by Proposition 1.5.3 and the computation above.
The general case: By Theorem 1.4.5, σ can be written as a product of m elementary transpositions, and we'll prove (1.5.5) by induction on m.

We have just dealt with the case $m = 1$.
The induction step: the "$m - 1$" case implies the "m" case. Let $\sigma = \tau\beta$ where β is a product of $m - 1$ elementary transpositions and τ is an elementary transposition. Then

$$\begin{aligned} T^\sigma = (T^\beta)^\tau &= (-1)^\tau T^\beta + \cdots \\ &= (-1)^\tau (-1)^\beta T + \cdots \\ &= (-1)^\sigma T + \cdots, \end{aligned}$$

where the "dots" are elements of $\mathcal{I}^k(V)$, and the induction hypothesis was used in the second line. □

Corollary 1.5.6. *If $T \in \mathcal{L}^k(V)$, then*

$$\operatorname{Alt}(T) = k!\, T + W, \tag{1.5.7}$$

where W is in $\mathcal{I}^k(V)$.

Proof. By definition $\operatorname{Alt}(T) = \sum_{\sigma \in S_k} (-1)^\sigma T^\sigma$, and by Proposition 1.5.4, $T^\sigma = (-1)^\sigma T + W_\sigma$, with $W_\sigma \in \mathcal{I}^k(V)$. Thus

$$\operatorname{Alt}(T) = \sum_{\sigma \in S_k} (-1)^\sigma (-1)^\sigma T + \sum_{\sigma \in S_k} (-1)^\sigma W_\sigma = k!\, T + W,$$

where $W = \sum_{\sigma \in S_k} (-1)^\sigma W_\sigma$. □

Corollary 1.5.8. *Let V be a vector space and $k \geq 1$. Then*

$$\mathcal{I}^k(V) = \ker(\mathrm{Alt}\colon \mathcal{L}^k(V) \to \mathcal{A}^k(V)).$$

Proof. We have already proved that if $T \in \mathcal{I}^k(V)$, then $\mathrm{Alt}(T) = 0$. To prove the converse assertion we note that if $\mathrm{Alt}(T) = 0$, then by (1.5.7)

$$T = -\frac{1}{k!}W$$

with $W \in \mathcal{I}^k(V)$. □

Putting these results together we conclude the following.

Theorem 1.5.9. *Every element $T \in \mathcal{L}^k(V)$ can be written uniquely as a sum $T = T_1 + T_2$, where $T_1 \in \mathcal{A}^k(V)$ and $T_2 \in \mathcal{I}^k(V)$.*

Proof. By (1.5.7), $T = T_1 + T_2$ with

$$T_1 = \frac{1}{k!}\,\mathrm{Alt}(T)$$

and

$$T_2 = -\frac{1}{k!}W.$$

To prove that this decomposition is unique, suppose $T_1 + T_2 = 0$, with $T_1 \in \mathcal{A}^k(V)$ and $T_2 \in \mathcal{I}^k(V)$. Then

$$0 = \mathrm{Alt}(T_1 + T_2) = k!\,T_1$$

so $T_1 = 0$, and hence $T_2 = 0$. □

Definition 1.5.10. Let V be a finite-dimensional vector space and $k \geq 0$. Define

(1.5.11) $$\Lambda^k(V^\star) := \mathcal{L}^k(V)/\mathcal{I}^k(V),$$

i.e., let $\Lambda^k(V^\star)$ be the quotient of the vector space $\mathcal{L}^k(V)$ by the subspace $\mathcal{I}^k(V)$. By (1.2.7) one has a linear map:

(1.5.12) $$\pi\colon \mathcal{L}^k(V) \to \Lambda^k(V^\star), \quad T \mapsto T + \mathcal{I}^k(V),$$

which is onto and has $\mathcal{I}^k(V)$ as kernel.

Theorem 1.5.13. *The map $\pi\colon \mathcal{L}^k(V) \to \Lambda^k(V^\star)$ maps $\mathcal{A}^k(V)$ bijectively onto $\Lambda^k(V^\star)$.*

Proof. By Theorem 1.5.9 every $\mathcal{I}^k(V)$ coset $T + \mathcal{I}^k(V)$ contains a unique element T_1 of $\mathcal{A}^k(V)$. Hence for every element of $\Lambda^k(V^\star)$ there is a unique element of $\mathcal{A}^k(V)$ which gets mapped onto it by π. □

Remark 1.5.14. Since $\Lambda^k(V^\star)$ and $\mathcal{A}^k(V)$ are isomorphic as vector spaces many treatments of multilinear algebra avoid mentioning $\Lambda^k(V^\star)$, reasoning that $\mathcal{A}^k(V)$ is a perfectly good substitute for it and that one should, if possible, not make two different definitions for what is essentially the same object. This is a justifiable point of view (and is the point of view taken by in [4, 10, 12]). There are, however, some advantages to distinguishing between $\mathcal{A}^k(V)$ and $\Lambda^k(V^\star)$, as we shall see in §1.6.

Exercises for §1.5

Exercise 1.5.i. A k-tensor $T \in \mathcal{L}^k(V)$ is *symmetric* if $T^\sigma = T$ for all $\sigma \in S_k$. Show that the set $\mathcal{S}^k(V)$ of symmetric k tensors is a vector subspace of $\mathcal{L}^k(V)$.

Exercise 1.5.ii. Let $e_1, \ldots, e_n$ be a basis of V. Show that every symmetric 2-tensor is of the form

$$\sum_{1 \le i,j \le n} a_{i,j} e_i^\star \otimes e_j^\star,$$

where $a_{i,j} = a_{j,i}$ and $e_1^\star, \ldots, e_n^\star$ are the dual basis vectors of $V^\star$.

Exercise 1.5.iii. Show that if T is a symmetric k-tensor, then, for $k \ge 2$, T is in $\mathcal{I}^k(V)$.

Hint: Let σ be a transposition and deduce from the identity, $T^\sigma = T$, that T has to be in the kernel of Alt.

Exercise 1.5.iv (a warning). In general $\mathcal{S}^k(V) \neq \mathcal{I}^k(V)$. Show, however, that if $k = 2$ these two spaces are equal.

Exercise 1.5.v. Show that if $\ell \in V^\star$ and $T \in \mathcal{L}^{k-2}(V)$, then $\ell \otimes T \otimes \ell$ is in $\mathcal{I}^k(V)$.

Exercise 1.5.vi. Show that if ℓ_1 and ℓ_2 are in $V^\star$ and $T \in \mathcal{L}^{k-2}(V)$, then

$$\ell_1 \otimes T \otimes \ell_2 + \ell_2 \otimes T \otimes \ell_1$$

is in $\mathcal{I}^k(V)$.

Exercise 1.5.vii. Given a permutation $\sigma \in S_k$ and $T \in \mathcal{I}^k(V)$, show that $T^\sigma \in \mathcal{I}^k(V)$.

Exercise 1.5.viii. Let $\mathcal{W}(V)$ be a subspace of $\mathcal{L}^k(V)$ having the following two properties.

(1) For $S \in \mathcal{S}^2(V)$ and $T \in \mathcal{L}^{k-2}(V)$, $S \otimes T$ is in $\mathcal{W}(V)$.
(2) For T in $\mathcal{W}(V)$ and $\sigma \in S_k$, T^σ is in $\mathcal{W}(V)$.

Show that $\mathcal{W}(V)$ has to contain $\mathcal{I}^k(V)$ and conclude that $\mathcal{I}^k(V)$ is the smallest subspace of $\mathcal{L}^k(V)$ having properties (1) and (2).

Exercise 1.5.ix. Show that there is a bijective linear map

$$\alpha \colon \Lambda^k(V^\star) \xrightarrow{\sim} \mathcal{A}^k(V)$$

with the property

$$\alpha\pi(T) = \frac{1}{k!}\,\mathrm{Alt}(T) \tag{1.5.15}$$

for all $T \in \mathcal{L}^k(V)$, and show that α is the inverse of the map of $\mathcal{A}^k(V)$ onto $\Lambda^k(V^\star)$ described in Theorem 1.5.13.

Hint: Exercise 1.2.viii.

Exercise 1.5.x. Let V be an n-dimensional vector space. Compute the dimension of $\mathcal{S}^k(V)$.

Hints:

(1) Introduce the following symmetrization operation on tensors $T \in \mathcal{L}^k(V)$:

$$\mathrm{Sym}(T) = \sum_{\tau \in S_k} T^\tau.$$

Prove that this operation has properties (2)–(4) of Proposition 1.4.17 and, as a substitute for (1), has the property: $\mathrm{Sym}(T)^\sigma = \mathrm{Sym}(T)$.

(2) Let $\phi_I = \mathrm{Sym}(e_I^\star)$, $e_I^\star = e_{i_1}^\star \otimes \cdots \otimes e_{i_n}^\star$. Prove that $\{\phi_I \mid I \text{ is non-decreasing}\}$ form a basis of $S^k(V)$.

(3) Conclude that $\dim(\mathcal{S}^k(V))$ is equal to the number of non-decreasing multi-indices of length k: $1 \le i_1 \le i_2 \le \cdots \le \ell_k \le n$.

(4) Compute this number by noticing that the assignment

$$(i_1, \ldots, i_k) \mapsto (i_1 + 0, i_2 + 1, \ldots, i_k + k - 1)$$

is a bijection between the set of these non-decreasing multi-indices and the set of increasing multi-indices $1 \le j_1 < \cdots < j_k \le n + k - 1$.

1.6. The wedge product

The tensor algebra operations on the spaces $\mathcal{L}^k(V)$ which we discussed in §§ 1.2 and 1.3, i.e., the "tensor product operation" and the "pullback" operation, give rise to similar operations on the spaces, $\Lambda^k(V^\star)$. We will discuss in this section the analogue of the tensor product operation.

Definition 1.6.1. Given $\omega_i \in \Lambda^{k_i}(V^\star)$, $i = 1, 2$ we can, by equation (1.5.12), find a $T_i \in \mathcal{L}^{k_i}(V)$ with $\omega_i = \pi(T_i)$. Then $T_1 \otimes T_2 \in \mathcal{L}^{k_1+k_2}(V)$. The ***wedge product*** $\omega_1 \wedge \omega_2$ is defined by

$$\omega_1 \wedge \omega_2 := \pi(T_1 \otimes T_2) \in \Lambda^{k_1+k_2}(V^\star). \tag{1.6.2}$$

Claim 1.6.3. *This wedge product is well defined, i.e., does not depend on our choices of T_1 and T_2.*

Proof. Let $\pi(T_1) = \pi(T_1') = \omega_1$. Then $T_1' = T_1 + W_1$ for some $W_1 \in \mathcal{I}^{k_1}(V)$, so

$$T_1' \otimes T_2 = T_1 \otimes T_2 + W_1 \otimes T_2.$$

But $W_1 \in \mathcal{I}^{k_1}(V)$ implies $W_1 \otimes T_2 \in \mathcal{I}^{k_1+k_2}(V)$ and this implies:

$$\pi(T_1' \otimes T_2) = \pi(T_1 \otimes T_2).$$

A symmetric argument shows that $\omega_1 \wedge \omega_2$ is well-defined, independent of the choice of T_2. □

More generally let $\omega_i \in \Lambda^{k_i}(V^\star)$, for $i = 1, 2, 3$, and let $\omega_i = \pi(T_i)$, $T_i \in \mathcal{L}^{k_i}(V)$. Define

$$\omega_1 \wedge \omega_2 \wedge \omega_3 \in \Lambda^{k_1+k_2+k_3}(V^\star)$$

by setting

$$\omega_1 \wedge \omega_2 \wedge \omega_3 = \pi(T_1 \otimes T_2 \otimes T_3).$$

As above it is easy to see that this is well-defined independent of the choice of T_1, T_2 and T_3. It is also easy to see that this triple wedge product is just the wedge product of $\omega_1 \wedge \omega_2$ with ω_3 or, alternatively, the wedge product of ω_1 with $\omega_2 \wedge \omega_3$, i.e.,

$$\omega_1 \wedge \omega_2 \wedge \omega_3 = (\omega_1 \wedge \omega_2) \wedge \omega_3 = \omega_1 \wedge (\omega_2 \wedge \omega_3).$$

We leave for you to check: for $\lambda \in \mathbf{R}$

$$\lambda(\omega_1 \wedge \omega_2) = (\lambda\omega_1) \wedge \omega_2 = \omega_1 \wedge (\lambda\omega_2) \tag{1.6.4}$$

and verify the two distributive laws:

$$(\omega_1 + \omega_2) \wedge \omega_3 = \omega_1 \wedge \omega_3 + \omega_2 \wedge \omega_3 \tag{1.6.5}$$

and

$$\omega_1 \wedge (\omega_2 + \omega_3) = \omega_1 \wedge \omega_2 + \omega_1 \wedge \omega_3. \tag{1.6.6}$$

As we noted in §1.4, $\mathcal{I}^k(V) = 0$ for $k = 1$, i.e., there are no non-zero "redundant" k tensors in degree $k = 1$. Thus

$$\Lambda^1(V^\star) = V^\star = \mathcal{L}^1(V).$$

A particularly interesting example of a wedge product is the following. Let $\ell_1, \ldots, \ell_k \in V^\star = \Lambda^1(V^\star)$. If $T = \ell_1 \otimes \cdots \otimes \ell_k$, then

$$\ell_1 \wedge \cdots \wedge \ell_k = \pi(T) \in \Lambda^k(V^\star). \tag{1.6.7}$$

We will call (1.6.7) a ***decomposable element*** of $\Lambda^k(V^\star)$.

We will prove that these elements satisfy the following wedge product identity. For $\sigma \in S_k$:

$$\ell_{\sigma(1)} \wedge \cdots \wedge \ell_{\sigma(k)} = (-1)^\sigma \ell_1 \wedge \cdots \wedge \ell_k. \tag{1.6.8}$$

Proof. For every $T \in \mathcal{L}^k(V)$, $T = (-1)^\sigma T + W$ for some $W \in \mathcal{I}^k(V)$ by Proposition 1.5.4. Therefore since $\pi(W) = 0$

$$\pi(T^\sigma) = (-1)^\sigma \pi(T).$$

In particular, if $T = \ell_1 \otimes \cdots \otimes \ell_k$, $T^\sigma = \ell_{\sigma(1)} \otimes \cdots \otimes \ell_{\sigma(k)}$, so

$$\begin{aligned}\pi(T^\sigma) &= \ell_{\sigma(1)} \wedge \cdots \wedge \ell_{\sigma(k)} = (-1)^\sigma \pi(T) \\ &= (-1)^\sigma \ell_1 \wedge \cdots \wedge \ell_k.\end{aligned}$$

□

In particular, for ℓ_1 and $\ell_2 \in V^\star$

$$\ell_1 \wedge \ell_2 = -\ell_2 \wedge \ell_1 \tag{1.6.9}$$

and for ℓ_1, ℓ_2 and $\ell_3 \in V^\star$

$$\ell_1 \wedge \ell_2 \wedge \ell_3 = -\ell_2 \wedge \ell_1 \wedge \ell_3 = \ell_2 \wedge \ell_3 \wedge \ell_1.$$

More generally, it is easy to deduce from equation (1.6.8) the following result (which we'll leave as an exercise).

Theorem 1.6.10. *If $\omega_1 \in \Lambda^r(V^\star)$ and $\omega_2 \in \Lambda^s(V^\star)$, then*

$$\omega_1 \wedge \omega_2 = (-1)^{rs} \omega_2 \wedge \omega_1. \tag{1.6.11}$$

Hint: It suffices to prove this for decomposable elements i.e., for $\omega_1 = \ell_1 \wedge \cdots \wedge \ell_r$ and $\omega_2 = \ell_1' \wedge \cdots \wedge \ell_s'$. Now make rs applications of (1.6.9).

Let $e_1, \ldots, e_n$ be a basis of V and let $e_1^\star, \ldots, e_n^\star$ be the dual basis of $V^\star$. For every multi-index I of length k,

$$e_{i_1}^\star \wedge \cdots \wedge e_{i_k}^\star = \pi(e_I^\star) = \pi(e_{i_1}^\star \otimes \cdots \otimes e_{i_k}^\star). \tag{1.6.12}$$

Theorem 1.6.13. *The elements* (1.6.12), *with I strictly increasing, are basis vectors of $\Lambda^k(V^\star)$.*

Proof. The elements $\psi_I = \mathrm{Alt}(e_I^\star)$, for I strictly increasing, are basis vectors of $\mathcal{A}^k(V)$ by Proposition 1.4.26; so their images, $\pi(\psi_I)$, are a basis of $\Lambda^k(V^\star)$. But

$$\begin{aligned}\pi(\psi_I) &= \pi\left(\sum_{\sigma\in S_k}(-1)^\sigma (e_I^\star)^\sigma\right) = \sum_{\sigma\in S_k}(-1)^\sigma \pi(e_I^\star)^\sigma \\ &= \sum_{\sigma\in S_k}(-1)^\sigma(-1)^\sigma \pi(e_I^\star) = k!\,\pi(e_I^\star).\end{aligned}$$ □

Exercises for §1.6

Exercise 1.6.i. Prove the assertions (1.6.4), (1.6.5), and (1.6.6).

Exercise 1.6.ii. Verify the multiplication law in equation (1.6.11) for the wedge product.

Exercise 1.6.iii. Given $\omega \in \Lambda^r(V^\star)$ let ω^k be the k-fold wedge product of ω with itself, i.e., let $\omega^2 = \omega \wedge \omega$, $\omega^3 = \omega \wedge \omega \wedge \omega$, etc.

(1) Show that if r is odd then, for $k > 1$, $\omega^k = 0$.
(2) Show that if ω is decomposable, then, for $k > 1$, $\omega^k = 0$.

Exercise 1.6.iv. If ω and μ are in $\Lambda^r(V^\star)$ prove:

$$(\omega + \mu)^k = \sum_{\ell=0}^{k} \binom{k}{\ell} \omega^\ell \wedge \mu^{k-\ell}.$$

Hint: As in freshman calculus, prove this binomial theorem by induction using the identity: $\binom{k}{\ell} = \binom{k-1}{\ell-1} + \binom{k-1}{\ell}$.

Exercise 1.6.v. Let ω be an element of $\Lambda^2(V^\star)$. By definition the ***rank*** of ω is k if $\omega^k \neq 0$ and $\omega^{k+1} = 0$. Show that if

$$\omega = e_1 \wedge f_1 + \cdots + e_k \wedge f_k$$

with $e_i, f_i \in V^\star$, then ω is of rank $\leq k$.

Hint: Show that

$$\omega^k = k!\, e_1 \wedge f_1 \wedge \cdots \wedge e_k \wedge f_k.$$

Exercise 1.6.vi. Given $e_i \in V^\star$, $i = 1, \ldots, k$ show that $e_1 \wedge \cdots \wedge e_k \neq 0$ if and only if the e_i's are linearly independent.

Hint: Induction on k.

1.7. The interior product

We'll describe in this section another basic product operation on the spaces $\Lambda^k(V^\star)$. As above we'll begin by defining this operator on the $\mathcal{L}^k(V)$'s.

Definition 1.7.1. Let V be a vector space and k a non-negative integer. Given $T \in \mathcal{L}^k(V)$ and $v \in V$, let $\iota_v T$ be the $(k-1)$-tensor which takes the value

$$(\iota_v T)(v_1, \dots, v_{k-1}) := \sum_{r=1}^{k} (-1)^{r-1} T(v_1, \dots, v_{r-1}, v, v_r, \dots, v_{k-1})$$

on the $k-1$-tuple of vectors, $v_1, \dots, v_{k-1}$, i.e., in the rth summand on the right, v gets inserted between v_{r-1} and v_r. (In particular, the first summand is $T(v, v_1, \dots, v_{k-1})$ and the last summand is $(-1)^{k-1} T(v_1, \dots, v_{k-1}, v)$.)

It is clear from the definition that if $v = v_1 + v_2$

$$\iota_v T = \iota_{v_1} T + \iota_{v_2} T, \tag{1.7.2}$$

and if $T = T_1 + T_2$

$$\iota_v T = \iota_v T_1 + \iota_v T_2, \tag{1.7.3}$$

and we will leave for you to verify by inspection the following two lemmas.

Lemma 1.7.4. *If T is the decomposable k-tensor $\ell_1 \otimes \cdots \otimes \ell_k$, then*

$$\iota_v T = \sum_{r=1}^{k} (-1)^{r-1} \ell_r(v) \ell_1 \otimes \cdots \otimes \hat{\ell}_r \otimes \cdots \otimes \ell_k, \tag{1.7.5}$$

where the "hat" over ℓ_r means that ℓ_r is deleted from the tensor product.

Lemma 1.7.6. *If $T_1 \in \mathcal{L}^p(V)$ and $T_2 \in \mathcal{L}^q(V)$, then*

$$\iota_v(T_1 \otimes T_2) = \iota_v T_1 \otimes T_2 + (-1)^p T_1 \otimes \iota_v T_2. \tag{1.7.7}$$

We will next prove the important identity.

Lemma 1.7.8. *Let V be a vector space and $T \in \mathcal{L}^k(V)$. Then for all $v \in V$ we have*

$$\iota_v(\iota_v T) = 0. \tag{1.7.9}$$

Proof. It suffices by linearity to prove this for decomposable tensors and since (1.7.9) is trivially true for $T \in \mathcal{L}^1(V)$, we can by induction assume (1.7.9) is true for decomposable tensors of degree $k-1$. Let $\ell_1 \otimes \cdots \otimes \ell_k$ be a decomposable tensor of degree k. Setting $T := \ell_1 \otimes \cdots \otimes \ell_{k-1}$ and $\ell = \ell_k$ we have

$$\iota_v(\ell_1 \otimes \cdots \otimes \ell_k) = \iota_v(T \otimes \ell) = \iota_v T \otimes \ell + (-1)^{k-1} \ell(v) T$$

by (1.7.7). Hence

$$\iota_v(\iota_v(T \otimes \ell)) = \iota_v(\iota_v T) \otimes \ell + (-1)^{k-2} \ell(v) \iota_v T + (-1)^{k-1} \ell(v) \iota_v T.$$

But by induction the first summand on the right is zero and the two remaining summands cancel each other out. ☐

From (1.7.9) we can deduce a slightly stronger result: For $v_1, v_2 \in V$

$$\iota_{v_1}\iota_{v_2} = -\iota_{v_2}\iota_{v_1}. \tag{1.7.10}$$

Proof. Let $v = v_1 + v_2$. Then $\iota_v = \iota_{v_1} + \iota_{v_2}$ so

$$\begin{aligned} 0 = \iota_v\iota_v &= (\iota_{v_1} + \iota_{v_2})(\iota_{v_1} + \iota_{v_2}) \\ &= \iota_{v_1}\iota_{v_1} + \iota_{v_1}\iota_{v_2} + \iota_{v_2}\iota_{v_1} + \iota_{v_2}\iota_{v_2} \\ &= \iota_{v_1}\iota_{v_2} + \iota_{v_2}\iota_{v_1} \end{aligned}$$

since the first and last summands are zero by (1.7.9). □

We'll now show how to define the operation ι_v on $\Lambda^k(V^\star)$. We'll first prove the following lemma.

Lemma 1.7.11. *If $T \in \mathcal{L}^k(V)$ is redundant, then so is $\iota_v T$.*

Proof. Let $T = T_1 \otimes \ell \otimes \ell \otimes T_2$ where ℓ is in $V^\star$, T_1 is in $\mathcal{L}^p(V)$ and T_2 is in $\mathcal{L}^q(V)$. Then by equation (1.7.7)

$$\iota_v T = \iota_v T_1 \otimes \ell \otimes \ell \otimes T_2 + (-1)^p T_1 \otimes \iota_v(\ell \otimes \ell) \otimes T_2 + (-1)^{p+2} T_1 \otimes \ell \otimes \ell \otimes \iota_v T_2.$$

However, the first and the third terms on the right are redundant and

$$\iota_v(\ell \otimes \ell) = \ell(v)\ell - \ell(v)\ell$$

by equation (1.7.5). □

Now let π be the projection (1.5.12) of $\mathcal{L}^k(V)$ onto $\Lambda^k(V^\star)$ and for $\omega = \pi(T) \in \Lambda^k(V^\star)$ define

$$\iota_v\omega = \pi(\iota_v T). \tag{1.7.12}$$

To show that this definition is legitimate we note that if $\omega = \pi(T_1) = \pi(T_2)$, then $T_1 - T_2 \in \mathcal{I}^k(V)$, so by Lemma 1.7.11 $\iota_v T_1 - \iota_v T_2 \in \mathcal{I}^{k-1}$ and hence

$$\pi(\iota_v T_1) = \pi(\iota_v T_2).$$

Therefore, (1.7.12) does not depend on the choice of T.

By definition, ι_v is a linear map $\Lambda^k(V^\star) \to \Lambda^{k-1}(V^\star)$. We call this the ***interior product operation***. From the identities (1.7.2)–(1.7.12) one gets, for $v, v_1, v_2 \in V$, $\omega \in \Lambda^k(V^\star)$, $\omega_1 \in \Lambda^p(V^\star)$, and $\omega_2 \in \Lambda^2(V^\star)$

$$\iota_{(v_1+v_2)}\omega = \iota_{v_1}\omega + \iota_{v_2}\omega, \tag{1.7.13}$$

$$\iota_v(\omega_1 \wedge \omega_2) = \iota_v\omega_1 \wedge \omega_2 + (-1)^p \omega_1 \wedge \iota_v\omega_2, \tag{1.7.14}$$

$$\iota_v(\iota_v\omega) = 0 \tag{1.7.15}$$

and

$$\iota_{v_1}\iota_{v_2}\omega = -\iota_{v_2}\iota_{v_1}\omega.$$

Moreover if $\omega = \ell_1 \wedge \cdots \wedge \ell_k$ is a decomposable element of $\Lambda^k(V^\star)$ one gets from (1.7.5)

$$\iota_v \omega = \sum_{r=1}^{k} (-1)^{r-1} \ell_r(v) \ell_1 \wedge \cdots \wedge \hat{\ell}_r \wedge \cdots \wedge \ell_k.$$

In particular if $e_1, \ldots, e_n$ is a basis of V, $e_1^\star, \ldots, e_n^\star$ the dual basis of $V^\star$ and $\omega_I = e_{i_1}^\star \wedge \cdots \wedge e_{i_k}^\star$, $1 \le i_1 < \cdots < i_k \le n$, then $\iota_{e_j} \omega_I = 0$ if $j \notin I$ and if $j = i_r$

$$\iota_{e_j} \omega_I = (-1)^{r-1} \omega_{I_r}, \tag{1.7.16}$$

where $I_r = (i_1, \ldots, \hat{\imath}_r, \ldots, i_k)$ (i.e., I_r is obtained from the multi-index I by deleting i_r).

Exercises for §1.7

Exercise 1.7.i. Prove Lemma 1.7.4.

Exercise 1.7.ii. Prove Lemma 1.7.6.

Exercise 1.7.iii. Show that if $T \in \mathcal{A}^k$, $\iota_v T = k T_v$ where T_v is the tensor (1.3.21). In particular conclude that $\iota_v T \in \mathcal{A}^{k-1}(V)$. (See Exercise 1.4.viii.)

Exercise 1.7.iv. Assume the dimension of V is n and let Ω be a non-zero element of the one-dimensional vector space $\Lambda^n(V^\star)$. Show that the map

$$\rho \colon V \to \Lambda^{n-1}(V^\star), \quad v \mapsto \iota_v \Omega, \tag{1.7.17}$$

is a bijective linear map.

Hint: One can assume $\Omega = e_1^\star \wedge \cdots \wedge e_n^\star$ where $e_1, \ldots, e_n$ is a basis of V. Now use equation (1.7.16) to compute this map on basis elements.

Exercise 1.7.v (cross-product). Let V be a three-dimensional vector space, B an inner product on V and Ω a non-zero element of $\Lambda^3(V^\star)$. Define a map

$$- \times - \colon V \times V \to V$$

by setting

$$v_1 \times v_2 \coloneqq \rho^{-1}(Lv_1 \wedge Lv_2)$$

where ρ is the map (1.7.17) and $L \colon V \to V^\star$ the map (1.2.17). Show that this map is linear in v_1, with v_2 fixed and linear in v_2 with v_1 fixed, and show that $v_1 \times v_2 = -v_2 \times v_1$.

Exercise 1.7.vi. For $V = \mathbf{R}^3$ let e_1, e_2 and e_3 be the standard basis vectors and B the standard inner product. (See §1.1.) Show that if $\Omega = e_1^\star \wedge e_2^\star \wedge e_3^\star$ the cross-product above is the standard cross-product:

$$e_1 \times e_2 = e_3,$$

$$e_2 \times e_3 = e_1,$$

$$e_3 \times e_1 = e_2.$$

Hint: If B is the standard inner product, then $Le_i = e_i^\star$.

Remark 1.7.18. One can make this standard cross-product look even more standard by using the calculus notation: $e_1 = \hat{\imath}$, $e_2 = \hat{\jmath}$, and $e_3 = \widehat{k}$.

1.8. The pullback operation on $\Lambda^k(V^\star)$

Let V and W be vector spaces and let A be a linear map of V into W. Given a k-tensor $T \in \mathcal{L}^k(W)$, recall that the *pullback* $A^\star T$ is the k-tensor

$$(A^\star T)(v_1, \dots, v_k) = T(Av_1, \dots, Av_k)$$

in $\mathcal{L}^k(V)$. (See §1.3 and equation (1.3.17).) In this section we'll show how to define a similar pullback operation on $\Lambda^k(V^\star)$.

Lemma 1.8.1. *If* $T \in \mathcal{I}^k(W)$, *then* $A^\star T \in \mathcal{I}^k(V)$.

Proof. It suffices to verify this when T is a redundant k-tensor, i.e., a tensor of the form

$$T = \ell_1 \otimes \cdots \otimes \ell_k,$$

where $\ell_r \in W^\star$ and $\ell_i = \ell_{i+1}$ for some index, i. But by equation (1.3.19),

$$A^\star T = A^\star \ell_1 \otimes \cdots \otimes A^\star \ell_k$$

and the tensor on the right is redundant since $A^\star \ell_i = A^\star \ell_{i+1}$. □

Now let ω be an element of $\Lambda^k(W^\star)$ and let $\omega = \pi(T)$ where T is in $\mathcal{L}^k(W)$. We define

$$A^\star \omega = \pi(A^\star T). \tag{1.8.2}$$

Claim 1.8.3. *The left-hand side of equation* (1.8.2) *is well-defined.*

Proof. If $\omega = \pi(T) = \pi(T')$, then $T = T' + S$ for some $S \in \mathcal{I}^k(W)$, and $A^\star T = A^\star T' + A^\star S$. But $A^\star S \in \mathcal{I}^k(V)$, so

$$\pi(A^\star T') = \pi(A^\star T).$$ □

Proposition 1.8.4. *The map* $A^\star : \Lambda^k(W^\star) \to \Lambda^k(V^\star)$ *sending* $\omega \mapsto A^\star \omega$ *is linear. Moreover,*

(1) *if* $\omega_i \in \Lambda^{k_i}(W^\star)$, $i = 1, 2$, *then*

$$A^\star(\omega_1 \wedge \omega_2) = A^\star(\omega_1) \wedge A^\star(\omega_2);$$

(2) *if* U *is a vector space and* $B : U \to V$ *a linear map, then for* $\omega \in \Lambda^k(W^\star)$,

$$B^\star A^\star \omega = (AB)^\star \omega.$$

We'll leave the proof of these three assertions as exercises. As a hint, they follow immediately from the analogous assertions for the pullback operation on tensors. (See equations (1.3.19) and (1.3.20).)

As an application of the pullback operation we'll show how to use it to define the notion of *determinant* for a linear mapping.

Definition 1.8.5. Let V be an n-dimensional vector space. Then $\dim \Lambda^n(V^\star) = \binom{n}{n} = 1$; i.e., $\Lambda^n(V^\star)$ is a *one-dimensional* vector space. Thus if $A\colon V \to V$ is a linear mapping, the induced pullback mapping:

$$A^\star : \Lambda^n(V^\star) \to \Lambda^n(V^\star),$$

is just "multiplication by a constant". We denote this constant by $\det(A)$ and call it the ***determinant*** of A, Hence, by definition,

$$A^\star \omega = \det(A)\omega \tag{1.8.6}$$

for all $\omega \in \Lambda^n(V^\star)$.

From equation (1.8.6) it is easy to derive a number of basic facts about determinants.

Proposition 1.8.7. *If A and B are linear mappings of V into V, then*

$$\det(AB) = \det(A)\det(B).$$

Proof. By (2) and

$$\begin{aligned}(AB)^\star\omega = \det(AB)\omega &= B^\star(A^\star\omega)\\ &= \det(B)A^\star\omega = \det(B)\det(A)\omega,\end{aligned}$$

so $\det(AB) = \det(A)\det(B)$. □

Proposition 1.8.8. *Write* $\mathrm{id}_V\colon V \to V$ *for the identity map. Then* $\det(\mathrm{id}_V) = 1$.

We'll leave the proof as an exercise. As a hint, note that $\mathrm{id}_V^\star$ is the identity map on $\Lambda^n(V^\star)$.

Proposition 1.8.9. *If* $A\colon V \to V$ *is not surjective, then* $\det(A) = 0$.

Proof. Let W be the image of A. Then if A is not onto, the dimension of W is less than n, so $\Lambda^n(W^\star) = 0$. Now let $A = i_W B$ where i_W is the inclusion map of W into V and B is the mapping, A, regarded as a mapping from V to W. Thus if ω is in $\Lambda^n(V^\star)$, then by (2)

$$A^\star\omega = B^\star i_W^\star \omega$$

and since $i_W^\star\omega$ is in $\Lambda^n(W^\star)$ it is zero. □

We will derive by wedge product arguments the familiar "matrix formula" for the determinant. Let V and W be n-dimensional vector spaces and let $e_1, \dots, e_n$ be a basis for V and $f_1, \dots, f_n$ a basis for W. From these bases we get dual bases, $e_1^\star, \dots, e_n^\star$ and $f_1^\star, \dots, f_n^\star$, for $V^\star$ and $W^\star$. Moreover, if A is a linear map of V into W and $(a_{i,j})$ the $n \times n$ matrix describing A in terms of these bases, then the transpose map, $A^\star : W^\star \to V^\star$, is described in terms of these dual bases by the $n \times n$ transpose matrix, i.e., if

$$Ae_j = \sum_{i=1}^n a_{i,j} f_i,$$

then

$$A^\star f_j^\star = \sum_{i=1}^{n} a_{j,i} e_i^\star .$$

(See §1.2.) Consider now $A^\star(f_1^\star \wedge \cdots \wedge f_n^\star)$. By (1),

$$\begin{aligned} A^\star(f_1^\star \wedge \cdots \wedge f_n^\star) &= A^\star f_1^\star \wedge \cdots \wedge A^\star f_n^\star \\ &= \sum_{1\le k_1,\ldots,k_n\le n} (a_{1,k_1} e_{k_1}^\star) \wedge \cdots \wedge (a_{n,k_n} e_{k_n}^\star). \end{aligned}$$

Thus,

$$A^\star(f_1^\star \wedge \cdots \wedge f_n^\star) = \sum a_{1,k_1} \ldots a_{n,k_n}\, e_{k_1}^\star \wedge \cdots \wedge e_{k_n}^\star .$$

If the multi-index, $k_1, \ldots, k_n$, is repeating, then $e_{k_1}^\star \wedge \cdots \wedge e_{k_n}^\star$ is zero, and if it is not repeating then we can write

$$k_i = \sigma(i) \quad i = 1, \ldots, n$$

for some permutation, σ, and hence we can rewrite $A^\star(f_1^\star \wedge \cdots \wedge f_n^\star)$ as

$$A^\star(f_1^\star \wedge \cdots \wedge f_n^\star) = \sum_{\sigma\in S_n} a_{1,\sigma(1)} \cdots a_{n,\sigma(n)} \quad (e_1^\star \wedge \cdots \wedge e_n^\star)^\sigma .$$

But

$$(e_1^\star \wedge \cdots \wedge e_n^\star)^\sigma = (-1)^\sigma e_1^\star \wedge \cdots \wedge e_n^\star$$

so we get finally the formula

(1.8.10) $$A^\star(f_1^\star \wedge \cdots \wedge f_n^\star) = \det((a_{i,j})) e_1^\star \wedge \cdots \wedge e_n^\star ,$$

where

(1.8.11) $$\det((a_{i,j})) = \sum_{\sigma\in S_n} (-1)^\sigma a_{1,\sigma(1)} \cdots a_{n,\sigma(n)}$$

summed over $\sigma \in S_n$. The sum on the right is (as most of you know) the *determinant* of the matrix $(a_{i,j})$.

Notice that if $V = W$ and $e_i = f_i$, $i = 1, \ldots, n$, then $\omega = e_1^\star \wedge \cdots \wedge e_n^\star = f_1^\star \wedge \cdots \wedge f_n^\star$, hence by (1.8.6) and (1.8.10),

$$\det(A) = \det((a_{i,j})).$$

Exercises for §1.8

Exercise 1.8.i. Verify the three assertions of Proposition 1.8.4.

Exercise 1.8.ii. Deduce from Proposition 1.8.9 a well-known fact about determinants of $n \times n$ matrices: If two columns are equal, the determinant is zero.

Exercise 1.8.iii. Deduce from Proposition 1.8.7 another well-known fact about determinants of $n \times n$ matrices: If one interchanges two columns, then one changes the sign of the determinant.

Hint: Let $e_1, \ldots, e_n$ be a basis of V and let $B\colon V \to V$ be the linear mapping: $Be_i = e_j$, $Be_j = e_i$ and $Be_\ell = e_\ell$, $\ell \neq i, j$. What is $B^\star(e_1^\star \wedge \cdots \wedge e_n^\star)$?

Exercise 1.8.iv. Deduce from Propositions 1.8.7 and 1.8.8 another well-known fact about determinants of $n \times n$ matrix: If $(b_{i,j})$ is the inverse of $(a_{i,j})$, its determinant is the inverse of the determinant of $(a_{i,j})$.

Exercise 1.8.v. Extract from (1.8.11) a well-known formula for determinants of 2×2 matrices:

$$\det \begin{pmatrix} a_{1,1} & a_{1,2} \\ a_{2,1}, & a_{2,2} \end{pmatrix} = a_{1,1}a_{2,2} - a_{1,2}a_{2,1}.$$

Exercise 1.8.vi. Show that if $A = (a_{i,j})$ is an $n \times n$ matrix and $A^\intercal = (a_{j,i})$ is its transpose $\det A = \det A^\intercal$.

Hint: You are required to show that the sums

$$\sum_{\sigma \in S_n} (-1)^\sigma a_{1,\sigma(1)} \dots a_{n,\sigma(n)}$$

and

$$\sum_{\sigma \in S_n} (-1)^\sigma a_{\sigma(1),1} \dots a_{\sigma(n),n}$$

are the same. Show that the second sum is identical with

$$\sum_{\sigma \in S_n} (-1)^{\sigma^{-1}} a_{\sigma^{-1}(1),1} \dots a_{\sigma^{-1}(n),n}.$$

Exercise 1.8.vii. Let A be an $n \times n$ matrix of the form

$$A = \begin{pmatrix} B & * \\ 0 & C \end{pmatrix},$$

where B is a $k \times k$ matrix and C an $\ell \times \ell$ matrix and the bottom $\ell \times k$ block is zero. Show that

$$\det(A) = \det(B)\det(C).$$

Hint: Show that in equation (1.8.11) every non-zero term is of the form

$$(-1)^{\sigma\tau} b_{1,\sigma(1)} \dots b_{k,\sigma(k)} c_{1,\tau(1)} \dots c_{\ell,\tau(\ell)},$$

where $\sigma \in S_k$ and $\tau \in S_\ell$.

Exercise 1.8.viii. Let V and W be vector spaces and let $A \colon V \to W$ be a linear map. Show that if $Av = w$ then for $\omega \in \Lambda^p(W^\star)$,

$$A^\star \iota_w \omega = \iota_v A^\star \omega.$$

Hint: By equation (1.7.14) and Proposition 1.8.4 it suffices to prove this for $\omega \in \Lambda^1(W^\star)$, i.e., for $\omega \in W^\star$.

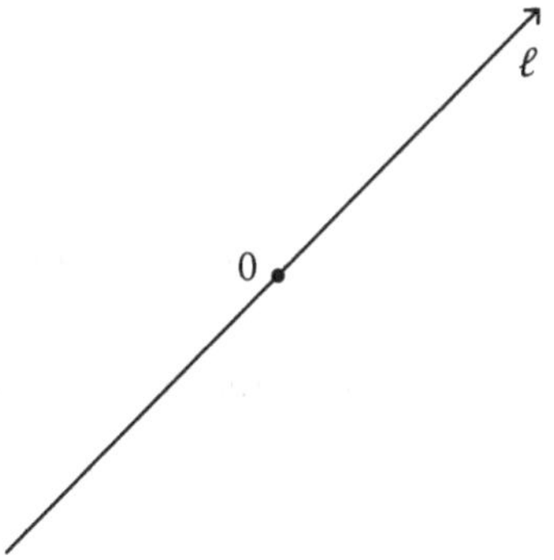

Figure 1.9.1. A line in $\mathbf{R}^2$.

1.9. Orientations

Definition 1.9.1. We recall from freshman calculus that if $\ell \subset \mathbf{R}^2$ is a line through the origin, then $\ell \setminus \{0\}$ has two connected components and an ***orientation*** of ℓ is a choice of one of these components (as in Figure 1.9.1).

More generally, if L is a one-dimensional vector space then $L \setminus \{0\}$ consists of two components: namely if v is an element of $L \setminus \{0\}$, then these two components are

$$L_1 = \{\lambda v \mid \lambda > 0\}$$

and

$$L_2 = \{\lambda v \mid \lambda < 0\}.$$

Definition 1.9.2. Let L be a one-dimensional vector space. An ***orientation*** of L is a choice of one of a connected component of $L \setminus \{0\}$. Usually the component chosen is denoted L_+, and called the ***positive component*** of $L \setminus \{0\}$ and the other component L_- the ***negative component*** of $L \setminus \{0\}$.

Definition 1.9.3. Let (L, L_+) be an oriented one-dimensional vector space. A vector $v \in L$ is ***positively oriented*** if $v \in L_+$.

Definition 1.9.4. Let V be an n-dimensional vector space. An ***orientation*** of V is an orientation of the one-dimensional vector space $\Lambda^n(V^\star)$.

One important way of assigning an orientation to V is to choose a basis, $e_1, \dots, e_n$ of V. Then, if $e_1^\star, \dots, e_n^\star$ is the dual basis, we can orient $\Lambda^n(V^\star)$ by requiring that $e_1^\star \wedge \cdots \wedge e_n^\star$ be in the positive component of $\Lambda^n(V^\star)$.

Definition 1.9.5. Let V be an oriented n-dimensional vector space. We say that an ordered basis $(e_1, \dots, e_n)$ of V is ***positively oriented*** if $e_1^\star \wedge \cdots \wedge e_n^\star$ is in the positive component of $\Lambda^n(V^\star)$.

Suppose that $e_1, \dots, e_n$ and $f_1, \dots, f_n$ are bases of V and that

$$e_j = \sum_{i=1}^{n} a_{i,j,} f_i. \tag{1.9.6}$$

Then by (1.7.10)

$$f_1^\star \wedge \cdots \wedge f_n^\star = \det((a_{i,j}))e_1^\star \wedge \cdots \wedge e_n^\star$$

so we conclude the following.

Proposition 1.9.7. *If $e_1, \ldots, e_n$ is positively oriented, then $f_1, \ldots, f_n$ is positively oriented if and only if $\det((a_{i,j}))$ is positive.*

Corollary 1.9.8. *If $e_1, \ldots, e_n$ is a positively oriented basis of V, then the basis*

$$e_1, \ldots, e_{i-1}, -e_i, e_{i+1}, \ldots, e_n$$

is negatively oriented.

Now let V be a vector space of dimension $n > 1$ and W a subspace of dimension $k < n$. We will use the result above to prove the following important theorem.

Theorem 1.9.9. *Given orientations on V and V/W, one gets from these orientations a natural orientation on W.*

Remark 1.9.10. What we mean by "natural" will be explained in the course of the proof.

Proof. Let $r = n - k$ and let π be the projection of V onto V/W. By Exercises 1.2.i and 1.2.ii we can choose a basis $e_1, \ldots, e_n$ of V such that $e_{r+1}, \ldots, e_n$ is a basis of W and $\pi(e_1), \ldots, \pi(e_r)$ a basis of V/W. Moreover, replacing e_1 by $-e_1$ if necessary we can assume by Corollary 1.9.8 that $\pi(e_1), \ldots, \pi(e_r)$ is a positively oriented basis of V/W and replacing e_n by $-e_n$ if necessary we can assume that $e_1, \ldots, e_n$ is a positively oriented basis of V. Now assign to W the orientation associated with the basis $e_{r+1}, \ldots, e_n$.

Let's show that this assignment is "natural" (i.e., does not depend on our choice of basis $e_1, \ldots, e_n$). To see this let $f_1, \ldots, f_n$ be another basis of V with the properties above and let $A = (a_{i,j})$ be the matrix (1.9.6) expressing the vectors $e_1, \ldots, e_n$ as linear combinations of the vectors $f_1, \ldots f_n$. This matrix has to have the form

$$A = \begin{pmatrix} B & C \\ 0 & D \end{pmatrix}, \tag{1.9.11}$$

where B is the $r \times r$ matrix expressing the basis vectors $\pi(e_1), \ldots, \pi(e_r)$ of V/W as linear combinations of $\pi(f_1), \ldots, \pi(f_r)$ and D the $k \times k$ matrix expressing the basis vectors $e_{r+1}, \ldots, e_n$ of W as linear combinations of $f_{r+1}, \ldots, f_n$. Thus

$$\det(A) = \det(B)\det(D).$$

However, by Proposition 1.9.7, $\det A$ and $\det B$ are positive, so $\det D$ is positive, and hence if $e_{r+1}, \ldots, e_n$ is a positively oriented basis of W so is $f_{r+1}, \ldots, f_n$. □

As a special case of this theorem suppose $\dim W = n - 1$. Then the choice of a vector $v \in V \setminus W$ gives one a basis vector $\pi(v)$ for the one-dimensional space V/W and hence if V is oriented, the choice of v gives one a natural orientation on W.

Definition 1.9.12. Let $A: V_1 \to V_2$ a bijective linear map of oriented n-dimensional vector spaces. We say that A is ***orientation-preserving*** if, for $\omega \in \Lambda^n(V_2^\star)_+$, we have that $A^\star\omega$ is in $\Lambda^n(V_1^\star)_+$.

Example 1.9.13. If $V_1 = V_2$ then $A^\star\omega = \det(A)\omega$ so A is orientation preserving if and only if $\det(A) > 0$.

The following proposition we'll leave as an exercise.

Proposition 1.9.14. *Let V_1, V_2, and V_3 be oriented n-dimensional vector spaces and $A_i: V_i \xrightarrow{\sim} V_{i+1}$ for $i = 1, 2$ be bijective linear maps. Then if A_1 and A_2 are orientation preserving, so is $A_2 \circ A_1$.*

Exercises for §1.9

Exercise 1.9.i. Prove Corollary 1.9.8.

Exercise 1.9.ii. Show that the argument in the proof of Theorem 1.9.9 can be modified to prove that if V and W are oriented then these orientations induce a natural orientation on V/W.

Exercise 1.9.iii. Similarly show that if W and V/W are oriented these orientations induce a natural orientation on V.

Exercise 1.9.iv. Let V be an n-dimensional vector space and $W \subset V$ a k-dimensional subspace. Let $U = V/W$ and let $\iota: W \to V$ and $\pi: V \to U$ be the inclusion and projection maps. Suppose V and U are oriented. Let μ be in $\Lambda^{n-k}(U^\star)_+$ and let ω be in $\Lambda^n(V^\star)_+$. Show that there exists a ν in $\Lambda^k(V^\star)$ such that $\pi^\star\mu \wedge \nu = \omega$. Moreover show that $\iota^\star\nu$ is *intrinsically* defined (i.e., does not depend on how we choose ν) and sits in the positive part $\Lambda^k(W^\star)_+$ of $\Lambda^k(W^\star)$.

Exercise 1.9.v. Let $e_1, \dots, e_n$ be the standard basis vectors of $\mathbf{R}^n$. The ***standard*** orientation of $\mathbf{R}^n$ is, by definition, the orientation associated with this basis. Show that if W is the subspace of $\mathbf{R}^n$ defined by the equation $x_1 = 0$, and $v = e_1 \notin W$ then the natural orientation of W associated with v and the standard orientation of $\mathbf{R}^n$ coincide with the orientation given by the basis vectors $e_2, \dots, e_n$ of W.

Exercise 1.9.vi. Let V be an oriented n-dimensional vector space and W an $(n-1)$-dimensional subspace. Show that if v and v' are in $V \smallsetminus W$ then $v' = \lambda v + w$, where w is in W and $\lambda \in \mathbf{R} \smallsetminus \{0\}$. Show that v and v' give rise to the same orientation of W if and only if λ is positive.

Exercise 1.9.vii. Prove Proposition 1.9.14.

Exercise 1.9.viii. A key step in the proof of Theorem 1.9.9 was the assertion that the matrix A expressing the vectors, e_i as linear combinations of the vectors f_j, had to have the form (1.9.11). Why is this the case?

Exercise 1.9.ix.

(1) Let V be a vector space, W a subspace of V and $A: V \to V$ a bijective linear map which maps W onto W. Show that one gets from A a bijective linear map
$$B: V/W \to V/W$$
with the property
$$\pi A = B\pi,$$
where $\pi: V \to V/W$ is the quotient map.

(2) Assume that V, W and V/W are compatibly oriented. Show that if A is orientation preserving and its restriction to W is orientation preserving then B is orientation preserving.

Exercise 1.9.x. Let V be an oriented n-dimensional vector space, W an $(n-1)$-dimensional subspace of V and $i: W \to V$ the inclusion map. Given $\omega \in \Lambda^n(V^\star)_+$ and $v \in V \setminus W$ show that for the orientation of W described in Exercise 1.9.v, $i^\star(\iota_v\omega) \in \Lambda^{n-1}(W^\star)_+$.

Exercise 1.9.xi. Let V be an n-dimensional vector space, $B: V \times V \to \mathbf{R}$ an inner product and $e_1, \dots, e_n$ a basis of V which is positively oriented and orthonormal. Show that the ***volume element***
$$\text{vol} := e_1^\star \wedge \cdots \wedge e_n^\star \in \Lambda^n(V^\star)$$
is intrinsically defined, independent of the choice of this basis.

Hint: Equations (1.2.20) and (1.8.10).

Exercise 1.9.xii.

(1) Let V be an oriented n-dimensional vector space and B an inner product on V. Fix an oriented orthonormal basis, $e_1, \dots, e_n$, of V and let $A: V \to V$ be a linear map. Show that if
$$Ae_i = v_i = \sum_{j=1}^{n} a_{j,i}e_j$$
and $b_{i,j} = B(v_i, v_j)$, the matrices $M_A = (a_{i,j})$ and $M_B = (b_{i,j})$ are related by: $M_B = M_A^\intercal M_A$.

(2) Show that if vol is the volume form $e_1^\star \wedge \cdots \wedge e_n^\star$, and A is orientation preserving
$$A^\star \text{vol} = \det(M_B)^{\frac{1}{2}} \text{vol}.$$

(3) By Theorem 1.5.13 one has a bijective map
$$\Lambda^n(V^\star) \cong \mathcal{A}^n(V).$$
Show that the element, Ω, of $\mathcal{A}^n(V)$ corresponding to the form vol has the property
$$|\Omega(v_1, \dots, v_n)|^2 = \det((b_{i,j})),$$
where $v_1, \dots, v_n$ are any n-tuple of vectors in V and $b_{i,j} = B(v_i, v_j)$.

Chapter 2

The Concept of a Differential Form

The goal of this chapter is to generalize to n dimensions the basic operations of three-dimensional vector calculus: divergence, curl, and gradient. The divergence and gradient operations have fairly straightforward generalizations, but the curl operation is more subtle. For vector fields it does not have any obvious generalization, however, if one replaces vector fields by a closely related class of objects, *differential forms,* then not only does it have a natural generalization but it turns out that divergence, curl, and gradient are all special cases of a general operation on differential forms called *exterior differentiation.*

2.1. Vector fields and 1-forms

In this section we will review some basic facts about vector fields in n variables and introduce their dual objects: *1-forms*. We will then take up in § 2.3 the theory of k-forms for $k > 1$. We begin by fixing some notation.

Definition 2.1.1. Let $p \in \mathbf{R}^n$. The ***tangent space to*** $\mathbf{R}^n$ ***at*** p is the set of pairs

$$T_p\mathbf{R}^n := \{(p, v) \mid v \in \mathbf{R}^n\}.$$

The identification

$$T_p\mathbf{R}^n \xrightarrow{\sim} \mathbf{R}^n, \quad (p, v) \mapsto v \tag{2.1.2}$$

makes $T_p\mathbf{R}^n$ into a vector space. More explicitly, for $v, v_1, v_2 \in \mathbf{R}^n$ and $\lambda \in \mathbf{R}$ we define the addition and scalar multiplication operations on $T_p\mathbf{R}^n$ by setting

$$(p, v_1) + (p, v_2) := (p, v_1 + v_2)$$

and

$$\lambda(p, v) := (p, \lambda v).$$

Let U be an open subset of $\mathbf{R}^n$ and $f: U \to \mathbf{R}^m$ a C^1 map. We recall that the derivative

$$Df(p): \mathbf{R}^n \to \mathbf{R}^m$$

of f at p is the linear map associated with the $m \times n$ matrix

$$\left(\frac{\partial f_i}{\partial x_j}(p)\right).$$

It will be useful to have a "base-pointed" version of this definition as well. Namely, if $q = f(p)$ we will define

$$df_p \colon T_p\mathbf{R}^n \to T_q\mathbf{R}^m$$

to be the map

$$df_p(p, v) := (q, Df(p)v).$$

It is clear from the way we have defined vector space structures on $T_p\mathbf{R}^n$ and $T_q\mathbf{R}^m$ that this map is linear.

Suppose that the image of f is contained in an open set V, and suppose $g \colon V \to \mathbf{R}^k$ is a C^1 map. Then the "base-pointed" version of the chain rule asserts that

$$dg_q \circ df_p = d(f \circ g)_p.$$

(This is just an alternative way of writing $Dg(q)Df(p) = D(g \circ f)(p)$.)

In three-dimensional vector calculus a ***vector field*** is a function which attaches to each point p of $\mathbf{R}^3$ a base-pointed arrow $(p, v) \in T_p\mathbf{R}^3$. The n-dimensional version of this definition is essentially the same.

Definition 2.1.3. Let U be an open subset of $\mathbf{R}^n$. A ***vector field*** $\boldsymbol{v}$ on U is a function which assigns to each point $p \in U$ a vector $\boldsymbol{v}(p) \in T_p\mathbf{R}^n$.

Thus a vector field is a vector-valued function, but its value at p is an element of a vector space $T_p\mathbf{R}^n$ that itself depends on p.

Examples 2.1.4.

(1) Given a fixed vector $v \in \mathbf{R}^n$, the function

$$p \mapsto (p, v) \tag{2.1.5}$$

is a vector field on $\mathbf{R}^n$. Vector fields of this type are called ***constant*** vector fields.

(2) In particular let $e_1, \dots, e_n$ be the standard basis vectors of $\mathbf{R}^n$. If $v = e_i$ we will denote the vector field (2.1.5) by $\partial/\partial x_i$. (The reason for this "derivation notation" will be explained below.)

(3) Given a vector field $\boldsymbol{v}$ on U and a function $f \colon U \to \mathbf{R}$, we denote by $f\boldsymbol{v}$ the vector field on U defined by

$$p \mapsto f(p)\boldsymbol{v}(p).$$

(4) Given vector fields $\boldsymbol{v}_1$ and $\boldsymbol{v}_2$ on U, we denote by $\boldsymbol{v}_1 + \boldsymbol{v}_2$ the vector field on U defined by

$$p \mapsto \boldsymbol{v}_1(p) + \boldsymbol{v}_2(p).$$

(5) The vectors, (p, e_i), $i = 1, \dots, n$, are a basis of $T_p\mathbf{R}^n$, so if $\boldsymbol{v}$ is a vector field on U, $\boldsymbol{v}(p)$ can be written uniquely as a linear combination of these vectors with real numbers $g_1(p), \dots, g_n(p)$ as coefficients. In other words, using the notation in example (2) above, $\boldsymbol{v}$ can be written uniquely as a sum

$$\boldsymbol{v} = \sum_{i=1}^{n} g_i \frac{\partial}{\partial x_i}, \tag{2.1.6}$$

where $g_i \colon U \to \mathbf{R}$ is the function $p \mapsto g_i(p)$.

Definition 2.1.7. We say that $\boldsymbol{v}$ is a C^∞ vector field if the g_i's are in $C^\infty(U)$.

Definition 2.1.8. A basic vector field operation is ***Lie differentiation***. If $f \in C^1(U)$ we define $L_{\boldsymbol{v}} f$ to be the function on U whose value at p is given by

$$L_{\boldsymbol{v}} f(p) := Df(p)v, \tag{2.1.9}$$

where $\boldsymbol{v}(p) = (p, v)$.

Example 2.1.10. If $\boldsymbol{v}$ is the vector field (2.1.6) then

$$L_{\boldsymbol{v}} f = \sum_{i=1}^{n} g_i \frac{\partial f}{\partial x_i}$$

(motivating our "derivation notation" for $\boldsymbol{v}$).

We leave the following generalization of the product rule for differentiation as an exercise.

Lemma 2.1.11. *Let U be an open subset of $\mathbf{R}^n$, $\boldsymbol{v}$ a vector field on U, and $f_1, f_2 \in C^1(U)$. Then*

$$L_{\boldsymbol{v}}(f_1 \cdot f_2) = L_{\boldsymbol{v}}(f_1) \cdot f_2 + f_1 \cdot L_{\boldsymbol{v}}(f_2).$$

We now discuss a class of objects which are in some sense "dual objects" to vector fields.

Definition 2.1.12. Let $p \in \mathbf{R}^n$. The ***cotangent space to $\mathbf{R}^n$ at p*** is the dual vector space

$$T_p^\star \mathbf{R}^n := (T_p \mathbf{R})^\star.$$

An element of $T_p^\star \mathbf{R}^n$ is called a ***cotangent vector*** to $\mathbf{R}^n$ at p.

Definition 2.1.13. Let U be an open subset of $\mathbf{R}^n$. A ***differential 1-form***, or simply ***1-form*** on U, is a function ω which assigns to each point $p \in U$ a vector ω_p in $T_p^\star \mathbf{R}^n$.

Examples 2.1.14.

(1) Let $f : U \to \mathbf{R}$ be a C^1 function. Then for $p \in U$ and $q = f(p)$ one has a linear map

$$df_p \colon T_p \mathbf{R}^n \to T_q \mathbf{R} \tag{2.1.15}$$

and by making the identification $T_q \mathbf{R} \xrightarrow{\sim} \mathbf{R}$ by sending $(q, v) \mapsto v$, the differential df_p can be regarded as a linear map from $T_p \mathbf{R}^n$ to $\mathbf{R}$, i.e., as an element of $T_p^\star \mathbf{R}^n$. Hence the assignment

$$p \mapsto df_p$$

defines a 1-form on U which we denote by df.

(2) Given a 1-form ω and a function $\phi \colon U \to \mathbf{R}$, the pointwise product of ϕ with ω defines 1-form $\phi\omega$ on U by $(\phi\omega)_p := \phi(p)\omega_p$.

(3) Given two 1-forms ω_1 and ω_2 their pointwise sum defines a 1-form $\omega_1 + \omega_2$ by $(\omega_1 + \omega_2)_p := (\omega_1)_p + (\omega_2)_p$.

(4) The 1-forms $dx_1, \ldots, dx_n$ play a particularly important role. By equation (2.1.15)

$$(dx_i)\left(\frac{\partial}{\partial x_j}\right)_p = \delta_{i,j}, \tag{2.1.16}$$

i.e., it is equal to 1 if $i = j$ and zero if $i \neq j$. Thus $(dx_1)_p, \ldots, (dx_n)_p$ are the basis of $T_p^\star \mathbf{R}^n$ *dual* to the basis $(\partial/\partial x_i)_p$. Therefore, if ω is *any* 1-form on U, ω_p can be written uniquely as a sum

$$\omega_p = \sum_{i=1}^n f_i(p)(dx_i)_p, \quad f_i(p) \in \mathbf{R},$$

and ω can be written uniquely as a sum

$$\omega = \sum_{i=1}^n f_i dx_i, \tag{2.1.17}$$

where $f_i \colon U \to \mathbf{R}$ is the function $p \mapsto f_i(p)$. We say that ω is a C^∞ ***1-form*** or ***smooth 1-form*** if the functions $f_1, \ldots, f_n$ are C^∞.

We leave the following as an exercise.

Lemma 2.1.18. *Let U be an open subset of $\mathbf{R}^n$. If $f : U \to \mathbf{R}$ is a C^∞ function, then*

$$df = \sum_{i=1}^n \frac{\partial f}{\partial x_i} dx_i.$$

Suppose that $\boldsymbol{v}$ is a vector field and ω a 1-form on $U \subset \mathbf{R}^n$. Then for every $p \in U$ the vectors $\boldsymbol{v}(p) \in T_p\mathbf{R}^n$ and $\omega_p \in (T_p\mathbf{R}^n)^\star$ can be paired to give a number $\iota_{\boldsymbol{v}(p)}\omega_p \in \mathbf{R}$, and hence, as p varies, an $\mathbf{R}$-valued function $\iota_{\boldsymbol{v}}\omega$ which we will call the ***interior product*** of $\boldsymbol{v}$ with ω.

Example 2.1.19. For instance if $\boldsymbol{v}$ is the vector field (2.1.6) and ω the 1-form (2.1.17), then

$$\iota_{\boldsymbol{v}}\omega = \sum_{i=1}^n f_i g_i.$$

Thus if $\boldsymbol{v}$ and ω are C^∞, so is the function $\iota_{\boldsymbol{v}}\omega$. Also notice that if $\phi \in C^\infty(U)$, then as we observed above

$$d\phi = \sum_{i=1}^n \frac{\partial \phi}{\partial x_i} \frac{\partial}{\partial x_i}$$

so if $\boldsymbol{v}$ is the vector field (2.1.6) then we have

$$\iota_{\boldsymbol{v}} d\phi = \sum_{i=1}^n g_i \frac{\partial \phi}{\partial x_i} = L_{\boldsymbol{v}}\phi.$$

Exercises for §2.1

Exercise 2.1.i. Prove Lemma 2.1.18.

Exercise 2.1.ii. Prove Lemma 2.1.11.

Exercise 2.1.iii. Let U be an open subset of $\mathbf{R}^n$ and $\boldsymbol{v}_1$ and $\boldsymbol{v}_2$ vector fields on U. Show that there is a unique vector field $\boldsymbol{w}$, on U with the property

$$L_{\boldsymbol{w}}\phi = L_{\boldsymbol{v}_1}(L_{\boldsymbol{v}_2}\phi) - L_{\boldsymbol{v}_2}(L_{\boldsymbol{v}_1}\phi)$$

for all $\phi \in C^\infty(U)$.

Exercise 2.1.iv. The vector field $\boldsymbol{w}$ in Exercise 2.1.iii is called the ***Lie bracket*** of the vector fields $\boldsymbol{v}_1$ and $\boldsymbol{v}_2$ and is denoted by $[\boldsymbol{v}_1, \boldsymbol{v}_2]$. Verify that the Lie bracket is *skew-symmetric*, i.e.,

$$[\boldsymbol{v}_1, \boldsymbol{v}_2] = -[\boldsymbol{v}_2, \boldsymbol{v}_1],$$

and satisfies the *Jacobi identity*

$$[\boldsymbol{v}_1, [\boldsymbol{v}_2, \boldsymbol{v}_3]] + [\boldsymbol{v}_2, [\boldsymbol{v}_3, \boldsymbol{v}_1]] + [\boldsymbol{v}_3, [\boldsymbol{v}_1, \boldsymbol{v}_2]] = 0.$$

(And thus defines the structure of a *Lie algebra*.)

Hint: Prove analogous identities for $L_{\boldsymbol{v}_1}$, $L_{\boldsymbol{v}_2}$ and $L_{\boldsymbol{v}_3}$.

Exercise 2.1.v. Let $\boldsymbol{v}_1 = \partial/\partial x_i$ and $\boldsymbol{v}_2 = \sum_{j=1}^n g_j \partial/\partial x_j$. Show that

$$[\boldsymbol{v}_1, \boldsymbol{v}_2] = \sum_{j=1}^n \frac{\partial g_j}{\partial x_i}\frac{\partial}{\partial x_j}.$$

Exercise 2.1.vi. Let $\boldsymbol{v}_1$ and $\boldsymbol{v}_2$ be vector fields and f a C^∞ function. Show that

$$[\boldsymbol{v}_1, f\boldsymbol{v}_2] = L_{\boldsymbol{v}_1} f \boldsymbol{v}_2 + f[\boldsymbol{v}_1, \boldsymbol{v}_2].$$

Exercise 2.1.vii. Let U be an open subset of $\mathbf{R}^n$ and let $\gamma\colon [a,b] \to U$, $t \mapsto (\gamma_1(t), \dots, \gamma_n(t))$ be a C^1 curve. Given a C^∞ 1-form $\omega = \sum_{i=1}^n f_i dx_i$ on U, define the ***line integral*** of ω over γ to be the integral

$$\int_\gamma \omega := \sum_{i=1}^n \int_a^b f_i(\gamma(t)) \frac{d\gamma_i}{dt}\, dt.$$

Show that if $\omega = df$ for some $f \in C^\infty(U)$

$$\int_\gamma \omega = f(\gamma(b)) - f(\gamma(a)).$$

In particular conclude that if γ is a closed curve, i.e., $\gamma(a) = \gamma(b)$, this integral is zero.

Exercise 2.1.viii. Let ω be the C^∞ 1-form on $\mathbf{R}^2 \smallsetminus \{0\}$ defined by

$$\omega = \frac{x_1 dx_2 - x_2 dx_1}{x_1^2 + x_2^2},$$

and let $\gamma\colon [0,2\pi] \to \mathbf{R}^2 \smallsetminus \{0\}$ be the closed curve $t \mapsto (\cos t, \sin t)$. Compute the line integral $\int_\gamma \omega$, and note that $\int_\gamma \omega \neq 0$. Conclude that ω is not of the form df for $f \in C^\infty(\mathbf{R}^2 \smallsetminus \{0\})$.

Exercise 2.1.ix. Let f be the function

$$f(x_1,x_2) = \begin{cases} \arctan \dfrac{x_2}{x_1} & x_1 > 0, \\ \dfrac{\pi}{2}, & x_2 > 0 \text{ or } x_1 = 0, \\ \arctan \dfrac{x_2}{x_1} + \pi & x_1 < 0\,. \end{cases}$$

Recall that $-\frac{\pi}{2} < \arctan(t) < \frac{\pi}{2}$. Show that f is C^∞ and that df is the 1-form ω in Exercise 2.1.viii. Why does not this contradict what you proved in Exercise 2.1.viii?

2.2. Integral curves for vector fields

In this section we'll discuss some properties of vector fields which we'll need for the manifold segment of this text. We'll begin by generalizing to n-variables the calculus notion of an "integral curve" of a vector field.

Definition 2.2.1. Let $U \subset \mathbf{R}^n$ be open and $\boldsymbol{v}$ a vector field on U. A C^1 curve $\gamma\colon (a,b) \to U$ is an ***integral curve*** of $\boldsymbol{v}$ if for all $t \in (a,b)$ we have

$$\boldsymbol{v}(\gamma(t)) = \left(\gamma(t), \frac{d\gamma}{dt}(t)\right).$$

Remark 2.2.2. If $\boldsymbol{v}$ is the vector field (2.1.6) and $g\colon U \to \mathbf{R}^n$ is the function $(g_1,\dots,g_n)$, the condition for $\gamma(t)$ to be an integral curve of $\boldsymbol{v}$ is that it satisfies the system of differential equations

$$\frac{d\gamma}{dt}(t) = g(\gamma(t)). \tag{2.2.3}$$

We will quote without proof a number of basic facts about systems of ordinary differential equations of the type (2.2.3).[1]

Theorem 2.2.4 (existence of integral curves). *Let U be an open subset of $\mathbf{R}^n$ and $\boldsymbol{v}$ a vector field on U. Given a point $p_0 \in U$ and $a \in \mathbf{R}$, there exist an interval $I = (a-T, a+T)$, a neighborhood U_0 of p_0 in U, and for every $p \in U_0$ an integral curve $\gamma_p\colon I \to U$ for $\boldsymbol{v}$ such that $\gamma_p(a) = p$.*

Theorem 2.2.5 (uniqueness of integral curves). *Let U be an open subset of $\mathbf{R}^n$ and $\boldsymbol{v}$ a vector field on U. For $i = 1,2$, let $\gamma_i\colon I_i \to U$, $i = 1,2$ be integral curves for $\boldsymbol{v}$. If $a \in I_1 \cap I_2$ and $\gamma_1(a) = \gamma_2(a)$ then $\gamma_1|_{I_1\cap I_2} = \gamma_2|_{I_1\cap I_2}$ and the curve $\gamma\colon I_1 \cup I_2 \to U$ defined by*

$$\gamma(t) := \begin{cases} \gamma_1(t), & t \in I_1, \\ \gamma_2(t), & t \in I_2 \end{cases}$$

is an integral curve for $\boldsymbol{v}$.

---

[1] A source for these results that we highly recommend is [2, Chapter 6].

Theorem 2.2.6 (smooth dependence on initial data). *Let $V \subset U \subset \mathbf{R}^n$ be open subsets. Let $\boldsymbol{v}$ be a C^∞ vector field on V, $I \subset \mathbf{R}$ an open interval, $a \in I$, and $h\colon V \times I \to U$ a map with the following properties.*

(1) *$h(p, a) = p$.*

(2) *For all $p \in V$ the curve*

$$\gamma_p\colon I \to U, \qquad \gamma_p(t) := h(p, t)$$

is an integral curve of $\boldsymbol{v}$.

Then the map h is C^∞.

One important feature of the system (2.2.3) is that it is an *autonomous* system of differential equations: the function $g(x)$ is a function of x alone (it does *not* depend on t). One consequence of this is the following.

Theorem 2.2.7. *Let $I = (a, b)$ and for $c \in \mathbf{R}$ let $I_c = (a - c, b - c)$. Then if $\gamma\colon I \to U$ is an integral curve, the reparameterized curve*

$$\gamma_c\colon I_c \to U, \quad \gamma_c(t) := \gamma(t + c) \tag{2.2.8}$$

is an integral curve.

We recall that C^1 function $\phi\colon U \to \mathbf{R}$ is an ***integral*** of the system (2.2.3) if for every integral curve $\gamma(t)$, the function $t \mapsto \phi(\gamma(t))$ is constant. This is true if and only if for all t

$$0 = \frac{d}{dt}\phi(\gamma(t)) = (D\phi)_{\gamma(t)}\left(\frac{d\gamma}{dt}\right) = (D\phi)_{\gamma(t)}(v),$$

where $\boldsymbol{v}(p) = (p, v)$. By equation (2.1.6) the term on the right is $L_{\boldsymbol{v}}\phi(p)$. Hence we conclude the following.

Theorem 2.2.9. *Let U be an open subset of $\mathbf{R}^n$ and $\phi \in C^1(U)$. Then ϕ is an integral of the system (2.2.3) if and only if $L_{\boldsymbol{v}}\phi = 0$.*

Definition 2.2.10. Let U be an open subset of $\mathbf{R}^n$ and $\boldsymbol{v}$ a vector field on U. We say that $\boldsymbol{v}$ is ***complete*** if for every $p \in U$ there exists an integral curve $\gamma\colon \mathbf{R} \to U$ with $\gamma(0) = p$, i.e., for every p there exists an integral curve that starts at p and *exists for all time.*

To see what completeness involves, we recall that an integral curve

$$\gamma\colon [0, b) \to U,$$

with $\gamma(0) = p$ is called ***maximal*** if it cannot be extended to an interval $[0, b')$, where $b' > b$. (See for instance [2, § 6.11].) For such curves it is known that either

(1) $b = +\infty$,

(2) $|\gamma(t)| \to +\infty$ as $t \to b$,

(3) or the limit set of

$$\{\gamma(t) \mid 0 \le t < b\}$$

contains points on the boundary of U.

Hence if we can exclude (2) and (3), we have shown that an integral curve with $\gamma(0) = p$ exists for all positive time. A simple criterion for excluding (2) and (3) is the following.

Lemma 2.2.11. *The scenarios (2) and (3) cannot happen if there exists a proper C^1 function $\phi: U \to \mathbf{R}$ with $L_{\boldsymbol{v}}\phi = 0$.*

Proof. The identity $L_{\boldsymbol{v}}\phi = 0$ implies that ϕ is constant on $\gamma(t)$, but if $\phi(p) = c$ this implies that the curve $\gamma(t)$ lies on the compact subset $\phi^{-1}(c) \subset U$, hence it cannot "run off to infinity" as in scenario (2) or "run off the boundary" as in scenario (3). □

Applying a similar argument to the interval $(-b, 0]$ we conclude the following.

Theorem 2.2.12. *Suppose there exists a proper C^1 function $\phi: U \to \mathbf{R}$ with the property $L_{\boldsymbol{v}}\phi = 0$. Then the vector field $\boldsymbol{v}$ is complete.*

Example 2.2.13. Let $U = \mathbf{R}^2$ and let v be the vector field

$$v = x^3 \frac{\partial}{\partial y} - y\frac{\partial}{\partial x}.$$

Then $\phi(x, y) = 2y^2 + x^4$ is a proper function with the property above.

Another hypothesis on $\boldsymbol{v}$ which excludes (2) and (3) is the following.

Definition 2.2.14. The ***support*** of $\boldsymbol{v}$ is set

$$\operatorname{supp}(\boldsymbol{v}) = \overline{\{q \in U \mid \boldsymbol{v}(q) \neq 0\}}.$$

We say that $\boldsymbol{v}$ is ***compactly supported*** if $\operatorname{supp}(\boldsymbol{v})$ is compact.

We now show that compactly supported vector fields are complete.

Theorem 2.2.15. *If a vector field $\boldsymbol{v}$ is compactly supported, then $\boldsymbol{v}$ is complete.*

Proof. Notice first that if $\boldsymbol{v}(p) = 0$, the constant curve, $\gamma_0(t) = p$, $-\infty < t < \infty$, satisfies the equation

$$\frac{d}{dt}\gamma_0(t) = 0 = \boldsymbol{v}(p),$$

so it is an integral curve of $\boldsymbol{v}$. Hence if $\gamma(t)$, $-a < t < b$, is any integral curve of $\boldsymbol{v}$ with the property $\gamma(t_0) = p$ for some t_0, it has to coincide with γ_0 on the interval $(-a, a)$, and hence has to be the constant curve $\gamma(t) = p$ on $(-a, a)$.

Now suppose the support $\operatorname{supp}(\boldsymbol{v})$ of $\boldsymbol{v}$ is compact. Then either $\gamma(t)$ is in $\operatorname{supp}(\boldsymbol{v})$ for all t or is in $U \smallsetminus \operatorname{supp}(\boldsymbol{v})$ for some t_0. But if this happens, and $p = \gamma(t_0)$ then $\boldsymbol{v}(p) = 0$, so $\gamma(t)$ has to coincide with the constant curve $\gamma_0(t) = p$, for all t. In neither case can it go off to ∞ or off to the boundary of U as $t \to b$. □

One useful application of this result is the following. Suppose $\boldsymbol{v}$ is a vector field on $U \subset \mathbf{R}^n$, and one wants to see what its integral curves look like on some compact set $A \subset U$. Let $\rho \in C_0^\infty(U)$ be a bump function which is equal to one on a neighborhood of A. Then the vector field $\boldsymbol{w} = \rho\boldsymbol{v}$ is compactly supported and hence complete, but it is identical with $\boldsymbol{v}$ on A, so its integral curves on A coincide with the integral curves of $\boldsymbol{v}$.

If $\boldsymbol{v}$ is complete then for every p, one has an integral curve $\gamma_p : \mathbf{R} \to U$ with $\gamma_p(0) = p$, so one can define a map

$$f_t : U \to U$$

by setting $f_t(p) := \gamma_p(t)$. If $\boldsymbol{v}$ is C^∞, this mapping is C^∞ by the smooth dependence on initial data theorem, and, by definition, $f_0 = \mathrm{id}_U$. We claim that the f_t's also have the property

$$f_t \circ f_a = f_{t+a}. \tag{2.2.16}$$

Indeed if $f_a(p) = q$, then by the reparameterization theorem, $\gamma_q(t)$ and $\gamma_p(t+a)$ are both integral curves of $\boldsymbol{v}$, and since $q = \gamma_q(0) = \gamma_p(a) = f_a(p)$, they have the same initial point, so

$$\begin{aligned}\gamma_q(t) &= f_t(q) = (f_t \circ f_a)(p)\\ &= \gamma_p(t+a) = f_{t+a}(p)\end{aligned}$$

for all t. Since f_0 is the identity it follows from (2.2.16) that $f_t \circ f_{-t}$ is the identity, i.e.,

$$f_{-t} = f_t^{-1},$$

so f_t is a C^∞ diffeomorphism. Hence if $\boldsymbol{v}$ is complete it generates a ***one-parameter group*** f_t, $-\infty < t < \infty$, of C^∞ diffeomorphisms of U.

If $\boldsymbol{v}$ is not complete there is an analogous result, but it is trickier to formulate precisely. Roughly speaking $\boldsymbol{v}$ generates a one-parameter group of diffeomorphisms f_t but these diffeomorphisms are not defined on all of U nor for all values of t. Moreover, the identity (2.2.16) only holds on the open subset of U where both sides are well-defined.

We devote the remainder of this section to discussing some "functorial" properties of vector fields and 1-forms.

Definition 2.2.17. Let $U \subset \mathbf{R}^n$ and $W \subset \mathbf{R}^m$ be open, and let $f : U \to W$ be a C^∞ map. If $\boldsymbol{v}$ is a C^∞ vector field on U and $\boldsymbol{w}$ a C^∞ vector field on W we say that $\boldsymbol{v}$ and $\boldsymbol{w}$ are ***f-related*** if, for all $p \in U$ we have

$$df_p(\boldsymbol{v}(p)) = \boldsymbol{w}(f(p)).$$

Writing

$$\boldsymbol{v} = \sum_{i=1}^{n} v_i \frac{\partial}{\partial x_i}, \quad v_i \in C^k(U)$$

and

$$\boldsymbol{w} = \sum_{j=1}^{m} w_j \frac{\partial}{\partial y_j}, \quad w_j \in C^k(V)$$

this equation reduces, in coordinates, to the equation

$$w_i(q) = \sum_{j=1}^{n} \frac{\partial f_i}{\partial x_j}(p) v_j(p).$$

In particular, if $m = n$ and f is a C^∞ diffeomorphism, the formula (3.2) defines a C^∞ vector field on W, i.e.,

$$\boldsymbol{w} = \sum_{j=1}^{n} w_i \frac{\partial}{\partial y_j}$$

is the vector field defined by the equation

$$w_i = \sum_{j=1}^{n} \left(\frac{\partial f_i}{\partial x_j} v_j \right) \circ f^{-1}.$$

Hence we have proved the following.

Theorem 2.2.18. *If $f : U \xrightarrow{\simeq} W$ is a C^∞ diffeomorphism and $\boldsymbol{v}$ a C^∞ vector field on U, there exists a* unique *C^∞ vector field $\boldsymbol{w}$ on W having the property that $\boldsymbol{v}$ and $\boldsymbol{w}$ are f-related.*

Definition 2.2.19. In the setting of Theorem 2.2.18, we denote the vector field $\boldsymbol{w}$ by $f_\star \boldsymbol{v}$, and call $f_\star \boldsymbol{v}$ the ***pushforward*** of $\boldsymbol{v}$ by f.

We leave the following assertions as easy exercises (but provide some hints).

Theorem 2.2.20. *For $i = 1, 2$, let U_i be an open subset of $\mathbf{R}^{n_i}$, $\boldsymbol{v}_i$ a vector field on U_i, and $f : U_1 \to U_2$ a C^∞ map. If $\boldsymbol{v}_1$ and $\boldsymbol{v}_2$ are f-related, every integral curve*

$$\gamma : I \to U_1$$

of $\boldsymbol{v}_1$ gets mapped by f onto an integral curve $f \circ \gamma : I \to U_2$ of $\boldsymbol{v}_2$.

Proof sketch. Theorem 2.2.20 follows from the chain rule: if $p = \gamma(t)$ and $q = f(p)$

$$df_p \left(\frac{d}{dt} \gamma(t) \right) = \frac{d}{dt} f(\gamma(t)). \qquad \square$$

Corollary 2.2.21. *In the setting of Theorem 2.2.20, suppose $\boldsymbol{v}_1$ and $\boldsymbol{v}_2$ are complete. Let $(f_{i,t})_{t \in \mathbf{R}} : U_i \xrightarrow{\simeq} U_i$ be the one-parameter group of diffeomorphisms generated by $\boldsymbol{v}_i$. Then $f \circ f_{1,t} = f_{2,t} \circ f$.*

Proof sketch. To deduce Corollary 2.2.21 from Theorem 2.2.20 note that for $p \in U$, $f_{1,t}(p)$ is just the integral curve $\gamma_p(t)$ of $\boldsymbol{v}_1$ with initial point $\gamma_p(0) = p$.

The notion of f-relatedness can be very succinctly expressed in terms of the Lie differentiation operation. For $\phi \in C^\infty(U_2)$ let $f^\star \phi$ be the composition, $\phi \circ f$, viewed as a C^∞ function on U_1, i.e., for $p \in U_1$ let $f^\star \phi(p) = \phi(f(p))$. Then

$$f^\star L_{\boldsymbol{v}_2} \phi = L_{\boldsymbol{v}_1} f^\star \phi.$$

To see this, note that if $f(p) = q$ then at the point p the right-hand side is

$$(d\phi)_q \circ df_p(\boldsymbol{v}_1(p))$$

by the chain rule and by definition the left-hand side is

$$d\phi_q(\boldsymbol{v}_2(q)).$$

Moreover, by definition

$$\boldsymbol{v}_2(q) = df_p(\boldsymbol{v}_1(p))$$

so the two sides are the same. □

Another easy consequence of the chain rule is the following theorem.

Theorem 2.2.22. *For $i = 1, 2, 3$, let U_i be an open subset of $\mathbf{R}^{n_i}$, $\boldsymbol{v}_i$ a vector field on U_i, and for $i = 1, 2$ let $f_i \colon U_i \to U_{i+1}$ be a C^∞ map. Suppose that $\boldsymbol{v}_1$ and $\boldsymbol{v}_2$ are f_1-related and that $\boldsymbol{v}_2$ and $\boldsymbol{v}_3$ are f_2-related. Then $\boldsymbol{v}_1$ and $\boldsymbol{v}_3$ are $(f_2 \circ f_1)$-related.*

In particular, if f_1 and f_2 are diffeomorphisms writing $\boldsymbol{v} \coloneqq \boldsymbol{v}_1$ we have

$$(f_2)_\star(f_1)_\star\boldsymbol{v} = (f_2 \circ f_1)_\star\boldsymbol{v}.$$

The results we described above have "dual" analogues for 1-forms.

Construction 2.2.23. Let $U \subset \mathbf{R}^n$ and $V \subset \mathbf{R}^m$ be open, and let $f \colon U \to V$ be a C^∞ map. Given a 1-form μ on V one can define a ***pullback*** 1-form $f^\star\mu$ on U by the following method. For $p \in U$, by definition $\mu_{f(p)}$ is a linear map

$$\mu_{f(p)} \colon T_{f(p)}\mathbf{R}^m \to \mathbf{R}$$

and by composing this map with the linear map

$$df_p \colon T_p\mathbf{R}^n \to T_{f(p)}\mathbf{R}^n$$

we get a linear map

$$\mu_{f(p)} \circ df_p \colon T_p\mathbf{R}^n \to \mathbf{R},$$

i.e., an element $\mu_{f(p)} \circ df_p$ of $T_p^\star\mathbf{R}^n$.

The pullback 1-form $f^\star\mu$ is defined by the assignment $p \mapsto (\mu_{f(p)} \circ df_p)$.

In particular, if $\phi \colon V \to \mathbf{R}$ is a C^∞ function and $\mu = d\phi$ then

$$\mu_q \circ df_p = d\phi_q \circ df_p = d(\phi \circ f)_p,$$

i.e.,

$$f^\star\mu = d\phi \circ f.$$

In particular if μ is a 1-form of the form $\mu = d\phi$, with $\phi \in C^\infty(V)$, $f^\star\mu$ is C^∞. From this it is easy to deduce the following theorem.

Theorem 2.2.24. *If μ is any C^∞ 1-form on V, its pullback $f^\star\omega$ is C^∞. (See Exercise 2.2.ii.)*

Notice also that the pullback operation on 1-forms and the pushforward operation on vector fields are somewhat different in character. The former is defined for *all* C^∞ maps, but the latter is only defined for diffeomorphisms.

Exercises for §2.2

Exercise 2.2.i. Prove the reparameterization result Theorem 2.2.7.

Exercise 2.2.ii. Let U be an open subset of $\mathbf{R}^n$, V an open subset of $\mathbf{R}^n$ and $f : U \to V$ a C^k map.

(1) Show that for $\phi \in C^\infty(V)$ (2.2) the chain the rule for f and ϕ

$$f^\star d\phi = df^\star\phi.$$

(2) Let μ be the 1-form

$$\mu = \sum_{i=1}^{m} \phi_i dx_i, \quad \phi_i \in C^\infty(V)$$

on V. Show that if $f = (f_1, \dots, f_m)$ then

$$f^\star\mu = \sum_{i=1}^{m} f^\star\phi_i df_i.$$

(3) Show that if μ is C^∞ and f is C^∞, $f^\star\mu$ is C^∞.

Exercise 2.2.iii. Let $\boldsymbol{v}$ be a complete vector field on U and $f_t : U \to U$ the one parameter group of diffeomorphisms generated by $\boldsymbol{v}$. Show that if $\phi \in C^1(U)$

$$L_{\boldsymbol{v}}\phi = \left.\frac{d}{dt} f_t^\star\phi\right|_{t=0}.$$

Exercise 2.2.iv.

(1) Let $U = \mathbf{R}^2$ and let $\boldsymbol{v}$ be the vector field, $x_1\partial/\partial x_2 - x_2\partial/\partial x_1$. Show that the curve

$$t \mapsto (r\cos(t+\theta), r\sin(t+\theta)),$$

for $t \in \mathbf{R}$, is the unique integral curve of $\boldsymbol{v}$ passing through the point, $(r\cos\theta, r\sin\theta)$, at $t = 0$.

(2) Let $U = \mathbf{R}^n$ and let $\boldsymbol{v}$ be the constant vector field: $\sum_{i=1}^n c_i\partial/\partial x_i$. Show that the curve

$$t \mapsto a + t(c_1, \dots, c_n),$$

for $t \in \mathbf{R}$, is the unique integral curve of $\boldsymbol{v}$ passing through $a \in \mathbf{R}^n$ at $t = 0$.

(3) Let $U = \mathbf{R}^n$ and let $\boldsymbol{v}$ be the vector field, $\sum_{i=1}^n x_i\partial/\partial x_i$. Show that the curve

$$t \mapsto e^t(a_1, \dots, a_n),$$

for $t \in \mathbf{R}$, is the unique integral curve of $\boldsymbol{v}$ passing through a at $t = 0$.

Exercise 2.2.v. Show that the following are one-parameter groups of diffeomorphisms:

(1) $f_t : \mathbf{R} \to \mathbf{R}$, $f_t(x) = x + t$;
(2) $f_t : \mathbf{R} \to \mathbf{R}$, $f_t(x) = e^t x$;
(3) $f_t : \mathbf{R}^2 \to \mathbf{R}^2$, $f_t(x, y) = (x\cos(t) - y\sin(t), x\sin(t) + y\cos(t))$.

Exercise 2.2.vi. Let $A\colon \mathbf{R}^n \to \mathbf{R}^n$ be a linear mapping. Show that the series

$$\exp(tA) := \sum_{n=0}^{\infty} \frac{(tA)^n}{n!} = \mathrm{id}_n + tA + \frac{t^2}{2!}A^2 + \frac{t^3}{3!}A^3 + \cdots$$

converges and defines a one-parameter group of diffeomorphisms of $\mathbf{R}^n$.

Exercise 2.2.vii.

(1) What are the infinitesimal generators of the one-parameter groups in Exercise 2.2.v?

(2) Show that the infinitesimal generator of the one-parameter group in Exercise 2.2.vi is the vector field

$$\sum_{1\le i,j\le n} a_{i,j}x_j \frac{\partial}{\partial x_i},$$

where $(a_{i,j})$ is the defining matrix of A.

Exercise 2.2.viii. Let $\boldsymbol{v}$ be the vector field on $\mathbf{R}$ given by $x^2\frac{d}{dx}$. Show that the curve

$$x(t) = \frac{a}{a - at}$$

is an integral curve of $\boldsymbol{v}$ with initial point $x(0) = a$. Conclude that for $a > 0$ the curve

$$x(t) = \frac{a}{1 - at}, \quad 0 < t < \frac{1}{a}$$

is a maximal integral curve. (In particular, conclude that $\boldsymbol{v}$ is not complete.)

Exercise 2.2.ix. Let U and V be open subsets of $\mathbf{R}^n$ and $f\colon U \xrightarrow{\sim} V$ a diffeomorphism. If $\boldsymbol{w}$ is a vector field on V, define the ***pullback*** of $\boldsymbol{w}$ to U to be the vector field

$$f^\star \boldsymbol{w} := (f_\star^{-1}\boldsymbol{w}).$$

Show that if ϕ is a C^∞ function on V

$$f^\star L_{\boldsymbol{w}}\phi = L_{f^\star\boldsymbol{w}} f^\star\phi.$$

Hint: Equation (2.2.25).

Exercise 2.2.x. Let U be an open subset of $\mathbf{R}^n$ and $\boldsymbol{v}$ and $\boldsymbol{w}$ vector fields on U. Suppose $\boldsymbol{v}$ is the infinitesimal generator of a one-parameter group of diffeomorphisms

$$f_t\colon U \xrightarrow{\sim} U, \quad -\infty < t < \infty.$$

Let $\boldsymbol{w}_t = f_t^\star \boldsymbol{w}$. Show that for $\phi \in C^\infty(U)$ we have

$$L_{[\boldsymbol{v},\boldsymbol{w}]}\phi = L_{\dot{\boldsymbol{w}}}\phi,$$

where

$$\dot{\boldsymbol{w}} = \frac{d}{dt} f_t^\star \boldsymbol{w}\Big|_{t=0}.$$

Hint: Differentiate the identity

$$f_t^\star L_w \phi = L_{w_t} f_t^\star \phi$$

with respect to t and show that at $t = 0$ the derivative of the left-hand side is $L_v L_w \phi$ by Exercise 2.2.iii, and the derivative of the right-hand side is

$$L_{\dot{w}} + L_w(L_v \phi).$$

Exercise 2.2.xi. Conclude from Exercise 2.2.x that

$$[v, w] = \frac{d}{dt} f_t^\star w \bigg|_{t=0} . \tag{2.2.25}$$

2.3. Differential k-forms

1-Forms are the bottom tier in a pyramid of objects whose kth tier is the space of *k-forms*. More explicitly, given $p \in \mathbf{R}^n$ we can, as in § 1.5, form the kth exterior powers

$$\Lambda^k(T_p^\star \mathbf{R}^n), \quad k = 1, 2, 3, \ldots, n$$

of the vector space $T_p^\star \mathbf{R}^n$, and since

$$\Lambda^1(T_p^\star \mathbf{R}^n) = T_p^\star \mathbf{R}^n \tag{2.3.1}$$

one can think of a 1-form as a function which takes its value at p in the space (2.3.1). This leads to an obvious generalization.

Definition 2.3.2. Let U be an open subset of $\mathbf{R}^n$. A *k-form* ω on U is a function which assigns to each point $p \in U$ an element $\omega_p \in \Lambda^k(T_p^\star \mathbf{R}^n)$.

The wedge product operation gives us a way to construct lots of examples of such objects.

Example 2.3.3. Let $\omega_1, \ldots, \omega_k$ be 1-forms. Then $\omega_1 \wedge \cdots \wedge \omega_k$ is the k-form whose value at p is the wedge product

$$(\omega_1 \wedge \cdots \wedge \omega_k)_p := (\omega_1)_p \wedge \cdots \wedge (\omega_k)_p. \tag{2.3.4}$$

Notice that since $(\omega_i)_p$ is in $\Lambda^1(T_p^\star \mathbf{R}^n)$ the wedge product (2.3.4) makes sense and is an element of $\Lambda^k(T_p^\star \mathbf{R}^n)$.

Example 2.3.5. Let $f_1, \ldots, f_k$ be a real-valued C^∞ functions on U. Letting $\omega_i = df_i$ we get from (2.3.4) a k-form

$$df_1 \wedge \cdots \wedge df_k$$

whose value at p is the wedge product

$$(df_1)_p \wedge \cdots \wedge (df_k)_p. \tag{2.3.6}$$

Since $(dx_1)_p, \ldots, (dx_n)_p$ are a basis of $T_p^\star \mathbf{R}^n$, the wedge products

$$(dx_{i_1})_p \wedge \cdots \wedge (dx_{1_k})_p, \quad 1 \le i_1 < \cdots < i_k \le n \tag{2.3.7}$$

are a basis of $\Lambda^k(T_p^\star)$. To keep our multi-index notation from getting out of hand, we denote these basis vectors by $(dx_I)_p$, where $I = (i_1, \ldots, i_k)$ and the multi-indices

I range over multi-indices of length k which are *strictly increasing*. Since these wedge products are a basis of $\Lambda^k(T_p^*\mathbf{R}^n)$, every element of $\Lambda^k(T_p^*\mathbf{R}^n)$ can be written uniquely as a sum

$$\sum_I c_I(dx_I)_p, \quad c_I \in \mathbf{R}$$

and every k-form ω on U can be written uniquely as a sum

$$\omega = \sum_I f_I dx_I, \tag{2.3.8}$$

where dx_I is the k-form $dx_{i_1} \wedge \cdots \wedge dx_{i_k}$, and f_I is a real-valued function $f_I : U \to \mathbf{R}$.

Definition 2.3.9. The k-form (2.3.8) is ***of class*** C^r if each function f_I is in $C^r(U)$.

Henceforth we assume, unless otherwise stated, that *all the k-forms we consider are of class* C^∞. We denote the set of k-forms of class C^∞ on U by $\Omega^k(U)$.

We will conclude this section by discussing a few simple operations on k-forms. Let U be an open subset of $\mathbf{R}^n$.

(1) Given a function $f \in C^\infty(U)$ and a k-form $\omega \in \Omega^k(U)$, we define $f\omega \in \Omega^k(U)$ to be the k-form defined by

$$p \mapsto f(p)\omega_p.$$

(2) Given $\omega_1, \omega_2 \in \Omega^k(U)$, we define $\omega_1 + \omega_2 \in \Omega^k(U)$ to be the k-form

$$p \mapsto (\omega_1)_p + (\omega_2)_p.$$

(Notice that this sum makes sense since each summand is in $\Lambda^k(T_p^*\mathbf{R}^n)$.)

(3) Given $\omega_1 \in \Omega^{k_1}(U)$ and $\omega_2 \in \Omega^{k_2}(U)$ we define their ***wedge product***, $\omega_1 \wedge \omega_2 \in \Omega^{k_1+k_2}(U)$, to be the $(k_1 + k_2)$-form

$$p \mapsto (\omega_1)_p \wedge (\omega_2)_p.$$

We recall that $\Lambda^0(T_p^*\mathbf{R}^n) = \mathbf{R}$, so a 0-form is an $\mathbf{R}$-valued function and a 0-form of class C^∞ is a C^∞ function, i.e., $\Omega^0(U) = C^\infty(U)$.

The addition and multiplication by functions operations naturally give the sets $\Omega^k(U)$ the structures of $\mathbf{R}$-vector spaces — we always regard $\Omega^k(U)$ as a vector space in this manner.

A fundamental operation on forms is the exterior differentiation operation which associates to a function $f \in C^\infty(U)$ the 1-form df. It is clear from the identity (2.2.3) that df is a 1-form of class C^∞, so the d-operation can be viewed as a map

$$d : \Omega^0(U) \to \Omega^1(U).$$

In the next section we show that an analogue of this map exists for every $\Omega^k(U)$.

Exercises for §2.3

Exercise 2.3.i. Let $\omega \in \Omega^2(\mathbf{R}^4)$ be the 2-form $dx_1 \wedge dx_2 + dx_3 \wedge dx_4$. Compute $\omega \wedge \omega$.

Exercise 2.3.ii. Let $\omega_1, \omega_2, \omega_3 \in \Omega^1(\mathbf{R}^3)$ be the 1-forms

$$\omega_1 = x_2 dx_3 - x_3 dx_2,$$
$$\omega_2 = x_3 dx_1 - x_1 dx_3,$$
$$\omega_3 = x_1 dx_2 - x_2 dx_1.$$

Compute the following.

(1) $\omega_1 \wedge \omega_2$.
(2) $\omega_2 \wedge \omega_3$.
(3) $\omega_3 \wedge \omega_1$.
(4) $\omega_1 \wedge \omega_2 \wedge \omega_3$.

Exercise 2.3.iii. Let U be an open subset of $\mathbf{R}^n$ and $f_1, \ldots, f_n \in C^\infty(U)$. Show that

$$df_1 \wedge \cdots \wedge df_n = \det\left[\frac{\partial f_i}{\partial x_j}\right] dx_1 \wedge \cdots \wedge dx_n.$$

Exercise 2.3.iv. Let U be an open subset of $\mathbf{R}^n$. Show that every $(n-1)$-form $\omega \in \Omega^{n-1}(U)$ can be written uniquely as a sum

$$\sum_{i=1}^{n} f_i dx_1 \wedge \cdots \wedge \widehat{dx_i} \wedge \cdots \wedge dx_n,$$

where $f_i \in C^\infty(U)$ and $\widehat{dx_i}$ indicates that dx_i is to be omitted from the wedge product $dx_1 \wedge \cdots \wedge dx_n$.

Exercise 2.3.v. Let $\mu = \sum_{i=1}^{n} x_i dx_i$. Show that there exists an $(n-1)$-form, $\omega \in \Omega^{n-1}(\mathbf{R}^n \smallsetminus \{0\})$ with the property

$$\mu \wedge \omega = dx_1 \wedge \cdots \wedge dx_n.$$

Exercise 2.3.vi. Let J be the multi-index $(j_1, \ldots, j_k)$ and let $dx_J = dx_{j_1} \wedge \cdots \wedge dx_{j_k}$. Show that $dx_J = 0$ if $j_r = j_s$ for some $r \neq s$ and show that if the numbers $j_1, \ldots, j_k$ are all distinct, then

$$dx_J = (-1)^\sigma dx_I,$$

where $I = (i_1, \ldots, i_k)$ is the strictly increasing rearrangement of $(j_1, \ldots, j_k)$ and σ is the permutation

$$(j_1, \ldots, j_k) \mapsto (i_1, \ldots, i_k).$$

Exercise 2.3.vii. Let I be a strictly increasing multi-index of length k and J a strictly increasing multi-index of length ℓ. What can one say about the wedge product $dx_I \wedge dx_J$?

2.4. Exterior differentiation

Let U be an open subset of $\mathbf{R}^n$. In this section we define an *exterior differentiation* operation

(2.4.1) $$d\colon \Omega^k(U) \to \Omega^{k+1}(U).$$

This operation is the fundamental operation in n-dimensional vector calculus.

For $k = 0$ we already defined the operation (2.4.1) in §2.1. Before defining it for $k > 0$, we list some properties that we require to this operation to satisfy.

Properties 2.4.2 (desired properties of exterior differentiation). Let U be an open subset of $\mathbf{R}^n$.

(1) For ω_1 and ω_2 in $\Omega^k(U)$, we have

$$d(\omega_1 + \omega_2) = d\omega_1 + d\omega_2.$$

(2) For $\omega_1 \in \Omega^k(U)$ and $\omega_2 \in \Omega^\ell(U)$, we have

(2.4.3) $$d(\omega_1 \wedge \omega_2) = d\omega_1 \wedge \omega_2 + (-1)^k \omega_1 \wedge d\omega_2.$$

(3) For all $\omega \in \Omega^k(U)$, we have

(2.4.4) $$d(d\omega) = 0.$$

Let us point out a few consequences of these properties.

Lemma 2.4.5. *Let U be an open subset of $\mathbf{R}^n$. For any functions $f_1, \dots, f_k \in C^\infty(U)$, we have*

$$d(df_1 \wedge \cdots \wedge df_k) = 0.$$

Proof. We prove this by induction on k. For the base case note that by (3)

(2.4.6) $$d(df_1) = 0$$

for every function $f_1 \in C^\infty(U)$.

For the induction step, suppose that we know the result for $k-1$ functions and that we are given functions $f_1, \dots, f_k \in C^\infty(U)$. Let $\mu = df_2 \wedge \cdots \wedge df_k$. Then, by the induction hypothesis $d\mu = 0$, combining (2.4.3) with (2.4.6) we see that

$$\begin{aligned} d(df_1 \wedge df_2 \wedge \cdots \wedge df_k) &= d(df_1 \wedge \mu) \\ &= d(df_1) \wedge \mu + (-1) df_1 \wedge d\mu \\ &= 0. \end{aligned}$$

□

Example 2.4.7. As a special case of Lemma 2.4.5, given a multi-index $I = (i_1, \dots, i_k)$ with $1 \le i_r \le n$ we have

(2.4.8) $$d(dx_I) = d(dx_{i_1} \wedge \cdots \wedge dx_{i_k}) = 0.$$

Recall that every k-form $\omega \in \Omega^k(U)$ can be written uniquely as a sum

$$\omega = \sum_I f_I dx_I\,, \quad f_I \in C^\infty(U),$$

where the multi-indices I are strictly increasing. Thus by (2.4.3) and (2.4.8)

$$d\omega = \sum_I df_I \wedge dx_I. \tag{2.4.9}$$

This shows that if there exists an operator $d\colon \Omega^*(U) \to \Omega^{*+1}(U)$ with Properties 2.4.2, then d is necessarily given by the formula (2.4.9). Hence all we have to show is that the operator defined by this equation (2.4.9) has these properties.

Proposition 2.4.10. *Let U be an open subset of $\mathbf{R}^n$. There is a unique operator $d\colon \Omega^*(U) \to \Omega^{*+1}(U)$ satisfying Properties 2.4.2.*

Proof. The property (1) is obvious.

To verify (2) we first note that for I strictly increasing equation (2.4.8) is a special case of equation (2.4.9) (take $f_I = 1$ and $f_J = 0$ for $J \neq I$). Moreover, if I is not strictly increasing it is either repeating, in which case $dx_I = 0$, or non-repeating in which case there exists a permutation $\sigma \in S_k$ such that I^σ is strictly increasing. Moreover,

$$dx_I = (-1)^\sigma dx_{I^\sigma}. \tag{2.4.11}$$

Hence (2.4.9) implies (2.4.8) for *all* multi-indices I. The same argument shows that for any sum $\sum_I f_I dx_I$ over multi-indices I of length k, we have

$$d(\textstyle\sum_I f_I dx_I) = \displaystyle\sum_I df_I \wedge dx_I. \tag{2.4.12}$$

(As above we can ignore the repeating multi-indices $dx_I = 0$ if I is repeating, and by (2.4.11) we can replace the non-repeating multi-indices by strictly increasing multi-indices.)

Suppose now that $\omega_1 \in \Omega^k(U)$ and $\omega_2 \in \Omega^\ell(U)$. Write $\omega_1 \coloneqq \sum_I f_I dx_I$ and $\omega_2 \coloneqq \sum_J g_J dx_J$, where $f_I, g_J \in C^\infty(U)$. We see that the wedge product $\omega_1 \wedge \omega_2$ is given by

$$\omega_1 \wedge \omega_2 = \sum_{I,J} f_I g_J dx_I \wedge dx_J, \tag{2.4.13}$$

and by equation (2.4.12)

$$d(\omega_1 \wedge \omega_2) = \sum_{I,J} d(f_I g_J) \wedge dx_I \wedge dx_J. \tag{2.4.14}$$

(Notice that if $I = (i_1, \ldots, i_k)$ and $J = (j_i, \ldots, i_\ell)$, we have $dx_I \wedge dx_J = dx_K$, where $K \coloneqq (i_1, \ldots, i_k, j_1, \ldots, j_\ell)$. Even if I and J are strictly increasing, K is not necessarily strictly increasing. However, in deducing (2.4.14) from (2.4.13) we have observed that this does not matter.)

Now note that by equation (2.2.8)

$$d(f_I g_J) = g_J df_I + f_I dg_J,$$

and by the wedge product identities of §1.6,

$$dg_J \wedge dx_I = dg_J \wedge dx_{i_1} \wedge \cdots \wedge dx_{i_k} = (-1)^k dx_I \wedge dg_J,$$

so the sum (2.4.14) can be rewritten:

$$\sum_{I,J} df_I \wedge dx_I \wedge g_J dx_J + (-1)^k \sum_{I,J} f_I dx_I \wedge dg_J \wedge dx_J,$$

or

$$\left(\sum_I df_I \wedge dx_I\right) \wedge \left(\sum_J g_J dx_J\right) + (-1)^k \left(\sum_J dg_J \wedge dx_J\right) \wedge \left(\sum_I df_I \wedge dx_I\right),$$

or, finally,

$$d\omega_1 \wedge \omega_2 + (-1)^k \omega_1 \wedge d\omega_2.$$

Thus the operator d defined by equation (2.4.9) satisfies (2).

Let us now check that d satisfies (3). If $\omega = \sum_I f_I dx_I$, where $f_I \in C^\infty(U)$, then by definition, $d\omega = \sum_I df_I \wedge dx_I$ and by (2.4.8) and (2.4.3)

$$d(d\omega) = \sum_I d(df_I) \wedge dx_I,$$

so it suffices to check that $d(df_I) = 0$, i.e., it suffices to check (2.4.6) for 0-forms $f \in C^\infty(U)$. However, by (2.2.3) we have

$$df = \sum_{j=1}^n \frac{\partial f}{\partial x_j} dx_j$$

so by equation (2.4.9)

$$\begin{aligned} d(df) &= \sum_{j=1}^n d\left(\frac{\partial f}{\partial x_j}\right) dx_j = \sum_{j=1}^n \left(\sum_{i=1}^n \frac{\partial^2 f}{\partial x_i \partial x_j} dx_i\right) \wedge dx_j \\ &= \sum_{1 \le i,j \le n} \frac{\partial^2 f}{\partial x_i \partial x_j} dx_i \wedge dx_j. \end{aligned}$$

Notice, however, that in this sum, $dx_i \wedge dx_j = -dx_j \wedge dx_i$ and

$$\frac{\partial^2 f}{\partial x_i \partial x_j} = \frac{\partial^2 f}{\partial x_j \partial x_i}$$

so the (i, j) term cancels the (j, i) term, and the total sum is zero. □

Definition 2.4.15. Let U be an open subset of $\mathbf{R}^n$. A k-form $\omega \in \Omega^k(U)$ is ***closed*** if $d\omega = 0$ and is ***exact*** if $\omega = d\mu$ for some $\mu \in \Omega^{k-1}(U)$.

By (3) every exact form is closed, but the converse is not true even for 1-forms (see Exercise 2.1.iii). In fact it is a very interesting (and hard) question to determine if an open set U has the following property: *For $k > 0$ every closed k-form is exact.*[2]

Some examples of spaces with this property are described in the exercises at the end of §2.6. We also sketch below a proof of the following result (and ask you to fill in the details).

---

[2]For $k = 0$, $df = 0$ *does not* imply that f is exact. In fact "exactness" does not make much sense for zero forms since there are not any "(−1)-forms". However, if $f \in C^\infty(U)$ and $df = 0$ then f is constant on connected components of U (see Exercise 2.2.iii).

Lemma 2.4.16 (Poincaré lemma). *If ω is a closed form on U of degree $k > 0$, then for every point $p \in U$, there exists a neighborhood of p on which ω is exact.*

Proof. See Exercises 2.4.v and 2.4.vi. ☐

Exercises for §2.4

Exercise 2.4.i. Compute the exterior derivatives of the following differential forms.

(1) $x_1 dx_2 \wedge dx_3$;
(2) $x_1 dx_2 - x_2 dx_1$;
(3) $e^{-f} df$ where $f = \sum_{i=1}^n x_i^2$;
(4) $\sum_{i=1}^n x_i dx_i$;
(5) $\sum_{i=1}^n (-1)^i x_i dx_1 \wedge \cdots \wedge \widehat{dx_i} \wedge \cdots \wedge dx_n$.

Exercise 2.4.ii. Solve the equation $d\mu = \omega$ for $\mu \in \Omega^1(\mathbf{R}^3)$, where ω is the 2-form:

(1) $dx_2 \wedge dx_3$;
(2) $x_2 dx_2 \wedge dx_3$;
(3) $(x_1^2 + x_2^2) dx_1 \wedge dx_2$;
(4) $\cos(x_1) dx_1 \wedge dx_3$.

Exercise 2.4.iii. Let U be an open subset of $\mathbf{R}^n$.

(1) Show that if $\mu \in \Omega^k(U)$ is exact and $\omega \in \Omega^\ell(U)$ is closed then $\mu \wedge \omega$ is exact.
Hint: Equation (2.4.3).
(2) In particular, dx_1 is exact, so if $\omega \in \Omega^\ell(U)$ is closed $dx_1 \wedge \omega = d\mu$. What is μ?

Exercise 2.4.iv. Let Q be the rectangle $(a_1, b_1) \times \cdots \times (a_n, b_n)$. Show that if ω is in $\Omega^n(Q)$, then ω is exact.

Hint: Let $\omega = f dx_1 \wedge \cdots \wedge dx_n$ with $f \in C^\infty(Q)$ and let g be the function

$$g(x_1, \ldots, x_n) = \int_{a_1}^{x_1} f(t, x_2, \ldots, x_n) dt.$$

Show that $\omega = d(g dx_2 \wedge \cdots \wedge dx_n)$.

Exercise 2.4.v. Let U be an open subset of $\mathbf{R}^{n-1}$, $A \subset \mathbf{R}$ an open interval and (x, t) product coordinates on $U \times A$. We say that a form $\mu \in \Omega^\ell(U \times A)$ is ***reduced*** if μ can be written as a sum

$$\mu = \sum_I f_I(x, t) dx_I \tag{2.4.17}$$

(i.e., with no terms involving dt).

(1) Show that every form, $\omega \in \Omega^k(U \times A)$ can be written uniquely as a sum:

$$\omega = dt \wedge \alpha + \beta, \tag{2.4.18}$$

where α and β are reduced.

(2) Let μ be the reduced form (2.4.17) and let

$$\frac{d\mu}{dt} = \sum \frac{d}{dt} f_I(x, t) dx_I$$

and

$$d_U\mu = \sum_I \left(\sum_{i=1}^n \frac{\partial}{\partial x_i} f_I(x,t) dx_i \right) \wedge dx_I .$$

Show that

$$d\mu = dt \wedge \frac{d\mu}{dt} + d_U\mu .$$

(3) Let ω be the form (2.4.18). Show that

$$d\omega = -dt \wedge d_U\alpha + dt \wedge \frac{d\beta}{dt} + d_U\beta$$

and conclude that ω is closed if and only if

(2.4.19)
$$\begin{cases} \dfrac{d\beta}{dt} = d_U\alpha, \\ d_U\beta = 0. \end{cases}$$

(4) Let α be a reduced $(k-1)$-form. Show that there exists a reduced $(k-1)$-form ν such that

(2.4.20)
$$\frac{d\nu}{dt} = \alpha .$$

Hint: Let $\alpha = \sum_I f_I(x,t)dx_I$ and $\nu = \sum g_I(x,t)dx_I$. Equation (2.4.20) reduces to the system of equations

(2.4.21)
$$\frac{d}{dt} g_I(x,t) = f_I(x,t).$$

Let c be a point on the interval, A, and using calculus show that equation (2.4.21) has a unique solution, $g_I(x,t)$, with $g_I(x,c) = 0$.

(5) Show that if ω is the form (2.4.18) and ν a solution of (2.4.20) then the form

(2.4.22)
$$\omega - d\nu$$

is reduced.

(6) Let

$$\gamma = \sum h_I(x,t)dx_I$$

be a reduced k-form. Deduce from (2.4.19) that if γ is closed then $\frac{d\gamma}{dt} = 0$ and $d_U\gamma = 0$. Conclude that $h_I(x,t) = h_I(x)$ and that

$$\gamma = \sum_I h_I(x)dx_I$$

is effectively a closed k-form on U. Prove that if every closed k-form on U is exact, then every closed k-form on $U \times A$ is exact.

Hint: Let ω be a closed k-form on $U \times A$ and let γ be the form (2.4.22).

Exercise 2.4.vi. Let $Q \subset \mathbf{R}^n$ be an open rectangle. Show that every closed form on Q of degree $k > 0$ is exact.

Hint: Let $Q = (a_1, b_1) \times \cdots \times (a_n, b_n)$. Prove this assertion by induction, at the nth stage of the induction letting $U = (a_1, b_1) \times \cdots \times (a_{n-1}, b_{n-1})$ and $A = (a_n, b_n)$.

2.5. The interior product operation

In § 2.1 we explained how to pair a 1-form ω and a vector field $\boldsymbol{v}$ to get a function $\iota_v\omega$. This pairing operation generalizes.

Definition 2.5.1. Let U be an open subset of $\mathbf{R}^n$, $\boldsymbol{v}$ a vector field on U, and $\omega \in \Omega^k(U)$. The ***interior product*** of $\boldsymbol{v}$ with ω is $(k-1)$-form $\iota_v\omega$ on U defined by declaring the value of $\iota_v\omega$ value at $p \in U$ to be the interior product

$$\iota_{v(p)}\omega_p. \tag{2.5.2}$$

Note that $\boldsymbol{v}(p)$ is in $T_p\mathbf{R}^n$ and ω_p in $\Lambda^k(T_p^\star\mathbf{R}^n)$, so by definition of interior product (see § 1.7), the expression (2.5.2) is an element of $\Lambda^{k-1}(T_p^\star\mathbf{R}^n)$.

From the properties of interior product on vector spaces which we discussed in § 1.7, one gets analogous properties for this interior product on forms. We list these properties, leaving their verification as an easy exercise.

Properties 2.5.3. Let U be an open subset of $\mathbf{R}^n$, $\boldsymbol{v}$ and $\boldsymbol{w}$ vector fields on U, ω_1 and ω_2 both k-forms on U, and ω a k-form and μ an ℓ-form on U.

(1) *Linearity in the form:* we have

$$\iota_v(\omega_1 + \omega_2) = \iota_v\omega_1 + \iota_v\omega_2.$$

(2) *Linearity in the vector field:* we have

$$\iota_{v+w}\omega = \iota_v\omega + \iota_w\omega.$$

(3) *Derivation property:* we have

$$\iota_v(\omega \wedge \mu) = \iota_v\omega \wedge \mu + (-1)^k \omega \wedge \iota_v\mu.$$

(4) The interior product satisfies the identity

$$\iota_v(\iota_w\omega) = -\iota_w(\iota_v\omega).$$

(5) As a special case of (4), the interior product satisfies the identity

$$\iota_v(\iota_v\omega) = 0.$$

(6) Moreover, if ω is decomposable, i.e., is a wedge product of 1-forms

$$\omega = \mu_1 \wedge \cdots \wedge \mu_k,$$

then

$$\iota_v\omega = \sum_{r=1}^{k} (-1)^{r-1} \iota_v(\mu_r)\mu_1 \wedge \cdots \wedge \hat{\mu}_r \wedge \cdots \wedge \mu_k.$$

We also leave for you to prove the following two assertions, both of which are special cases of (6).

Example 2.5.4. If $\boldsymbol{v} = \partial/\partial x_r$ and $\omega = dx_I = dx_{i_1} \wedge \cdots \wedge dx_{i_k}$, then

$$\iota_v\omega = \sum_{i=1}^{k} (-1)^{i-1} \delta_{i,i_r} dx_{I_r}, \tag{2.5.5}$$

where

$$\delta_{i,i_r} := \begin{cases} 1, & i = i_r, \\ 0, & i \neq i_r, \end{cases}$$

and $I_r = (i_1, \ldots, \hat{i}_r, \ldots, i_k)$.

Example 2.5.6. If $\boldsymbol{v} = \sum_{i=1}^n f_i \partial/\partial x_i$ and $\omega = dx_1 \wedge \cdots \wedge dx_n$, then

$$\iota_v \omega = \sum_{r=1}^{n} (-1)^{r-1} f_r \, dx_1 \wedge \cdots \widehat{dx_r} \cdots \wedge dx_n. \tag{2.5.7}$$

By combining exterior differentiation with the interior product operation one gets another basic operation of vector fields on forms: the *Lie differentiation* operation. For 0-forms, i.e., for C^∞ functions, we defined this operation by the formula (2.1.16). For k-forms we define it by a slightly more complicated formula.

Definition 2.5.8. Let U be an open subset of $\mathbf{R}^n$, $\boldsymbol{v}$ a vector field on U, and $\omega \in \Omega^k(U)$. The ***Lie derivative*** of ω with respect to $\boldsymbol{v}$ is the k-form

$$L_v \omega := \iota_v(d\omega) + d(\iota_v \omega). \tag{2.5.9}$$

Notice that for 0-forms the second summand is zero, so (2.5.9) and (2.1.16) agree.

Properties 2.5.10. Let U be an open subset of $\mathbf{R}^n$, $\boldsymbol{v}$ a vector field on U, $\omega \in \Omega^k(U)$, and $\mu \in \Omega^\ell(U)$. The Lie derivative enjoys the following properties:

(1) *Commutativity with exterior differentiation:* we have

$$d(L_v \omega) = L_v(d\omega). \tag{2.5.11}$$

(2) *Interaction with wedge products:* we have

$$L_v(\omega \wedge \mu) = L_v \omega \wedge \mu + \omega \wedge L_v \mu. \tag{2.5.12}$$

From Properties 2.5.10 it is fairly easy to get an explicit formula for $L_v \omega$. Namely let ω be the k-form

$$\omega = \sum_I f_I dx_I, \quad f_I \in C^\infty(U)$$

and $\boldsymbol{v}$ the vector field

$$\boldsymbol{v} = \sum_{i=1}^{n} g_i \frac{\partial}{\partial x_i}, \quad g_i \in C^\infty(U).$$

By equation (2.5.12)

$$L_v(f_I dx_I) = (L_v f_I) dx_I + f_I (L_v dx_I)$$

and

$$L_v dx_I = \sum_{r=1}^{k} dx_{i_1} \wedge \cdots \wedge L_v dx_{i_r} \wedge \cdots \wedge dx_{i_k},$$

and by equation (2.5.11)

$$L_v dx_{i_r} = d L_v x_{i_r}$$

so to compute $L_v \omega$ one is reduced to computing $L_v x_{i_r}$ and $L_v f_I$.

However by equation (2.5.12)

$$L_v x_{i_r} = g_{i_r}$$

and

$$L_v f_I = \sum_{i=1}^{n} g_i \frac{\partial f_I}{\partial x_i}.$$

We leave the verification of Properties 2.5.10 as exercises, and also ask you to prove (by the method of computation that we have just sketched) the following *divergence formula*.

Lemma 2.5.13. *Let $U \subset \mathbf{R}^n$ be open, let $g_1, \ldots, g_n \in C^\infty(U)$, and set $\boldsymbol{v} := \sum_{i=1}^n g_i \, \partial/\partial x_i$. Then*

$$L_v(dx_1 \wedge \cdots \wedge dx_n) = \sum_{i=1}^{n} \left(\frac{\partial g_i}{\partial x_i} \right) dx_1 \wedge \cdots \wedge dx_n. \tag{2.5.14}$$

Exercises for §2.5

Exercise 2.5.i. Verify Properties 2.5.3(1)–(6).

Exercise 2.5.ii. Show that if ω is the k-form dx_I and $\boldsymbol{v}$ the vector field $\partial/\partial x_r$, then $\iota_v \omega$ is given by equation (2.5.5).

Exercise 2.5.iii. Show that if ω is the n-form $dx_1 \wedge \cdots \wedge dx_n$ and $\boldsymbol{v}$ the vector field $\sum_{i=1}^n f_i \partial/\partial x_i$, then $\iota_v \omega$ is given by equation (2.5.7).

Exercise 2.5.iv. Let U be an open subset of $\mathbf{R}^n$ and $\boldsymbol{v}$ a C^∞ vector field on U. Show that for $\omega \in \Omega^k(U)$

$$dL_v \omega = L_v d\omega$$

and

$$\iota_v(L_v \omega) = L_v(\iota_v \omega).$$

Hint: Deduce the first of these identities from the identity $d(d\omega) = 0$ and the second from the identity $\iota_v(\iota_v \omega) = 0$.

Exercise 2.5.v. Given $\omega_i \in \Omega^{k_i}(U)$, for $i = 1, 2$ show that

$$L_v(\omega_1 \wedge \omega_2) = L_v \omega_1 \wedge \omega_2 + \omega_1 \wedge L_v \omega_2.$$

Hint: Plug $\omega = \omega_1 \wedge \omega_2$ into equation (2.5.9) and use equation (2.4.3) and the derivation property of the interior product to evaluate the resulting expression.

Exercise 2.5.vi. Let $\boldsymbol{v}_1$ and $\boldsymbol{v}_2$ be vector fields on U and let $\boldsymbol{w}$ be their Lie bracket. Show that for $\omega \in \Omega^k(U)$

$$L_w \omega = L_{v_1}(L_{v_2}\omega) - L_{v_2}(L_{v_1}\omega).$$

Hint: By definition this is true for 0-forms and by equation (2.5.11) for exact 1-forms. Now use the fact that every form is a sum of wedge products of 0-forms and 1-forms and the fact that L_v satisfies the product identity (2.5.12).

Exercise 2.5.vii. Prove the divergence formula (2.5.14).

Exercise 2.5.viii.

(1) Let $\omega = \Omega^k(\mathbf{R}^n)$ be the form

$$\omega = \sum_I f_I(x_1, \dots, x_n) dx_I$$

and $\boldsymbol{v}$ the vector field $\partial/\partial x_n$. Show that

$$L_{\boldsymbol{v}}\omega = \sum \frac{\partial}{\partial x_n} f_I(x_1, \dots, x_n) dx_I.$$

(2) Suppose $\iota_{\boldsymbol{v}}\omega = L_{\boldsymbol{v}}\omega = 0$. Show that ω only depends on $x_1, \dots, x_{n-1}$ and $dx_1, \dots, dx_{n-1}$, i.e., is effectively a k-form on $\mathbf{R}^{n-1}$.

(3) Suppose $\iota_{\boldsymbol{v}}\omega = d\omega = 0$. Show that ω is effectively a closed k-form on $\mathbf{R}^{n-1}$.

(4) Use these results to give another proof of the Poincaré lemma for $\mathbf{R}^n$. Prove by induction on n that every closed form on $\mathbf{R}^n$ is exact.

Hints:

› Let ω be the form in part (1) and let

$$g_I(x_1, \dots, x_n) = \int_0^{x_n} f_I(x_1, \dots, x_{n-1}, t) dt.$$

Show that if $\nu = \sum_I g_I dx_I$, then $L_{\boldsymbol{v}}\nu = \omega$.

› Conclude that

(2.5.15)
$$\omega - d\iota_{\boldsymbol{v}}\nu = \iota_{\boldsymbol{v}} d\nu.$$

› Suppose $d\omega = 0$. Conclude from equation (2.5.15) and (5) of Properties 2.5.3 that the form $\beta = \iota_{\boldsymbol{v}} d\nu$ satisfies $d\beta = \iota(\boldsymbol{v})\beta = 0$.

› By part (3), β is effectively a closed form on $\mathbf{R}^{n-1}$, and by induction, $\beta = d\alpha$. Thus by equation (2.5.15)

$$\omega = d\iota_{\boldsymbol{v}}\nu + d\alpha.$$

2.6. The pullback operation on forms

Let U be an open subset of $\mathbf{R}^n$, V an open subset of $\mathbf{R}^m$, and $f : U \to V$ a C^∞ map. Then for $p \in U$ and the derivative of f at p

$$df_p \colon T_p\mathbf{R}^n \to T_{f(p)}\mathbf{R}^m$$

is a linear map, so (as explained in § 1.7) we get from df_p a pullback map

(2.6.1)
$$df_p^\star := (df_p)^\star : \Lambda^k(T^\star_{f(p)}\mathbf{R}^m) \to \Lambda^k(T^\star_p\mathbf{R}^n).$$

In particular, let ω be a k-form on V. Then at $f(p) \in V$, ω takes the value

$$\omega_{f(p)} \in \Lambda^k(T^\star_q\mathbf{R}^m),$$

so we can apply to it the operation (2.6.1), and this gives us an element:

(2.6.2)
$$df_p^\star \omega_{f(p)} \in \Lambda^k(T^\star_p\mathbf{R}^n).$$

In fact we can do this for every point $p \in U$, and the assignment

$$p \mapsto (df_p)^* \omega_{f(p)} \tag{2.6.3}$$

defines a k-form on U, which we denote by $f^*\omega$. We call $f^*\omega$ the ***pullback*** of ω along the map f. A few of its basic properties are described below.

Properties 2.6.4. Let U be an open subset of $\mathbf{R}^n$, V an open subset of $\mathbf{R}^m$, and $f : U \to V$ a C^∞ map.

(1) Let ϕ be a 0-form, i.e., a function $\phi \in C^\infty(V)$. Since

$$\Lambda^0(T_p^*) = \Lambda^0(T_{f(p)}^*) = \mathbf{R}$$

the map (2.6.1) is just the identity map of $\mathbf{R}$ onto $\mathbf{R}$ when $k = 0$. Hence for 0-forms

$$(f^*\phi)(p) = \phi(f(p)), \tag{2.6.5}$$

that is, $f^*\phi$ is just the composite function, $\phi \circ f \in C^\infty(U)$.

(2) Let $\phi \in \Omega^0(U)$ and let $\mu \in \Omega^1(V)$ be the 1-form $\mu = d\phi$. By the chain rule (2.6.2) unwinds to:

$$(df_p)^* d\phi_q = (d\phi)_q \circ df_p = d(\phi \circ f)_p \tag{2.6.6}$$

and hence by (2.6.5) we have

$$f^* \, d\phi = df^*\phi. \tag{2.6.7}$$

(3) If $\omega_1, \omega_2 \in \Omega^k(V)$ from equation (2.6.2) we see that

$$(df_p)^*(\omega_1 + \omega_2)_q = (df_p)^*(\omega_1)_q + (df_p)^*(\omega_2)_q,$$

and hence by (2.6.3) we have

$$f^*(\omega_1 + \omega_2) = f^*\omega_1 + f^*\omega_2.$$

(4) We observed in §1.7 that the operation (2.6.1) commutes with wedge product, hence if $\omega_1 \in \Omega^k(V)$ and $\omega_2 \in \Omega^\ell(V)$

$$df_p^*(\omega_1)_q \wedge (\omega_2)_q = df_p^*(\omega_1)_q \wedge df_p^*(\omega_2)_q.$$

In other words

$$f^*\omega_1 \wedge \omega_2 = f^*\omega_1 \wedge f^*\omega_2. \tag{2.6.8}$$

(5) Let W be an open subset of $\mathbf{R}^k$ and $g : V \to W$ a C^∞ map. Given a point $p \in U$, let $q = f(p)$ and $w = g(q)$. Then the composition of the map

$$(df_p)^* : \Lambda^k(T_q^*) \to \Lambda^k(T_p^*)$$

and the map

$$(dg_q)^* : \Lambda^k(T_w^*) \to \Lambda^k(T_q^*)$$

is the map

$$(dg_q \circ df_p)^* : \Lambda^k(T_w^*) \to \Lambda^k(T_p^*)$$

by formula (1.7.5). However, by the chain rule

$$(dg_q) \circ (df)_p = d(g \circ f)_p$$

so this composition is the map

$$d(g \circ f)_p^\star : \Lambda^k(T_w^\star) \to \Lambda^k(T_p^\star).$$

Thus if $\omega \in \Omega^k(W)$

$$f^\star(g^\star \omega) = (g \circ f)^\star \omega.$$

Let us see what the pullback operation looks like in coordinates. Using multi-index notation we can express every k-form $\omega \in \Omega^k(V)$ as a sum over multi-indices of length k

$$\omega = \sum_I \phi_I dx_I, \tag{2.6.9}$$

the coefficient ϕ_I of dx_I being in $C^\infty(V)$. Hence by (2.6.5)

$$f^\star \omega = \sum f^\star \phi_I f^\star(dx_I),$$

where $f^\star \phi_I$ is the function of $\phi \circ f$.

What about $f^\star dx_I$? If I is the multi-index, $(i_1, \dots, i_k)$, then by definition

$$dx_I = dx_{i_1} \wedge \cdots \wedge dx_{i_k}$$

so

$$f^\star(dx_I) = f^\star(dx_{i_1}) \wedge \cdots \wedge f^\star(dx_{i_k})$$

by (2.6.8), and by (2.6.7), we have

$$f^\star dx_i = d f^\star x_i = d f_i,$$

where f_i is the ith coordinate function of the map f. Thus, setting

$$d f_I \coloneqq d f_{i_1} \wedge \cdots \wedge d f_{i_k},$$

we see that for each multi-index I we have

$$f^\star(dx_I) = d f_I, \tag{2.6.10}$$

and for the pullback of the form (2.6.9)

$$f^\star \omega = \sum_I f^\star \phi_I \, d f_I. \tag{2.6.11}$$

We will use this formula to prove that pullback commutes with exterior differentiation:

$$d(f^\star \omega) = f^\star(d\omega). \tag{2.6.12}$$

To prove this we recall that by (2.3.6) we have $d(d f_I) = 0$, hence by (2.3.1) and (2.6.10) we have

$$\begin{aligned} d(f^\star \omega) &= \sum_I d(f^\star \phi_I) \wedge d f_I = \sum_I f^\star(d\phi_I) \wedge f^\star(dx_I) \\ &= f^\star \sum_I d\phi_I \wedge dx_I = f^\star(d\omega). \end{aligned}$$

A special case of formula (2.6.10) will be needed in Chapter 4: Let U and V be open subsets of $\mathbf{R}^n$ and let $\omega = dx_1 \wedge \cdots \wedge dx_n$. Then by (2.6.10) we have

$$f^\star \omega_p = (df_1)_p \wedge \cdots \wedge (df_n)_p$$

for all $p \in U$. However,

$$(df_i)_p = \sum \frac{\partial f_i}{\partial x_j}(p)(dx_j)_p$$

and hence by formula (1.7.10)

$$f^\star \omega_p = \det \left[\frac{\partial f_i}{\partial x_j}(p) \right] (dx_1 \wedge \cdots \wedge dx_n)_p.$$

In other words

$$f^\star(dx_1 \wedge \cdots \wedge dx_n) = \det \left[\frac{\partial f_i}{\partial x_j} \right] dx_1 \wedge \cdots \wedge dx_n. \tag{2.6.13}$$

In Exercise 2.6.iv and equation (2.6.6) we outline the proof of an important topological property of the pullback operation.

Definition 2.6.14. Let U be an open subset of $\mathbf{R}^n$, V an open subset of $\mathbf{R}^m$, $A \subset \mathbf{R}$ an open interval containing 0 and 1, and $f_0, f_1 : U \to V$ two C^∞ maps. A C^∞ map $F : U \times A \to V$ is a C^∞ ***homotopy*** between f_0 and f_1 if $F(x,0) = f_0(x)$ and $F(x,1) = f_1(x)$.

If there exists a homotopy between f_0 and f_1, we say that f_0 and f_1 are ***homotopic*** and write $f_0 \simeq f_1$.

Intuitively, f_0 and f_1 are homotopic if there exists a family of C^∞ maps $f_t : U \to V$, where $f_t(x) = F(x,t)$, which "smoothly deform f_0 into f_1". In Exercise 2.6.iv and equation (2.6.6) you will be asked to verify that for f_0 and f_1 to be homotopic they have to satisfy the following criteria.

Theorem 2.6.15. *Let U be an open subset of $\mathbf{R}^n$, V an open subset of $\mathbf{R}^m$, and $f_0, f_1 : U \to V$ two C^∞ maps. If f_0 and f_1 are homotopic then for every closed form $\omega \in \Omega^k(V)$ the form $f_1^\star \omega - f_0^\star \omega$ is exact.*

This theorem is closely related to the Poincaré lemma, and, in fact, one gets from it a slightly stronger version of the Poincaré lemma than that described in Exercise 2.3.iv and equation (2.3.7).

Definition 2.6.16. An open subset U of $\mathbf{R}^n$ is ***contractible*** if, for some point $p_0 \in U$, the identity map $\mathrm{id}_U : U \to U$ is homotopic to the constant map

$$f_0 : U \to U\,, \quad f_0(p) = p_0$$

at p_0.

From Theorem 2.6.15 it is easy to see that the Poincaré lemma holds for contractible open subsets of $\mathbf{R}^n$. If U is contractible, every closed k-form on U of degree $k > 0$ is exact. To see this, note that if $\omega \in \Omega^k(U)$, then for the identity map $\mathrm{id}_U^\star \omega = \omega$ and if f is the constant map at a point of U, then $f^\star \omega = 0$.

Exercises for §2.6

Exercise 2.6.i. Let $f : \mathbf{R}^3 \to \mathbf{R}^3$ be the map

$$f(x_1, x_2, x_3) = (x_1 x_2, x_2 x_3^2, x_3^3).$$

Compute the pullback, $f^\star \omega$ for the following forms:

(1) $\omega = x_2 dx_3$;
(2) $\omega = x_1 dx_1 \wedge dx_3$;
(3) $\omega = x_1 dx_1 \wedge dx_2 \wedge dx_3$.

Exercise 2.6.ii. Let $f : \mathbf{R}^2 \to \mathbf{R}^3$ be the map

$$f(x_1, x_2) = (x_1^2, x_2^2, x_1 x_2).$$

Complete the pullback, $f^\star \omega$, for the following forms:

(1) $\omega = x_2 dx_2 + x_3 dx_3$;
(2) $\omega = x_1 dx_2 \wedge dx_3$;
(3) $\omega = dx_1 \wedge dx_2 \wedge dx_3$.

Exercise 2.6.iii. Let U be an open subset of $\mathbf{R}^n$, V an open subset of $\mathbf{R}^m$, $f : U \to V$ a C^∞ map, and $\gamma : [a, b] \to U$ a C^∞ curve. Show that for $\omega \in \Omega^1(V)$

$$\int_\gamma f^\star \omega = \int_{\gamma_1} \omega,$$

where $\gamma_1 : [a, b] \to V$ is the curve, $\gamma_1(t) = f(\gamma(t))$. (See Exercise 2.2.viii.)

Exercise 2.6.iv. Let U be an open subset of $\mathbf{R}^n$, $A \subset \mathbf{R}$ an open interval containing the points, 0 and 1, and (x, t) product coordinates on $U \times A$. Recall from Exercise 2.3.v that a form $\mu \in \Omega^\ell(U \times A)$ is *reduced* if it can be written as a sum

$$\mu = \sum_I f_I(x, t) dx_I$$

(i.e., none of the summands involve dt). For a reduced form μ, let $Q\mu \in \Omega^\ell(U)$ be the form

$$Q\mu := \left(\sum_I \int_0^1 f_I(x, t) dt \right) dx_I \tag{2.6.17}$$

and let $\mu_1, \mu_2 \in \Omega^\ell(U)$ be the forms

$$\mu_0 = \sum_I f_I(x, 0) dx_I$$

and

$$\mu_1 = \sum_I f_I(x, 1) dx_I.$$

Now recall from Exercise 2.4.v that every form $\omega \in \Omega^k(U \times A)$ can be written uniquely as a sum

$$\omega = dt \wedge \alpha + \beta, \tag{2.6.18}$$

where α and β are reduced.

(1) Prove the following theorem.

Theorem 2.6.19. *If the form (2.6.18) is closed then*

$$\beta_1 - \beta_0 = dQ\alpha. \tag{2.6.20}$$

Hint: Equation (2.4.19).

(2) Let ι_0 and ι_1 be the maps of U into $U \times A$ defined by $\iota_0(x) = (x, 0)$ and $\iota_1(x) = (x, 1)$. Show that (2.6.20) can be rewritten

$$\iota_1^\star \omega - \iota_0^\star \omega = dQ\alpha. \tag{2.6.21}$$

Exercise 2.6.v. Let V be an open subset of $\mathbf{R}^m$ and let $f_0, f_1 : U \to V$ be C^∞ maps. Suppose f_0 and f_1 are homotopic. Show that for every closed form, $\mu \in \Omega^k(V)$, $f_1^\star \mu - f_0^\star \mu$ is exact.

Hint: Let $F : U \times A \to V$ be a homotopy between f_0 and f_1 and let $\omega = F^\star \mu$. Show that ω is closed and that $f_0^\star \mu = \iota_0^\star \omega$ and $f_1^\star \mu = \iota_1^\star \omega$. Conclude from equation (2.6.21) that

$$f_1^\star \mu - f_0^\star \mu = dQ\alpha,$$

where $\omega = dt \wedge \alpha + \beta$ and α and β are reduced.

Exercise 2.6.vi. Show that if $U \subset \mathbf{R}^n$ is a contractible open set, then the Poincaré lemma holds: every closed form of degree $k > 0$ is exact.

Exercise 2.6.vii. An open subset, U, of $\mathbf{R}^n$ is said to be ***star-shaped*** if there exists a point $p_0 \in U$, with the property that for every point $p \in U$, the line segment,

$$\{ tp + (1-t)p_0 \mid 0 \le t \le 1 \},$$

joining p to p_0 is contained in U. Show that if U is star-shaped it is contractible.

Exercise 2.6.viii. Show that the following open sets are star-shaped.

(1) The open unit ball

$$\{ x \in \mathbf{R}^n \mid |x| < 1 \}.$$

(2) The open rectangle, $I_1 \times \cdots \times I_n$, where each I_k is an open subinterval of $\mathbf{R}$.
(3) $\mathbf{R}^n$ itself.
(4) Product sets

$$U_1 \times U_2 \subset \mathbf{R}^n \cong \mathbf{R}^{n_1} \times \mathbf{R}^{n_2},$$

where U_i is a star-shaped open set in $\mathbf{R}^{n_i}$.

Exercise 2.6.ix. Let U be an open subset of $\mathbf{R}^n$, $f_t : U \to U$, $t \in \mathbf{R}$, a one-parameter group of diffeomorphisms and v its infinitesimal generator. Given $\omega \in \Omega^k(U)$ show that at $t = 0$

$$\frac{d}{dt} f_t^\star \omega = L_v \omega. \tag{2.6.22}$$

Here is a sketch of a proof:

(1) Let $\gamma(t)$ be the curve, $\gamma(t) = f_t(p)$, and let ϕ be a 0-form, i.e., an element of $C^\infty(U)$. Show that

$$f_t^\star \phi(p) = \phi(\gamma(t))$$

and by differentiating this identity at $t = 0$ conclude that (2.6.22) holds for 0-forms.

(2) Show that if (2.6.22) holds for ω it holds for $d\omega$.

Hint: Differentiate the identity

$$f_t^\star d\omega = d f_t^\star \omega$$

at $t = 0$.

(3) Show that if (2.6.22) holds for ω_1 and ω_2 it holds for $\omega_1 \wedge \omega_2$.

Hint: Differentiate the identity

$$f_t^\star(\omega_1 \wedge \omega_2) = f_t^\star \omega_1 \wedge f_t^\star \omega_2$$

at $t = 0$.

(4) Deduce (2.6.22) from (1)–(3).

Hint: Every k-form is a sum of wedge products of 0-forms and exact 1-forms.

Exercise 2.6.x. In Exercise 2.6.ix show that for *all* t

$$\frac{d}{dt} f_t^\star \omega = f_t^\star L_v \omega = L_v f_t^\star \omega. \tag{2.6.23}$$

Hint: By the definition of a one-parameter group, $f_{s+t} = f_s \circ f_t = f_r \circ f_s$, hence:

$$f_{s+t}^\star \omega = f_t^\star(f_s^\star \omega) = f_s^\star(f_t^\star \omega).$$

Prove the first assertion by differentiating the first of these identities with respect to s and then setting $s = 0$, and prove the second assertion by doing the same for the second of these identities.

In particular conclude that

$$f_t^\star L_v \omega = L_v f_t^\star \omega.$$

Exercise 2.6.xi.

(1) By massaging the result above show that

$$\frac{d}{dt} f_t^\star \omega = dQ_t \omega + Q_t d\omega,$$

where

$$Q_t \omega = f_t^\star \iota_v \omega.$$

Hint: Equation (2.5.9).

(2) Let

$$Q\omega = \int_0^1 f_t^\star \iota_v \omega dt.$$

Prove the homotopy identity

$$f_1^\star \omega - f_0^\star \omega = dQ\omega + Qd\omega.$$

Exercise 2.6.xii. Let U be an open subset of $\mathbf{R}^n$, V an open subset of $\mathbf{R}^m$, $\boldsymbol{v}$ a vector field on U, $\boldsymbol{w}$ a vector field on V, and $f : U \to V$ a C^∞ map. Show that if $\boldsymbol{v}$ and $\boldsymbol{w}$ are f-related

$$\iota_{\boldsymbol{v}} f^\star \omega = f^\star \iota_{\boldsymbol{w}} \omega.$$

2.7. Divergence, curl, and gradient

The basic operations in three-dimensional vector calculus: gradient, curl, and divergence are, by definition, operations on *vector fields*. As we see below these operations are closely related to the operations

$$d\colon \Omega^k(\mathbf{R}^3) \to \Omega^{k+1}(\mathbf{R}^3) \tag{2.7.1}$$

in degrees $k = 0, 1, 2$. However, only the gradient and divergence generalize to n dimensions. (They are essentially the d-operations in degrees zero and $n-1$.) And, unfortunately, there is no simple description in terms of vector fields for the other d-operations. This is one of the main reasons why an adequate theory of vector calculus in n-dimensions forces on one the differential form approach that we have developed in this chapter.

Even in three dimensions, however, there is a good reason for replacing the gradient, divergence, and curl by the three operations (2.7.1). A problem that physicists spend a lot of time worrying about is the problem of *general covariance*: formulating the laws of physics in such a way that they admit as large a set of symmetries as possible, and frequently these formulations involve differential forms. An example is Maxwell's equations, the fundamental laws of electromagnetism. These are usually expressed as identities involving div and curl. However, as we explain below, there is an alternative formulation of Maxwell's equations based on the operations (2.7.1), and from the point of view of general covariance, this formulation is much more satisfactory: the only symmetries of $\mathbf{R}^3$ which preserve div and curl are translations and rotations, whereas the operations (2.7.1) admit all diffeomorphisms of $\mathbf{R}^3$ as symmetries.

To describe how the gradient, divergence, and curl are related to the operations (2.7.1) we first note that there are two ways of converting vector fields into forms.

The first makes use of the natural inner product $B(v, w) = \sum v_i w_i$ on $\mathbf{R}^n$. From this inner product one gets by Exercise 1.2.ix a bijective linear map

$$L\colon \mathbf{R}^n \xrightarrow{\sim} (\mathbf{R}^n)^*$$

with the defining property: $L(v) = \ell$ if and only if $\ell(w) = B(v, w)$. Via the identification (2.1.2) B and L can be transferred to $T_p\mathbf{R}^n$, giving one an inner product B_p on $T_p\mathbf{R}^n$ and a bijective linear map

$$L_p\colon T_p\mathbf{R}^n \xrightarrow{\sim} T_p^*\mathbf{R}^n.$$

Hence if we are given a vector field $\boldsymbol{v}$ on U we can convert it into a 1-form $\boldsymbol{v}^\sharp$ by setting

$$\boldsymbol{v}^\sharp(p) := L_p \boldsymbol{v}(p) \tag{2.7.2}$$

and this sets up a bijective correspondence between vector fields and 1-forms.

For instance

$$\boldsymbol{v} = \frac{\partial}{\partial x_i} \iff \boldsymbol{v}^\sharp = dx_i$$

(see Exercise 2.7.iii) and, more generally,

$$v = \sum_{i=1}^{n} f_i \frac{\partial}{\partial x_i} \iff v^\sharp = \sum_{i=1}^{n} f_i dx_i. \tag{2.7.3}$$

Example 2.7.4. In particular if f is a C^∞ function on $U \subset \mathbf{R}^n$ the ***gradient*** of f is the vector field

$$\mathrm{grad}(f) := \sum_{i=1}^{n} \frac{\partial f}{\partial x_i},$$

and this gets converted by (2.7.5) into the 1-form df. Thus the gradient operation in vector calculus is basically just the exterior derivative operation $d\colon \Omega^0(U) \to \Omega^1(U)$.

The second way of converting vector fields into forms is via the interior product operation. Namely let Ω be the n-form $dx_1 \wedge \cdots \wedge dx_n$. Given an open subset U of $\mathbf{R}^n$ and a C^∞ vector field

$$v = \sum_{i=1}^{n} f_i \frac{\partial}{\partial x_i} \tag{2.7.5}$$

on U the interior product of v with Ω is the $(n-1)$-form

$$\iota_v \Omega = \sum_{r=1}^{n} (-1)^{r-1} f_r dx_1 \wedge \cdots \wedge \widehat{dx_r} \cdots \wedge dx_n. \tag{2.7.6}$$

Moreover, every $(n-1)$-form can be written uniquely as such a sum, so (2.7.5) and (2.7.6) set up a bijective correspondence between vector fields and $(n-1)$-forms. Under this correspondence the d-operation gets converted into an operation on vector fields

$$v \mapsto d\iota_v \Omega.$$

Moreover, by (2.5.9)

$$d\iota_v \Omega = L_v \Omega$$

and by (2.5.14)

$$L_v \Omega = \mathrm{div}(v)\Omega,$$

where

$$\mathrm{div}(v) = \sum_{i=1}^{n} \frac{\partial f_i}{\partial x_i}. \tag{2.7.7}$$

In other words, this correspondence between $(n-1)$-forms and vector fields converts the d-operation into the divergence operation (2.7.7) on vector fields.

Notice that divergence and gradient are well-defined as vector calculus operations in n dimensions, even though one usually thinks of them as operations in three-dimensional vector calculus. The curl operation, however, is intrinsically a three-dimensional vector calculus operation. To define it we note that by (2.7.6) every 2-form μ on an open subset $U \subset \mathbf{R}^3$ can be written uniquely as an interior product,

$$\mu = \iota_w dx_1 \wedge dx_2 \wedge dx_3, \tag{2.7.8}$$

for some vector field $\boldsymbol{w}$, and the left-hand side of this formula determines $\boldsymbol{w}$ uniquely.

Definition 2.7.9. Let U be an open subset of $\mathbf{R}^3$ and $\boldsymbol{v}$ a vector field on U. From $\boldsymbol{v}$ we get by (2.7.3) a 1-form $\boldsymbol{v}^\sharp$, and hence by (2.7.8) a vector field $\boldsymbol{w}$ satisfying

$$d\boldsymbol{v}^\sharp = \iota_w dx_1 \wedge dx_2 \wedge dx_3.$$

The *curl* of $\boldsymbol{v}$ is the vector field

$$\mathrm{curl}(\boldsymbol{v}) \coloneqq \boldsymbol{w}.$$

We leave for you to check that this definition coincides with the definition one finds in calculus books. More explicitly we leave for you to check that if $\boldsymbol{v}$ is the vector field

$$\boldsymbol{v} = f_1 \frac{\partial}{\partial x_1} + f_2 \frac{\partial}{\partial x_2} + f_3 \frac{\partial}{\partial x_3},$$

then

$$\mathrm{curl}\,(\boldsymbol{v}) = g_1 \frac{\partial}{\partial x_1} + g_2 \frac{\partial}{\partial x_2} + g_3 \frac{\partial}{\partial x_3},$$

where

$$\begin{aligned} g_1 &= \frac{\partial f_2}{\partial x_3} - \frac{\partial f_3}{\partial x_2}, \\ g_2 &= \frac{\partial f_3}{\partial x_1} - \frac{\partial f_1}{\partial x_3}, \\ g_3 &= \frac{\partial f_1}{\partial x_2} - \frac{\partial f_2}{\partial x_1}. \end{aligned} \tag{2.7.10}$$

To summarize: the gradient, curl, and divergence operations in three-dimensions are basically just the three operations (2.7.1). The gradient operation is the operation (2.7.1) in degree zero, curl is the operation (2.7.1) in degree one, and divergence is the operation (2.7.1) in degree two. However, to define gradient we had to assign an inner product B_p to the tangent space $T_p\mathbf{R}^n$ for each $p \in U$; to define divergence we had to equip U with the 3-form Ω and to define curl, the most complicated of these three operations, we needed the inner products B_p *and* the form Ω. This is why diffeomorphisms preserve the three operations (2.7.1) but do not preserve gradient, curl, and divergence. The additional structures which one needs to define grad, curl and div are only preserved by translations and rotations.

Maxwell's equations and differential forms

We conclude this section by showing how Maxwell's equations, which are usually formulated in terms of divergence and curl, can be reset into "form" language. The discussion below is an abbreviated version of [5, §1.20].

Maxwell's equations assert:

$$\operatorname{div}(\boldsymbol{v}_E) = q, \tag{2.7.11}$$

$$\operatorname{curl}(\boldsymbol{v}_E) = -\frac{\partial}{\partial t}\boldsymbol{v}_M, \tag{2.7.12}$$

$$\operatorname{div}(\boldsymbol{v}_M) = 0, \tag{2.7.13}$$

$$c^2 \operatorname{curl}(\boldsymbol{v}_M) = \boldsymbol{w} + \frac{\partial}{\partial t}\boldsymbol{v}_E, \tag{2.7.14}$$

where $\boldsymbol{v}_E$ and $\boldsymbol{v}_M$ are the ***electric*** and ***magnetic*** fields, q is the ***scalar charge density***, $\boldsymbol{w}$ is the ***current density*** and c is the velocity of light. (To simplify (2.7.15) slightly we assume that our units of space–time are chosen so that $c = 1$.) As above let $\Omega = dx_1 \wedge dx_2 \wedge dx_3$ and let

$$\mu_E = \iota(\boldsymbol{v}_E)\Omega$$

and

$$\mu_M = \iota(\boldsymbol{v}_M)\Omega.$$

We can then rewrite equations (2.7.11) and (2.7.13) in the form

$$d\mu_E = q\Omega \tag{2.7.11$'$}$$

and

$$d\mu_M = 0. \tag{2.7.13$'$}$$

What about (2.7.12) and (2.7.14)? We leave the following "form" versions of these equations as an exercise:

$$d\boldsymbol{v}_E^\sharp = -\frac{\partial}{\partial t}\mu_M \tag{2.7.12$'$}$$

and

$$d\boldsymbol{v}_M^\sharp = \iota_{\boldsymbol{w}}\Omega + \frac{\partial}{\partial t}\mu_E, \tag{2.7.14$'$}$$

where the 1-forms, $\boldsymbol{v}_E^\sharp$ and $\boldsymbol{v}_M^\sharp$, are obtained from $\boldsymbol{v}_E$ and $\boldsymbol{v}_M$ by the operation (2.7.2).

These equations can be written more compactly as differential form identities in $(3+1)$ dimensions. Let ω_M and ω_E be the 2-forms

$$\omega_M = \mu_M - \boldsymbol{v}_E^\sharp \wedge dt$$

and

$$\omega_E = \mu_E - \boldsymbol{v}_M^\sharp \wedge dt \tag{2.7.15}$$

and let Λ be the 3-form

$$\Lambda := q\Omega + \iota_{\boldsymbol{w}}\Omega \wedge dt.$$

We will leave for you to show that the four equations (2.7.11)–(2.7.14) are equivalent to two elegant and compact $(3+1)$-dimensional identities

$$d\omega_M = 0$$

and

$$d\omega_E = \Lambda.$$

Exercises for §2.7

Exercise 2.7.i. Verify that the curl operation is given in coordinates by equation (2.7.10).

Exercise 2.7.ii. Verify that Maxwell's equations (2.7.11) and (2.7.12) become equations (2.7.13) and (2.7.14) when rewritten in differential form notation.

Exercise 2.7.iii. Show that in $(3+1)$-dimensions Maxwell's equations take the form of equations (2.7.10) and (2.7.11).

Exercise 2.7.iv. Let U be an open subset of $\mathbf{R}^3$ and $\boldsymbol{v}$ a vector field on U. Show that if $\boldsymbol{v}$ is the gradient of a function, its curl has to be zero.

Exercise 2.7.v. If U is simply connected prove the converse: If the curl of $\boldsymbol{v}$ vanishes, $\boldsymbol{v}$ is the gradient of a function.

Exercise 2.7.vi. Let $\boldsymbol{w} = \operatorname{curl}(\boldsymbol{v})$. Show that the divergence of $\boldsymbol{w}$ is zero.

Exercise 2.7.vii. Is the converse to Exercise 2.7.vi true? Suppose the divergence of $\boldsymbol{w}$ is zero. Is $\boldsymbol{w} = \operatorname{curl}(\boldsymbol{v})$ for some vector field $\boldsymbol{v}$?

2.8. Symplectic geometry and classical mechanics

In this section we describe some other applications of the theory of differential forms to physics. Before describing these applications, however, we say a few words about the geometric ideas that are involved.

Definition 2.8.1. Let $x_1, \ldots, x_{2n}$ be the standard coordinate functions on $\mathbf{R}^{2n}$ and for $i = 1, \ldots, n$ let $y_i \coloneqq x_{n+i}$. The 2-form

$$\omega \coloneqq \sum_{i=1}^{n} dx_i \wedge dy_i = \sum_{i=1}^{n} dx_i \wedge dx_{n+i}$$

is called the ***Darboux form*** on $\mathbf{R}^{2n}$.

From the identity

$$\omega = -d\left(\sum_{i=1}^{n} y_i dx_i\right), \tag{2.8.2}$$

it follows that ω is exact. Moreover computing the n-fold wedge product of ω with itself we get

$$\omega^n = \left(\sum_{i_i=1}^{n} dx_{i_1} \wedge dy_{i_1}\right) \wedge \cdots \wedge \left(\sum_{i_n=1}^{n} dx_{i_n} \wedge dy_{i_n}\right)$$
$$= \sum_{i_1,\dots,i_n} dx_{i_1} \wedge dy_{i_1} \wedge \cdots \wedge dx_{i_n} \wedge dy_{i_n}.$$

We can simplify this sum by noting that if the multi-index $I = (i_1, \dots, i_n)$ is repeating, the wedge product

$$dx_{i_1} \wedge dy_{i_1} \wedge \cdots \wedge dx_{i_n} \wedge dx_{i_n} \tag{2.8.3}$$

involves two repeating dx_{i_j} and hence is zero, and if I is non-repeating we can permute the factors and rewrite (2.8.3) in the form

$$dx_1 \wedge dy_1 \wedge \cdots \wedge dx_n \wedge dy_n.$$

(See Exercise 1.6.v.)

Hence since these are exactly $n!$ non-repeating multi-indices

$$\omega^n = n!\, dx_1 \wedge dy_1 \wedge \cdots \wedge dx_n \wedge dy_n,$$

i.e.,

$$\frac{1}{n!}\omega^n = \Omega, \tag{2.8.4}$$

where

$$\Omega = dx_1 \wedge dy_1 \wedge \cdots \wedge dx_n \wedge dy_n \tag{2.8.5}$$

is the *symplectic volume form* on $\mathbf{R}^{2n}$.

Definition 2.8.6. Let U and V be open subsets of $\mathbf{R}^{2n}$. A diffeomorphism $f : U \xrightarrow{\sim} V$ is a ***symplectic diffeomorphism*** or ***symplectomorphism*** if $f^\star\omega = \omega$, where ω denotes the Darboux form on $\mathbf{R}^{2n}$.

Definition 2.8.7. Let U be an open subset of $\mathbf{R}^{2n}$, let

$$f_t \colon U \xrightarrow{\sim} U, \quad -\infty < t < \infty \tag{2.8.8}$$

be a one-parameter group of diffeomorphisms of U, and $\boldsymbol{v}$ be the vector field generating (2.8.8). We say that $\boldsymbol{v}$ is a ***symplectic vector field*** if the diffeomorphisms (2.8.8) are symplectomorphisms, i.e., for all t we have $f_t^\star\omega = \omega$.

Let us see what such vector fields have to look like. Note that by (2.6.23)

$$\frac{d}{dt} f_t^\star \omega = f_t^\star L_{\boldsymbol{v}}\omega, \tag{2.8.9}$$

hence if $f_t^\star\omega = \omega$ for all t, the left-hand side of (2.8.9) is zero, so

$$f_t^\star L_{\boldsymbol{v}}\omega = 0.$$

In particular, for $t = 0$, f_t is the identity map so $f_t^\star L_{\boldsymbol{v}}\omega = L_{\boldsymbol{v}}\omega = 0$. Conversely, if $L_{\boldsymbol{v}}\omega = 0$, then $f_t^\star L_{\boldsymbol{v}}\omega = 0$ so by (2.8.9) $f_t^\star\omega$ does not depend on t. However, since

$f_t^\star\omega = \omega$ for $t = 0$ we conclude that $f_t^\star\omega = \omega$ for all t. To summarize, we have proved the following.

Theorem 2.8.10. *Let $f_t : U \to U$ be a one-parameter group of diffeomorphisms and $\boldsymbol{v}$ the infinitesimal generator of this group. Then $\boldsymbol{v}$ is symplectic of and only if $L_{\boldsymbol{v}}\omega = 0$.*

There is an equivalent formulation of Theorem 2.8.10 in terms of the interior product $\iota_{\boldsymbol{v}}\omega$. By (2.5.9)

$$L_{\boldsymbol{v}}\omega = d\iota_{\boldsymbol{v}}\omega + \iota_{\boldsymbol{v}}\, d\omega.$$

But by (2.8.2) we have $d\omega = 0$, so

$$L_{\boldsymbol{v}}\omega = d\iota_{\boldsymbol{v}}\omega.$$

Thus we have shown the following.

Theorem 2.8.11. *A vector field $\boldsymbol{v}$ on an open subset of $\mathbf{R}^{2n}$ is symplectic if and only if the form $\iota_{\boldsymbol{v}}\omega$ is closed.*

Let U be an open subset of $\mathbf{R}^{2n}$ and $\boldsymbol{v}$ a vector field on U. If $\iota_{\boldsymbol{v}}\omega$ is not only closed but is exact we say that $\boldsymbol{v}$ is a ***Hamiltonian vector field***. In other words, $\boldsymbol{v}$ is Hamiltonian if

$$\iota_{\boldsymbol{v}}\omega = dH \tag{2.8.12}$$

for some C^∞ function $H \in C^\infty(U)$.

Let us see what this condition looks like in coordinates. Let

$$\boldsymbol{v} = \sum_{i=1}^{n}\left(f_i\frac{\partial}{\partial x_i} + g_i\frac{\partial}{\partial y_i}\right). \tag{2.8.13}$$

Then

$$\iota_{\boldsymbol{v}}\omega = \sum_{1\le i,j\le n} f_i\,\iota_{\partial/\partial x_i}(dx_j \wedge dy_j) + \sum_{1\le i,j\le n} g_i\,\iota_{\partial/\partial y_i}(dx_j \wedge dy_j).$$

But

$$\iota_{\partial/\partial x_i} dx_j = \begin{cases} 1, & i = j, \\ 0, & i \neq j \end{cases}$$

and

$$\iota_{\partial/\partial x_i} dy_j = 0$$

so the first summand above is $\sum_{i=1}^{n} f_i\, dy_i$, and a similar argument shows that the second summand is $-\sum_{i=1}^{n} g_i dx_i$.

Hence if $\boldsymbol{v}$ is the vector field (2.8.13), then

$$\iota_{\boldsymbol{v}}\omega = \sum_{i=1}^{n}(f_i\, dy_i - g_i dx_i). \tag{2.8.14}$$

Thus since

$$dH = \sum_{i=1}^{n}\left(\frac{\partial H}{\partial x_i}dx_i + \frac{\partial H}{\partial y_i}\, dy_i\right)$$

we get from (2.8.12)–(2.8.14)

$$f_i = \frac{\partial H}{\partial y_i} \quad \text{and} \quad g_i = -\frac{\partial H}{\partial x_i}$$

so $\boldsymbol{v}$ has the form:

$$\boldsymbol{v} = \sum_{i=1}^{n} \left(\frac{\partial H}{\partial y_i}\frac{\partial}{\partial x_i} - \frac{\partial H}{\partial x_i}\frac{\partial}{\partial y_i} \right). \tag{2.8.15}$$

In particular if $\gamma(t) = (x(t), y(t))$ is an integral curve of $\boldsymbol{v}$ it has to satisfy the system of differential equations

$$\begin{cases} \dfrac{dx_i}{dt} = \dfrac{\partial H}{\partial y_i}(x(t), y(t)), \\ \dfrac{dy_i}{dt} = -\dfrac{\partial H}{\partial x_i}(x(t), y(t)). \end{cases} \tag{2.8.16}$$

The formulas (2.8.13) and (2.8.14) exhibit an important property of the Darboux form ω. Every 1-form on U can be written uniquely as a sum

$$\sum_{i=1}^{n} (f_i dy_i - g_i dx_i)$$

with f_i and g_i in $C^\infty(U)$ and hence (2.8.13) and (2.8.14) imply the following theorem.

Theorem 2.8.17. *The map $\boldsymbol{v} \mapsto \iota_{\boldsymbol{v}}\omega$ defines a bijective between vector field and 1-forms.*

In particular, for every C^∞ function H, we get by correspondence a unique vector field $\boldsymbol{v} = \boldsymbol{v}_H$ with the property (2.8.12). We next note that by (1.7.9)

$$L_{\boldsymbol{v}} H = \iota_{\boldsymbol{v}} dH = \iota_{\boldsymbol{v}}(\iota_{\boldsymbol{v}}\omega) = 0.$$

Thus

$$L_{\boldsymbol{v}} H = 0 \tag{2.8.18}$$

i.e., H is an integral of motion of the vector field $\boldsymbol{v}$. In particular, if the function $H: U \to \mathbf{R}$ is proper, then by Theorem 2.2.12 the vector field, $\boldsymbol{v}$, is complete and hence by Theorem 2.8.10 generates a one-parameter group of symplectomorphisms.

Remark 2.8.19. If the one-parameter group (2.8.8) is a group of symplectomorphisms, then $f_t^\star \omega^n = f_t^\star \omega \wedge \cdots \wedge f_t^\star \omega = \omega^n$, so by (2.8.4)

$$f_t^\star \Omega = \Omega, \tag{2.8.20}$$

where Ω is the symplectic volume form (2.8.5).

Application 2.8.21. The application we want to make of these ideas concerns the description, in Newtonian mechanics, of a physical system consisting of N interacting point-masses. The ***configuration space*** of such a system is

$$\mathbf{R}^n = \mathbf{R}^3 \times \cdots \times \mathbf{R}^3,$$

where there are N copies of $\mathbf{R}^3$ in the product, with position coordinates $x_1, \ldots, x_n$ and the ***phase space*** is $\mathbf{R}^{2n}$ with position coordinates $x_1, \ldots, x_n$ and momentum coordinates, $y_1, \ldots, y_n$. The ***kinetic energy*** of this system is a quadratic function of the momentum coordinates

$$\frac{1}{2}\sum_{i=1}^{n} \frac{1}{m_i} y_i^2,$$

and for simplicity we assume that the potential energy is a function $V(x_1, \ldots, x_n)$ of the position coordinates alone, i.e., it does not depend on the momenta and is time independent as well. Let

$$H = \frac{1}{2}\sum_{i=1}^{n} \frac{1}{m_i} y_i^2 + V(x_1, \ldots, x_n) \tag{2.8.22}$$

be the ***total energy*** of the system.

We now show that Newton's second law of motion in classical mechanics reduces to the assertion.

Proposition 2.8.23. *The trajectories in phase space of the system above are just the integral curves of the Hamiltonian vector field* $\boldsymbol{v}_H$.

Proof. For the function (2.8.22), equations (2.8.16) become

$$\begin{cases} \dfrac{dx_i}{dt} = \dfrac{1}{m_i} y_i, \\[2ex] \dfrac{dy_i}{dt} = -\dfrac{\partial V}{\partial x_i}. \end{cases}$$

The first set of equation are essentially just the definitions of momentum, however, if we plug them into the second set of equations we get

$$m_i \frac{d^2 x_i}{dt^2} = -\frac{\partial V}{\partial x_i} \tag{2.8.24}$$

and interpreting the term on the right as the force exerted on the ith point-mass and the term on the left as mass times acceleration this equation becomes Newton's second law. □

In classical mechanics, equations (2.8.16) are known as the *Hamilton–Jacobi equations*. For a more detailed account of their role in classical mechanics, we highly recommend [1]. Historically, these equations came up for the first time, not in Newtonian mechanics, but in geometric optics and a brief description of their origins there and of their relation to Maxwell's equations can be found in [5].

We conclude this chapter by mentioning a few implications of the Hamiltonian description (2.8.16) of Newton's equations (2.8.24).

(1) *Conservation of energy:* By (2.8.18) the energy function (2.8.22) is constant along the integral curves of $\boldsymbol{v}$, hence the energy of the system (2.8.16) does not change in time.

(2) *Noether's principle:* Let $\gamma_t \colon \mathbf{R}^{2n} \to \mathbf{R}^{2n}$ be a one-parameter group of diffeomorphisms of phase space and $\boldsymbol{w}$ its infinitesimal generator. Then $(\gamma_t)_{t\in\mathbf{R}}$ is a ***symmetry*** of the system above if each γ_t preserves the function (2.8.22) and the vector field $\boldsymbol{w}$ is Hamiltonian. The Hamiltonian condition means that

$$\iota_{\boldsymbol{w}}\omega = dG$$

for some C^∞ function G, and what Noether's principle asserts is that *this function is an integral of motion of the system* (2.8.16), i.e., satisfies $L_{\boldsymbol{v}}G = 0$. In other words stated more succinctly: symmetries of the system (2.8.16) give rise to integrals of motion.

(3) *Poincaré recurrence:* An important theorem of Poincaré asserts that if the function

$$H \colon \mathbf{R}^{2n} \to \mathbf{R}$$

defined by (2.8.22) is proper then every trajectory of the system (2.8.16) returns arbitrarily close to its initial position at some positive time t_0, and, in fact, does this not just once but does so infinitely often. We sketch a proof of this theorem, using (2.8.20), in the next chapter.

Exercises for §2.8

Exercise 2.8.i. Let $\boldsymbol{v}_H$ be the vector field (2.8.15). Prove that $\operatorname{div}(\boldsymbol{v}_H) = 0$.

Exercise 2.8.ii. Let U be an open subset of $\mathbf{R}^m$, $f_t \colon U \to U$ a one-parameter group of diffeomorphisms of U and $\boldsymbol{v}$ the infinitesimal generator of this group. Show that if α is a k-form on U then $f_t^\star\alpha = \alpha$ for all t if and only if $L_{\boldsymbol{v}}\alpha = 0$ (i.e., generalize to arbitrary k-forms the result we proved above for the Darboux form).

Exercise 2.8.iii (the harmonic oscillator). Let H be the function $\sum_{i=1}^n m_i(x_i^2 + y_i^2)$ where the m_i's are positive constants.

(1) Compute the integral curves of $\boldsymbol{v}_H$.

(2) *Poincaré recurrence:* Show that if $(x(t), y(t))$ is an integral curve with initial point $(x_0, y_0) \coloneqq (x(0), y(0))$ and U an arbitrarily small neighborhood of (x_0, y_0), then for *every* $c > 0$ there exists a $t > c$ such that $(x(t), y(t)) \in U$.

Exercise 2.8.iv. Let U be an open subset of $\mathbf{R}^{2n}$ and let $H_1, H_2 \in C^\infty(U)$. Show that

$$[\boldsymbol{v}_{H_1}, \boldsymbol{v}_{H_2}] = \boldsymbol{v}_H,$$

where

$$H = \sum_{i=1}^n \frac{\partial H_1}{\partial x_i}\frac{\partial H_2}{\partial y_i} - \frac{\partial H_2}{\partial x_i}\frac{\partial H_1}{\partial y_i}. \tag{2.8.25}$$

Exercise 2.8.v. The expression (2.8.25) is known as the ***Poisson bracket*** of H_1 and H_2, denoted by $\{H_1, H_2\}$. Show that it is anti-symmetric

$$\{H_1, H_2\} = -\{H_2, H_1\}$$

and satisfies the *Jacobi identity*

$$0 = \{H_1, \{H_2, H_3\}\} + \{H_2, \{H_3, H_1\}\} + \{H_3, \{H_1, H_2\}\}.$$

Exercise 2.8.vi. Show that

$$\{H_1, H_2\} = L_{\boldsymbol{v}_{H_1}} H_2 = -L_{\boldsymbol{v}_{H_2}} H_1.$$

Exercise 2.8.vii. Prove that the following three properties are equivalent.

(1) $\{H_1, H_2\} = 0$.
(2) H_1 is an integral of motion of $\boldsymbol{v}_2$.
(3) H_2 is an integral of motion of $\boldsymbol{v}_1$.

Exercise 2.8.viii. Verify Noether's principle.

Exercise 2.8.ix (conservation of linear momentum). Suppose the potential V in equation (2.8.22) is invariant under the one-parameter group of translations

$$T_t(x_1, \dots, x_n) = (x_1 + t, \dots, x_n + t).$$

(1) Show that the function (2.8.22) is invariant under the group of diffeomorphisms

$$\gamma_t(x, y) = (T_t x, y).$$

(2) Show that the infinitesimal generator of this group is the Hamiltonian vector field $\boldsymbol{v}_G$ where $G = \sum_{i=1}^n y_i$.
(3) Conclude from Noether's principle that this function is an integral of the vector field $\boldsymbol{v}_H$, i.e., that "total linear momentum" is conserved.
(4) Show that "total linear momentum" is conserved if V is the Coulomb potential

$$\sum_{i \neq j} \frac{m_i}{|x_i - x_j|}.$$

Exercise 2.8.x. Let $R_t^i \colon \mathbf{R}^{2n} \to \mathbf{R}^{2n}$ be the rotation which fixes the variables, (x_k, y_k), $k \neq i$ and rotates (x_i, y_i) by the angle, t:

$$R_t^i(x_i, y_i) = (\cos t\, x_i + \sin t\, y_i, \, -\sin t\, x_i + \cos t\, y_i).$$

(1) Show that R_t^i, $-\infty < t < \infty$, is a one-parameter group of symplectomorphisms.
(2) Show that its generator is the Hamiltonian vector field, $\boldsymbol{v}_{H_i}$, where $H_i = (x_i^2 + y_i^2)/2$.
(3) Let H be the harmonic oscillator Hamiltonian from Exercise 2.8.iii. Show that the R_t^j's preserve H.
(4) What does Noether's principle tell one about the classical mechanical system with energy function H?

Exercise 2.8.xi. Show that if U is an open subset of $\mathbf{R}^{2n}$ and $\boldsymbol{v}$ is a symplectic vector field on U then for every point $p_0 \in U$ there exists a neighborhood U_0 of p_0 on which $\boldsymbol{v}$ is Hamiltonian.

Exercise 2.8.xii. Deduce from Exercises 2.8.iv and 2.8.xi that if $\boldsymbol{v}_1$ and $\boldsymbol{v}_2$ are symplectic vector fields on an open subset U of $\mathbf{R}^{2n}$ their Lie bracket $[\boldsymbol{v}_1, \boldsymbol{v}_2]$ is a Hamiltonian vector field.

Exercise 2.8.xiii. Let $\alpha = \sum_{i=1}^n y_i dx_i$.

(1) Show that $\omega = -d\alpha$.
(2) Show that if α_1 is any 1-form on $\mathbf{R}^{2n}$ with the property $\omega = -d\alpha_1$, then

$$\alpha = \alpha_1 + dF$$

for some C^∞ function F.
(3) Show that $\alpha = \iota_{\boldsymbol{w}}\omega$ where $\boldsymbol{w}$ is the vector field

$$\boldsymbol{w} := -\sum_{i=1}^n y_i \frac{\partial}{\partial y_i}.$$

Exercise 2.8.xiv. Let U be an open subset of $\mathbf{R}^{2n}$ and $\boldsymbol{v}$ a vector field on U. Show that $\boldsymbol{v}$ has the property, $L_{\boldsymbol{v}}\alpha = 0$, if and only if

$$\iota_{\boldsymbol{v}}\omega = d\iota_{\boldsymbol{v}}\alpha.$$

In particular conclude that if $L_{\boldsymbol{v}}\alpha = 0$ then $\boldsymbol{v}$ is Hamiltonian.

Hint: Equation (2.8.2).

Exercise 2.8.xv. Let H be the function

$$H(x, y) = \sum_{i=1}^n f_i(x) y_i, \tag{2.8.26}$$

where the f_i's are C^∞ functions on $\mathbf{R}^n$. Show that

$$L_{\boldsymbol{v}_H}\alpha = 0. \tag{2.8.27}$$

Exercise 2.8.xvi. Conversely show that if H is any C^∞ function on $\mathbf{R}^{2n}$ satisfying equation (2.8.27) it has to be a function of the form (2.8.26).

Hints:

(1) Let $\boldsymbol{v}$ be a vector field on $\mathbf{R}^{2n}$ satisfying $L_{\boldsymbol{v}}\alpha = 0$. By equation (2.8.16) we have $\boldsymbol{v} = \boldsymbol{v}_H$, where $H = \iota_{\boldsymbol{v}}\alpha$.
(2) Show that H has to satisfy the equation

$$\sum_{i=1}^n y_i \frac{\partial H}{\partial y_i} = H.$$

(3) Conclude that if $H_r = \frac{\partial H}{\partial y_r}$ then H_r has to satisfy the equation

$$\sum_{i=1}^n y_i \frac{\partial}{\partial y_i} H_r = 0.$$

(4) Conclude that H_r has to be constant along the rays (x, ty), for $0 \le t < \infty$.
(5) Conclude finally that H_r has to be a function of x alone, i.e., does not depend on y.

Exercise 2.8.xvii. Show that if $\boldsymbol{v}_{\mathbf{R}^n}$ is a vector field

$$\sum_{i=1}^{n} f_i(x)\frac{\partial}{\partial x_i}$$

on configuration space there is a unique lift of $\boldsymbol{v}_{\mathbf{R}^n}$ to phase space

$$\boldsymbol{v} = \sum_{i=1}^{n} f_i(x)\frac{\partial}{\partial x_i} + g_i(x, y)\frac{\partial}{\partial y_i}$$

satisfying $L_{\boldsymbol{v}}\alpha = 0$.

Chapter 3

Integration of Forms

3.1. Introduction

The change of variables formula asserts that if U and V are open subsets of $\mathbf{R}^n$ and $f : U \to V$ a C^1-diffeomorphism then, for every continuous function $\phi : V \to \mathbf{R}$, the integral

$$\int_V \phi(y)dy$$

exists if and only if the integral

$$\int_U (\phi \circ f)(x)\,|\det Df(x)|dx$$

exists, and if these integrals exist they are equal. Proofs of this can be found in [10] or [12]. This chapter contains an alternative proof of this result. This proof is due to Peter Lax. Our version of his proof in § 3.5 below makes use of the theory of differential forms; but, as Lax shows in the article [8] (which we strongly recommend as collateral reading for this course), references to differential forms can be avoided, and the proof described in § 3.5 can be couched entirely in the language of elementary multivariable calculus.

The virtue of Lax's proof allows one to prove a version of the change of variables theorem for other mappings besides diffeomorphisms, and involves a topological invariant, the *degree of a map*, which is itself quite interesting. Some properties of this invariant, and some topological applications of the change of variables formula will be discussed in § 3.6 of these notes.

Remark 3.1.1. The proof we are about to describe is somewhat simpler and more transparent if we assume that f is a C^∞ diffeomorphism. We'll henceforth make this assumption.

3.2. The Poincaré lemma for compactly supported forms on rectangles

Definition 3.2.1. Let ν be a k-form on $\mathbf{R}^n$. We define the ***support*** of ν by

$$\operatorname{supp}(\nu) := \overline{\{x \in \mathbf{R}^n \,|\, \nu_x \neq 0\}},$$

and we say that ν is ***compactly supported*** if $\operatorname{supp}(\nu)$ is compact.

We will denote by $\Omega_c^k(\mathbf{R}^n)$ the set of all C^∞ k-forms which are compactly supported, and if U is an open subset of $\mathbf{R}^n$, we will denote by $\Omega_c^k(U)$ the set of all compactly supported k-forms whose support is contained in U.

Let $\omega = f\,dx_1 \wedge \cdots \wedge dx_n$ be a compactly supported n-form with $f \in C_0^\infty(\mathbf{R}^n)$. We will define the integral of ω over $\mathbf{R}^n$:

$$\int_{\mathbf{R}^n} \omega$$

to be the usual integral of f over $\mathbf{R}^n$

$$\int_{\mathbf{R}^n} f\,dx.$$

(Since f is C^∞ and compactly supported this integral is well-defined.)

Now let Q be the rectangle

$$[a_1, b_1] \times \cdots \times [a_n, b_n].$$

The Poincaré lemma for rectangles asserts the following.

Theorem 3.2.2 (Poincaré lemma for rectangles). *Let ω be a compactly supported n-form with* $\operatorname{supp}(\omega) \subset \operatorname{int}(Q)$. *Then the following assertions are equivalent.*

(1) $\int \omega = 0$.
(2) *There exists a compactly supported $(n-1)$-form μ with* $\operatorname{supp}(\mu) \subset \operatorname{int}(Q)$ *satisfying $d\mu = \omega$.*

We will first prove that (2) $\Rightarrow$ (1).

Proof. (2) $\Rightarrow$ (1) Let

$$\mu = \sum_{i=1}^{n} f_i\,dx_1 \wedge \cdots \wedge \widehat{dx_i} \wedge \cdots \wedge dx_n$$

(the "hat" over the dx_i meaning that dx_i has to be omitted from the wedge product). Then

$$d\mu = \sum_{i=1}^{n} (-1)^{i-1} \frac{\partial f_i}{\partial x_i} dx_1 \wedge \cdots \wedge dx_n,$$

and to show that the integral of $d\mu$ is zero it suffices to show that each of the integrals

$$(3.2.2)_i \qquad \int_{\mathbf{R}^n} \frac{\partial f_i}{\partial x_i}\,dx$$

is zero. By Fubini we can compute $(3.2.2)_i$ by first integrating with respect to the variable, x_i, and then with respect to the remaining variables. But

$$\int \frac{\partial f_i}{\partial x_i}\,dx_i = f(x)\Big|_{x_i=a_i}^{x_i=b_i} = 0$$

since f_i is supported on U. □

We will prove that (1) $\Rightarrow$ (2) by proving a somewhat stronger result. Let U be an open subset of $\mathbf{R}^m$. We'll say that U has ***property*** P if for every form $\omega \in \Omega_c^m(U)$ such that $\int_U \omega = 0$ we have $\omega \in d\,\Omega_c^{m-1}(U)$. We will prove the following theorem.

Theorem 3.2.3. *Let U be an open subset of $\mathbf{R}^{n-1}$ and $A \subset \mathbf{R}$ an open interval. Then if U has property P, $U \times A$ does as well.*

Remark 3.2.4. It is very easy to see that the open interval A itself has property P. (See Exercise 3.2.i below.) Hence it follows by induction from Theorem 3.2.3 that

$$\operatorname{int} Q = A_1 \times \cdots \times A_n\,, \quad A_i = (a_i, b_i)$$

has property P, and this proves "(1) $\Rightarrow$ (2)".

Proof of Theorem 3.2.3. Let $(x,t) = (x_1, \dots, x_{n-1}, t)$ be product coordinates on $U \times A$. Given $\omega \in \Omega_c^n(U \times A)$ we can express ω as a wedge product, $dt \wedge \alpha$ with $\alpha = f(x,t)\,dx_1 \wedge \cdots \wedge dx_{n-1}$ and $f \in C_0^\infty(U \times A)$. Let $\theta \in \Omega_c^{n-1}(U)$ be the form

$$\theta = \left(\int_A f(x,t) dt \right) dx_1 \wedge \cdots \wedge dx_{n-1}. \tag{3.2.5}$$

Then

$$\int_{\mathbf{R}^{n-1}} \theta = \int_{\mathbf{R}^n} f(x,t) dx dt = \int_{\mathbf{R}^n} \omega$$

so if the integral of ω is zero, the integral of θ is zero. Hence since U has property P, $\theta = d\nu$ for some $\nu \in \Omega_c^{n-2}(U)$. Let $\rho \in C^\infty(\mathbf{R})$ be a bump function which is supported on A and whose integral over A is 1. Setting

$$\kappa = -\rho(t) dt \wedge \nu$$

we have

$$d\kappa = \rho(t) dt \wedge d\nu = \rho(t) dt \wedge \theta,$$

and hence

$$\begin{aligned} \omega - d\kappa &= dt \wedge (\alpha - \rho(t)\theta) \\ &= dt \wedge u(x,t) dx_1 \wedge \cdots \wedge dx_{n-1}, \end{aligned}$$

where

$$u(x,t) = f(x,t) - \rho(t) \int_A f(x,t) dt$$

by (3.2.5). Thus

$$\int u(x,t)\, dt = 0. \tag{3.2.6}$$

Let a and b be the endpoints of A and let

$$v(x,t) = \int_a^t u(x,s)\, ds. \tag{3.2.7}$$

By equation (3.2.6) $v(a,x) = v(b,x) = 0$, so v is in $C_0^\infty(U \times A)$ and by (3.2.7), $\partial v/\partial t = u$. Hence if we let γ be the form, $v(x,t)\,dx_1 \wedge \cdots \wedge dx_{n-1}$, we have:

$$d\gamma = dt \wedge u(x,t)\,dx_1 \wedge \cdots \wedge dx_{n-1} = \omega - d\kappa$$

and

$$\omega = d(\gamma + \kappa).$$

Since γ and κ are both in $\Omega_c^{n-1}(U \times A)$ this proves that ω is in $d\,\Omega_c^{n-1}(U \times A)$ and hence that $U \times A$ has property P. □

Exercises for §3.2

Exercise 3.2.i. Let $f : \mathbf{R} \to \mathbf{R}$ be a compactly supported function of class C^r with support on the interval (a,b). Show that the following conditions are equivalent.

(1) $\int_a^b f(x)dx = 0$.
(2) There exists a function $g : \mathbf{R} \to \mathbf{R}$ of class C^{r+1} with support on (a,b) with $\frac{dg}{dx} = f$.

Hint: Show that the function

$$g(x) = \int_a^x f(s)ds$$

is compactly supported.

Exercise 3.2.ii. Let $f = f(x,y)$ be a compactly supported function on $\mathbf{R}^k \times \mathbf{R}^\ell$ with the property that the partial derivatives

$$\frac{\partial f}{\partial x_i}(x,y), \quad \text{for } i = 1, \ldots, k,$$

exist and are continuous as functions of x and y. Prove the following "differentiation under the integral sign" theorem (which we implicitly used in our proof of Theorem 3.2.3).

Theorem 3.2.8. *The function* $g(x) := \int_{\mathbf{R}^\ell} f(x,y)dy$ *is of class* C^1 *and*

$$\frac{\partial g}{\partial x_i}(x) = \int \frac{\partial f}{\partial x_i}(x,y)dy.$$

Hints: For y fixed and $h \in \mathbf{R}^k$,

$$f(x+h,y) - f(x,y) = D_x f(c,y)h$$

for some point c on the line segment joining x to $x+h$. Using the fact that $D_x f$ is continuous as a function of x and y and compactly supported, we conclude the following.

Lemma 3.2.9. *Given* $\varepsilon > 0$ *there exists a* $\delta > 0$ *such that for* $|h| \le \delta$

$$|f(x+h,y) - f(x,y) - D_x f(x,y)h| \le \varepsilon |h|.$$

Now let $Q \subset \mathbf{R}^\ell$ be a rectangle with $\text{supp}(f) \subset \mathbf{R}^k \times Q$ and show that

$$|g(x+h) - g(x) - \left(\int D_x f(x,y)\,dy\right)h| \le \varepsilon \,\text{vol}(Q)|h|.$$

Conclude that g is differentiable at x and that its derivative is

$$\int D_x f(x,y)dy.$$

Exercise 3.2.iii. Let $f : \mathbf{R}^k \times \mathbf{R}^\ell \to \mathbf{R}$ be a compactly supported continuous function. Prove the following theorem.

Theorem 3.2.10. *If all the partial derivatives of $f(x,y)$ with respect to x of order $\le r$ exist and are continuous as functions of x and y the function*

$$g(x) = \int f(x,y)dy$$

is of class C^r.

Exercise 3.2.iv. Let U be an open subset of $\mathbf{R}^{n-1}$, $A \subset \mathbf{R}$ an open interval and (x,t) product coordinates on $U \times A$. Recall from Exercise 2.3.v that every form $\omega \in \Omega^k(U \times A)$ can be written uniquely as a sum $\omega = dt \wedge \alpha + \beta$ where α and β are *reduced*, i.e., do not contain a factor of dt.

(1) Show that if ω is compactly supported on $U \times A$ then so are α and β.
(2) Let $\alpha = \sum_I f_I(x,t)dx_I$. Show that the form

$$\theta = \sum_I \left(\int_A f_I(x,t)\,dt\right)\,dx_I \tag{3.2.11}$$

is in $\Omega_c^{k-1}(U)$.
(3) Show that if $d\omega = 0$, then $d\theta = 0$.
Hint: By equation (3.2.11),

$$d\theta = \sum_{I,i}\left(\int_A \frac{\partial f_I}{\partial x_i}(x,t)\,dt\right)dx_i \wedge dx_I = \int_A (d_U\alpha)\,dt$$

and by equation (2.4.19) we have $d_U\alpha = \frac{d\beta}{dt}$.

Exercise 3.2.v. In Exercise 3.2.iv, show that if θ is in $d\Omega_c^{k-2}(U)$ then ω is in $d\Omega_c^{k-1}(U \times A)$ as follows.

(1) Let $\theta = d\nu$, with $\nu \in \Omega_c^{k-2}(U)$ and let $\rho \in C^\infty(\mathbf{R})$ be a bump function which is supported on A and whose integral over A is one. Setting $k = -\rho(t)dt \wedge \nu$ show that

$$\begin{aligned}\omega - d\kappa &= dt \wedge (\alpha - \rho(t)\theta) + \beta \\ &= dt \wedge \left(\textstyle\sum_I u_I(x,t)dx_I\right) + \beta,\end{aligned}$$

where

$$u_I(x,t) = f_I(x,t) - \rho(t)\int_A f_I(x,t)dt.$$

(2) Let a and b be the endpoints of A and let

$$v_I(x,t) = \int_a^t u_I(x,t)\,dt.$$

Show that the form $\sum_I v_I(x,t)\,dx_I$ is in $\Omega_c^{k-1}(U \times A)$ and that

$$d\gamma = \omega - d\kappa - \beta + d_U\gamma.$$

(3) Conclude that the form $\omega - d(\kappa + \gamma)$ is reduced.

(4) Prove that if $\lambda \in \Omega_c^k(U \times A)$ is reduced and $d\lambda = 0$ then $\lambda = 0$.

Hint: Let $\lambda = \sum_I g_I(x,t)\,dx_I$. Show that $d\lambda = 0 \Rightarrow \frac{\partial}{\partial t} g_I(x,t) = 0$ and exploit the fact that for fixed x, $g_I(x,t)$ is compactly supported in t.

Exercise 3.2.vi. Let U be an open subset of $\mathbf{R}^m$. We say that U ***has property*** P_k, for $1 \le k < m$, if every closed k-form $\omega \in \Omega_c^k(U)$ is in $d\,\Omega_c^{k-1}(U)$. Prove that if the open set $U \subset \mathbf{R}^{n-1}$ in Exercise 3.2.iv has property P_{k-1} then $U \times A$ has property P_k.

Exercise 3.2.vii. Show that if Q is the rectangle $[a_1, b_1] \times \cdots \times [a_n, b_n]$ and $U = \operatorname{int} Q$ then U has property P_k.

Exercise 3.2.viii. Let $\mathbf{H}^n$ be the half-space

$$\mathbf{H}^n := \{(x_1, \dots, x_n) \in \mathbf{R}^n \mid x_1 \le 0\} \tag{3.2.12}$$

and let $\omega \in \Omega_c^n(\mathbf{R}^n)$ be the n-form $\omega := f\,dx_1 \wedge \cdots \wedge dx_n$ with $f \in C_0^\infty(\mathbf{R}^n)$. Define

$$\int_{\mathbf{H}^n} \omega := \int_{\mathbf{H}^n} f(x_1, \dots, x_n)\,dx_1 \cdots dx_n, \tag{3.2.13}$$

where the right-hand side is the usual Riemann integral of f over $\mathbf{H}^n$. (This integral makes sense since f is compactly supported.) Show that if $\omega = d\mu$ for some $\mu \in \Omega_c^{n-1}(\mathbf{R}^n)$ then

$$\int_{\mathbf{H}^n} \omega = \int_{\mathbf{R}^{n-1}} \iota^\star \mu, \tag{3.2.14}$$

where $\iota\colon \mathbf{R}^{n-1} \to \mathbf{R}^n$ is the inclusion map

$$(x_2, \dots, x_n) \mapsto (0, x_2, \dots, x_n).$$

Hint: Let $\mu = \sum_i f_i\,dx_1 \wedge \cdots \wedge \widehat{dx_i} \wedge \cdots \wedge dx_n$. Mimicking the "(2) $\Rightarrow$ (1)" part of the proof of Theorem 3.2.2 show that the integral (3.2.13) is the integral over $\mathbf{R}^{n-1}$ of the function

$$\int_{-\infty}^0 \frac{\partial f_1}{\partial x_1}(x_1, x_2, \dots, x_n)\,dx_1.$$

3.3. The Poincaré lemma for compactly supported forms on open subsets of $\mathbf{R}^n$

In this section we will generalize Theorem 3.2.2 to arbitrary connected open subsets of $\mathbf{R}^n$.

Theorem 3.3.1 (Poincaré lemma for compactly supported forms). *Let U be a connected open subset of $\mathbf{R}^n$ and let ω be a compactly supported n-form with* $\operatorname{supp}(\omega) \subset U$. *Then the following assertions are equivalent:*

(1) $\int_{\mathbf{R}^n} \omega = 0$.
(2) *There exists a compactly supported $(n-1)$-form μ with* $\operatorname{supp}\mu \subset U$ *and* $\omega = d\mu$.

Proof. (2) $\Rightarrow$ (1) The support of μ is contained in a large rectangle, so the integral of $d\mu$ is zero by Theorem 3.2.2. □

Proof. (1) $\Rightarrow$ (2) Let ω_1 and ω_2 be compactly supported n-forms with support in U. We will write

$$\omega_1 \sim \omega_2$$

as shorthand notation for the statement: *There exists a compactly supported $(n-1)$-form, μ, with support in U and with $\omega_1 - \omega_2 = d\mu$.* We will prove that (1) $\Rightarrow$ (2) by proving an equivalent statement: Fix a rectangle, $Q_0 \subset U$ and an n-form, ω_0, with $\operatorname{supp}\omega_0 \subset Q_0$ and integral equal to one.

Theorem 3.3.2. *If ω is a compactly supported n-form with* $\operatorname{supp}(\omega) \subset U$ *and* $c = \int \omega$, *then* $\omega \sim c\omega_0$.

Thus in particular if $c=0$, Theorem 3.3.2 says that $\omega \sim 0$ proving that (1) $\Rightarrow$ (2). □

To prove Theorem 3.3.2 let $Q_i \subset U$, $i = 1, 2, 3, \ldots$, be a collection of rectangles with $U = \bigcup_{i=1}^{\infty} \operatorname{int}(Q_i)$ and let ϕ_i be a partition of unity with $\operatorname{supp}(\phi_i) \subset \operatorname{int}(Q_i)$. Replacing ω by the finite sum $\sum_{i=1}^{m} \phi_i\omega$ for m large, it suffices to prove Theorem 3.3.2 for each of the summands $\phi_i\omega$. In other words we can assume that $\operatorname{supp}(\omega)$ is contained in one of the open rectangles $\operatorname{int}(Q_i)$. Denote this rectangle by Q. We claim that one can join Q_0 to Q by a sequence of rectangles as in Figure 3.3.1.

Lemma 3.3.3. *There exists a sequence of rectangles $R_0, \ldots, R_{N+1}$ such that $R_0 = Q_0$, $R_{N+1} = Q$ and* $\operatorname{int}(R_i) \cap \operatorname{int}(R_{i+1})$ *is non-empty.*

Proof. Denote by A the set of points, $x \in U$, for which there exists a sequence of rectangles, R_i, $i = 0, \ldots, N+1$ with $R_0 = Q_0$, with $x \in \operatorname{int} R_{N+1}$ and with $\operatorname{int} R_i \cap$

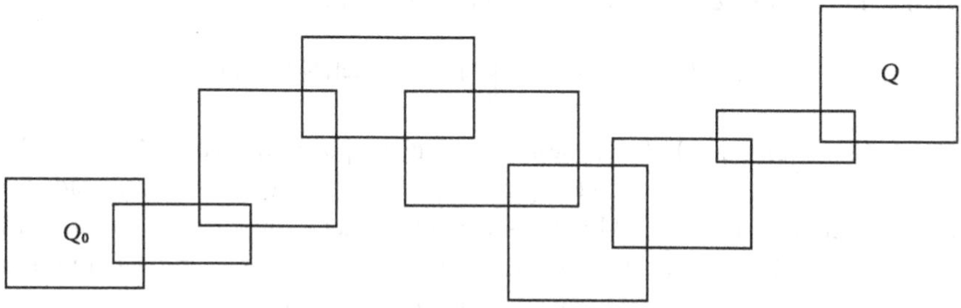

Figure 3.3.1. A sequence of rectangles joining the rectangles Q_0 and Q.

int R_{i+1} non-empty. It is clear that this set is open and that its complement is open; so, by the connectivity of U, $U = A$. □

To prove Theorem 3.3.2 with $\operatorname{supp}\omega \subset Q$, select, for each i, a compactly supported n-form ν_i with $\operatorname{supp}(\nu_i) \subset \operatorname{int}(R_i) \cap \operatorname{int}(R_{i+1})$ and with $\int \nu_i = 1$. The difference, $\nu_i - \nu_{i+1}$ is supported in int R_{i+1}, and its integral is zero. So by Theorem 3.2.2, $\nu_i \sim \nu_{i+1}$. Similarly, $\omega_0 \sim \nu_0$ and, setting $c := \int \omega$, we have $\omega \sim c\nu_N$. Thus

$$c\omega_0 \sim c\nu_0 \sim \cdots \sim c\nu_N = \omega$$

proving the theorem.

3.4. The degree of a differentiable mapping

Definition 3.4.1. Let U and V be open subsets of $\mathbf{R}^n$ and $\mathbf{R}^k$. A continuous map $f: U \to V$, is ***proper*** if for every compact subset $K \subset V$, the preimage $f^{-1}(K)$ is compact in U.

Proper mappings have a number of nice properties which will be investigated in the exercises below. One obvious property is that if f is a C^∞ mapping and ω is a compactly supported k-form with support on V, $f^\star\omega$ is a compactly supported k-form with support on U. Our goal in this section is to show that if U and V are connected open subsets of $\mathbf{R}^n$ and $f: U \to V$ is a proper C^∞ mapping then there exists a topological invariant of f, which we will call its *degree* (and denote by $\deg(f)$), such that the "change of variables" formula:

$$\int_U f^\star\omega = \deg(f)\int_V \omega \tag{3.4.2}$$

holds for all $\omega \in \Omega_c^n(V)$.

Before we prove this assertion let's see what this formula says in coordinates. If

$$\omega = \phi(y)dy_1 \wedge \cdots \wedge dy_n,$$

then at $x \in U$

$$f^\star\omega = (\phi \circ f)(x)\det(Df(x))dx_1 \wedge \cdots \wedge dx_n.$$

Hence, in coordinates, equation (3.4.2) takes the form

$$\int_V \phi(y)dy = \deg(f)\int_U \phi \circ f(x)\det(Df(x))dx. \tag{3.4.3}$$

Proof of equation (3.4.2). Let ω_0 be a compactly-supported n-form with $\operatorname{supp}(\omega_0) \subset V$ and with $\int \omega_0 = 1$. If we set $\deg f := \int_U f^\star\omega_0$ then (3.4.2) clearly holds for ω_0. We will prove that (3.4.2) holds for every compactly supported n-form ω with $\operatorname{supp}(\omega) \subset V$. Let $c := \int_V \omega$. Then by Theorem 3.3.1 $\omega - c\omega_0 = d\mu$, where μ is a compactly supported $(n-1)$-form with $\operatorname{supp}\mu \subset V$. Hence

$$f^\star\omega - cf^\star\omega_0 = f^\star d\mu = df^\star\mu,$$

and by part (1) of Theorem 3.3.1

$$\int_U f^\star \omega = c \int f^\star \omega_0 = \deg(f) \int_V \omega. \qquad \square$$

We will show in §3.6 that the degree of f is always an integer and explain why it is a "topological" invariant of f.

Proposition 3.4.4. *For the moment, however, we'll content ourselves with pointing out a simple but useful property of this invariant. Let U, V and W be connected open subsets of $\mathbf{R}^n$ and $f\colon U \to V$ and $g\colon V \to W$ proper C^∞ maps. Then*

$$\deg(g \circ f) = \deg(g)\deg(f). \tag{3.4.5}$$

Proof. Let ω be a compactly supported n-form with support on W. Then

$$(g \circ f)^\star \omega = f^\star g^\star \omega;$$

so

$$\begin{aligned} \int_U (g \circ f)^\star \omega = \int_U f^\star g^\star \omega &= \deg(f) \int_V g^\star \omega \\ &= \deg(f)\deg(g) \int_W \omega. \end{aligned} \qquad \square$$

From this multiplicative property it is easy to deduce the following result (which we will need in the next section).

Theorem 3.4.6. *Let A be a non-singular $n \times n$ matrix and $f_A\colon \mathbf{R}^n \to \mathbf{R}^n$ the linear mapping associated with A. Then $\deg(f_A) = +1$ if $\det A$ is positive and -1 if $\det A$ is negative.*

A proof of this result is outlined in Exercises 3.4.v to 3.4.ix.

Exercises for §3.4

Exercise 3.4.i. Let U be an open subset of $\mathbf{R}^n$ and let $(\phi_i)_{i \geq 1}$ be a partition of unity on U. Show that the mapping $f\colon U \to \mathbf{R}$ defined by

$$f := \sum_{k=1}^{\infty} k\phi_k$$

is a proper C^∞ mapping.

Exercise 3.4.ii. Let U and V be open subsets of $\mathbf{R}^n$ and $\mathbf{R}^k$ and let $f\colon U \to V$ be a proper continuous mapping. Prove the following theorem.

Theorem 3.4.7. *If B is a compact subset of V and $A = f^{-1}(B)$ then for every open subset U_0 with $A \subset U_0 \subset U$, there exists an open subset V_0 with $B \subset V_0 \subset V$ and $f^{-1}(V_0) \subset U_0$.*

Hint: Let C be a compact subset of V with $B \subset \operatorname{int} C$. Then the set $W = f^{-1}(C) \smallsetminus U_0$ is compact; so its image $f(W)$ is compact. Show that $f(W)$ and B are disjoint and let

$$V_0 = \operatorname{int} C \smallsetminus f(W).$$

Exercise 3.4.iii. Show that if $f: U \to V$ is a proper continuous mapping and X is a closed subset of U, then $f(X)$ is closed.

Hint: Let $U_0 = U - X$. Show that if p is in $V \setminus f(X)$, then $f^{-1}(p)$ is contained in U_0 and conclude from Exercise 3.4.ii that there exists a neighborhood V_0 of p such that $f^{-1}(V_0)$ is contained in U_0. Conclude that V_0 and $f(X)$ are disjoint.

Exercise 3.4.iv. Let $f: \mathbf{R}^n \to \mathbf{R}^n$ be the translation $f(x) = x + a$. Show that $\deg(f) = 1$.

Hint: Let $\psi: \mathbf{R} \to \mathbf{R}$ be a compactly supported C^∞ function. For $a \in \mathbf{R}$, the identity

$$\int_{\mathbf{R}} \psi(t)\,dt = \int_{\mathbf{R}} \psi(t-a)\,dt \tag{3.4.8}$$

is easy to prove by elementary calculus, and this identity proves the assertion above in dimension one. Now let

$$\phi(x) = \psi(x_1)\cdots\psi(x_n) \tag{3.4.9}$$

and compute the right and left sides of equation (3.4.3) by Fubini's theorem.

Exercise 3.4.v. Let σ be a permutation of the numbers, $1, \dots, n$ and let $f_\sigma: \mathbf{R}^n \to \mathbf{R}^n$ be the diffeomorphism, $f_\sigma(x_1, \dots, x_n) = (x_{\sigma(1)}, \dots, x_{\sigma(n)})$. Prove that $\deg f_\sigma = (-1)^\sigma$.

Hint: Let ϕ be the function (3.4.9). Show that if $\omega = \phi(x)\,dx_1 \wedge \cdots \wedge dx_n$, then we have $f^\star\omega = (-1)^\sigma\omega$.

Exercise 3.4.vi. Let $f: \mathbf{R}^n \to \mathbf{R}^n$ be the mapping

$$f(x_1, \dots, x_n) = (x_1 + \lambda x_2, x_2, \dots, x_n).$$

Prove that $\deg(f) = 1$.

Hint: Let $\omega = \phi(x_1, \dots, x_n)\,dx_1 \wedge \cdots \wedge dx_n$ where $\phi: \mathbf{R}^n \to \mathbf{R}$ is compactly supported and of class C^∞. Show that

$$\int f^\star\omega = \int \phi(x_1 + \lambda x_2, x_2, \dots, x_n)\,dx_1 \cdots dx_n$$

and evaluate the integral on the right by Fubini's theorem; i.e., by first integrating with respect to the x_1 variable and then with respect to the remaining variables. Note that by equation (3.4.8)

$$\int f(x_1 + \lambda x_2, x_2, \dots, x_n)\,dx_1 = \int f(x_1, x_2, \dots, x_n)\,dx_1.$$

Exercise 3.4.vii. Let $f: \mathbf{R}^n \to \mathbf{R}^n$ be the mapping

$$f(x_1, \dots, x_n) = (\lambda x_1, x_2, \dots, x_n)$$

with $\lambda \neq 0$. Show that $\deg f = +1$ if λ is positive and -1 if λ is negative.

Hint: In dimension one this is easy to prove by elementary calculus techniques. Prove it in d-dimensions by the same trick as in the previous exercise.

Exercise 3.4.viii.

(1) Let $e_1, \ldots, e_n$ be the standard basis vectors of $\mathbf{R}^n$ and A, B and C the linear mappings defined by

$$Ae_i = \begin{cases} e_1, & i = 1, \\ \sum_{j=1}^{n} a_{j,i} e_j, & i \neq 1, \end{cases}$$

$$Be_i = \begin{cases} \sum_{j=1}^{n} b_j e_j, & i = 1, \\ e_i, & i \neq 1, \end{cases} \tag{3.4.10}$$

$$Ce_i = \begin{cases} e_1, & i = 1, \\ e_i + c_i e_1, & i \neq 1. \end{cases}$$

Show that

$$BACe_i = \begin{cases} \sum_{j=1}^{m} b_j e_j, & i = 1, \\ \sum_{j=1}^{n} (a_{j,i} + c_i b_j) e_j + c_i b_1 e_1, & i \neq 1, \end{cases}$$

for $i > 1$.

(2) Let $L : \mathbf{R}^n \to \mathbf{R}^n$ be the linear mapping

$$Le_i = \sum_{j=1}^{n} \ell_{j,i} e_j, \quad i = 1, \ldots, n. \tag{3.4.11}$$

Show that if $\ell_{1,1} \neq 0$ one can write L as a product, $L = BAC$, where A, B and C are linear mappings of the form (3.4.10).

Hint: First solve the equations

$$\ell_{j,1} = b_j$$

for $j = 1, \ldots, n$. Next solve the equations

$$\ell_{1,i} = b_1 c_i$$

for $i > 1$. Finally, solve the equations

$$\ell_{j,i} = a_{j,i} + c_i b_j$$

for $i, j > 1$.

(3) Suppose L is invertible. Conclude that A, B and C are invertible and verify that Theorem 3.4.6 holds for B and C using the previous exercises in this section.

(4) Show by an inductive argument that Theorem 3.4.6 holds for A and conclude from (3.4.5) that it holds for L.

Exercise 3.4.ix. To show that Theorem 3.4.6 holds for an arbitrary linear mapping L of the form (3.4.11) we'll need to eliminate the assumption: $\ell_{1,1} \neq 0$. Show that for some j, $\ell_{j,1}$ is non-zero, and show how to eliminate this assumption by considering $f_{\tau_{1,j}} \circ L$ where $\tau_{1,j}$ is the transposition $1 \leftrightarrow j$.

Exercise 3.4.x. Here is an alternative proof of Theorem 3.4.6 which is shorter than the proof outlined in Exercise 3.4.ix but uses some slightly more sophisticated linear algebra.

(1) Prove Theorem 3.4.6 for linear mappings which are *orthogonal*, i.e., satisfy $L^{\intercal}L = \mathrm{id}_n$.

Hints:

- Show that $L^\star(x_1^2 + \cdots + x_n^2) = x_1^2 + \cdots + x_n^2$.
- Show that $L^\star(dx_1 \wedge \cdots \wedge dx_n)$ is equal to $dx_1 \wedge \cdots \wedge dx_n$ or $-dx_1 \wedge \cdots \wedge dx_n$ depending on whether L is orientation preserving or orientation reversing. (See Exercise 1.2.x.)
- Let ψ be as in Exercise 3.4.iv and let ω be the form

$$\omega = \psi(x_1^2 + \cdots + x_n^2)\, dx_1 \wedge \cdots \wedge dx_n.$$

Show that $L^\star\omega = \omega$ if L is orientation preserving and $L^\star\omega = -\omega$ if L is orientation reversing.

(2) Prove Theorem 3.4.6 for linear mappings which are *self-adjoint* (satisfy $L^{\intercal} = L$).

Hint: A self-adjoint linear mapping is diagonizable: there exists an intervertible linear mapping, $M : \mathbf{R}^n \to \mathbf{R}^n$ such that

$$M^{-1}LMe_i = \lambda_i e_i, \quad i = 1, \dots, n. \tag{3.4.12}$$

(3) Prove that every invertible linear mapping, L, can be written as a product, $L = BC$ where B is orthogonal and C is self-adjoint.

Hints:

- Show that the mapping, $A = L^{\intercal}L$, is self-adjoint and its eigenvalues (the λ_i's in equation (3.4.12)) are positive.
- Show that there exists an invertible self-adjoint linear mapping, C, such that $A = C^2$ and $AC = CA$.
- Show that the mapping $B = LC^{-1}$ is orthogonal.

3.5. The change of variables formula

Let U and V be connected open subsets of $\mathbf{R}^n$. If $f : U \xrightarrow{\sim} V$ is a diffeomorphism, the determinant of $Df(x)$ at $x \in U$ is non-zero, and hence, since $Df(x)$ is a continuous function of x, its sign is the same at every point. We will say that f is ***orientation preserving*** if this sign is positive and ***orientation reversing*** if it is negative. We will prove the following theorem.

Theorem 3.5.1. *The degree of f is $+1$ if f is orientation preserving and -1 if f is orientation reversing.*

We will then use this result to prove the following change of variables formula for diffeomorphisms.

Theorem 3.5.2. *Let $\phi\colon V \to \mathbf{R}$ be a compactly supported continuous function. Then*

$$\int_U (\phi \circ f)(x)|\det(Df(x))| = \int_V \phi(y)\,dy. \tag{3.5.3}$$

Proof of Theorem 3.5.1. Given a point $a_1 \in U$, let $a_2 = -f(a_1)$ and for $i = 1,2$ let $g_i\colon \mathbf{R}^n \to \mathbf{R}^n$ be the translation, $g_i(x) = x + a_i$. By equation (3.4.2) and Exercise 3.4.iv the composite diffeomorphism

$$g_2 \circ f \circ g_1 \tag{3.5.4}$$

has the same degree as f, so it suffices to prove the theorem for this mapping. Notice however that this mapping maps the origin onto the origin. Hence, replacing f by this mapping, we can, without loss of generality, assume that 0 is in the domain of f and that $f(0) = 0$.

Next notice that if $A\colon \mathbf{R}^n \xrightarrow{\sim} \mathbf{R}^n$ is a bijective linear mapping the theorem is true for A (by Exercise 3.4.ix), and hence if we can prove the theorem for $A^{-1} \circ f$, equation (3.4.2) will tell us that the theorem is true for f. In particular, letting $A = Df(0)$, we have

$$D(A^{-1} \circ f)(0) = A^{-1}Df(0) = \mathrm{id}_n,$$

where id_n is the identity mapping. Therefore, replacing f by $A^{-1} \circ f$, we can assume that the mapping f (for which we are attempting to prove Theorem 3.5.1) has the properties: $f(0) = 0$ and $Df(0) = \mathrm{id}_n$. Let $g(x) = f(x) - x$. Then these properties imply that $g(0) = 0$ and $Dg(0) = 0$. □

Lemma 3.5.5. *There exists a $\delta > 0$ such that $|g(x)| \le \frac{1}{2}|x|$ for $|x| \le \delta$.*

Proof. Let $g(x) = (g_1(x), \dots, g_n(x))$. Then

$$\frac{\partial g_i}{\partial x_j}(0) = 0;$$

so there exists a $\delta > 0$ such that

$$\left|\frac{\partial g_i}{\partial x_j}(x)\right| \le \frac{1}{2}$$

for $|x| \le \delta$. However, by the mean value theorem,

$$g_i(x) = \sum_{j=1}^{n} \frac{\partial g_i}{\partial x_j}(c)x_j$$

for $c = t_0 x$, $0 < t_0 < 1$. Thus, for $|x| < \delta$,

$$|g_i(x)| \le \frac{1}{2}\sup|x_i| = \frac{1}{2}|x|,$$

so

$$|g(x)| = \sup|g_i(x)| \le \frac{1}{2}|x|.$$

□

Let ρ be a compactly supported C^∞ function with $0 \le \rho \le 1$ and with $\rho(x) = 0$ for $|x| \ge \delta$ and $\rho(x) = 1$ for $|x| \le \frac{\delta}{2}$ and let $\tilde{f} : \mathbf{R}^n \to \mathbf{R}^n$ be the mapping

$$\tilde{f}(x) := x + \rho(x)g(x).$$

It is clear that

$$\tilde{f}(x) = x \quad \text{for } |x| \ge \delta \tag{3.5.6}$$

and, since $f(x) = x + g(x)$,

$$\tilde{f}(x) = f(x) \quad \text{for } |x| \le \frac{\delta}{2}. \tag{3.5.7}$$

In addition, for all $x \in \mathbf{R}^n$:

$$|\tilde{f}(x)| \ge \frac{1}{2}|x|. \tag{3.5.8}$$

Indeed, by (3.5.6), $|\tilde{f}(x)| \ge |x|$ for $|x| \ge \delta$, and for $|x| \le \delta$

$$\begin{aligned} |\tilde{f}(x)| &\ge |x| - \rho(x)|g(x)| \\ &\ge |x| - |g(x)| \ge |x| - \frac{1}{2}|x| \\ &= \frac{1}{2}|x| \end{aligned}$$

by Lemma 3.5.5.

Now let Q_r be the cube $Q_r := \{x \in \mathbf{R}^n \,|\, |x| \le r\}$, and let $Q_r^c := \mathbf{R}^n \smallsetminus Q_r$. From (3.5.8) we easily deduce that

$$\tilde{f}^{-1}(Q_r) \subset Q_{2r} \tag{3.5.9}$$

for all r, and hence that $\tilde{f}$ is *proper*. Also notice that for $x \in Q_\delta$,

$$|\tilde{f}(x)| \le |x| + |g(x)| \le \frac{3}{2}|x|$$

by Lemma 3.5.5 and hence

$$\tilde{f}^{-1}(Q^c_{\frac{3}{2}\delta}) \subset Q^c_\delta. \tag{3.5.10}$$

We will now prove Theorem 3.5.1.

Proof of Theorem 3.5.1. Since f is a diffeomorphism mapping 0-to-0, it maps a neighborhood U_0 of 0 in U diffeomorphically onto a neighborhood V_0 of 0 in V, and, by shrinking U_0 if necessary, we can assume that U_0 is contained in $Q_{\delta/2}$ and V_0 contained in $Q_{\delta/4}$. Let ω be an n-form with support in V_0 whose integral over $\mathbf{R}^n$ is equal to one. Then $f^\star\omega$ is supported in U_0 and hence in $Q_{\delta/2}$. Also by (3.5.9) $\tilde{f}^\star\omega$ is supported in $Q_{\delta/2}$. Thus both of these forms are zero outside $Q_{\delta/2}$. However, on $Q_{\delta/2}$, $\tilde{f} = f$ by (3.5.7), so these forms are equal everywhere, and hence

$$\deg(f) = \int f^\star\omega = \int \tilde{f}^\star\omega = \deg(\tilde{f}).$$

Next let ω be a compactly supported n-form with support in $Q^c_{3\delta/2}$ and with integral equal to one. Then $\tilde{f}^\star\omega$ is supported in Q^c_δ by (3.5.10), and hence since $f(x) = x$ on Q^c_δ, we have $\tilde{f}^\star\omega = \omega$. Thus

$$\deg(\tilde{f}) = \int \tilde{f}^\star\omega = \int \omega = 1.$$

Putting these two identities together we conclude that $\deg(f) = 1$. □

If the function, ϕ, in equation (3.5.4) is a C^∞ function, the identity (3.5.3) is an immediate consequence of the result above and the identity (3.4.3). If ϕ is not C^∞, but is just continuous, we will deduce equation (3.5.4) from the following result.

Theorem 3.5.11. *Let V be an open subset of $\mathbf{R}^n$. If $\phi\colon \mathbf{R}^n \to \mathbf{R}$ is a continuous function of compact support with* $\operatorname{supp}\phi \subset V$, *then for every $\varepsilon > 0$ there exists a C^∞ function of compact support, $\psi\colon \mathbf{R}^n \to \mathbf{R}$ with* $\operatorname{supp}\psi \subset V$ *and*

$$\sup|\psi(x) - \phi(x)| < \varepsilon.$$

Proof. Let A be the support of ϕ and let d be the distance in the sup norm from A to the complement of V. Since ϕ is continuous and compactly supported it is uniformly continuous; so for every $\varepsilon > 0$ there exists a $\delta > 0$ with $\delta < \frac{d}{2}$ such that $|\phi(x) - \phi(y)| < \varepsilon$ when $|x - y| \leq \delta$. Now let Q be the cube: $|x| < \delta$ and let $\rho\colon \mathbf{R}^n \to \mathbf{R}$ be a non-negative C^∞ function with $\operatorname{supp}\rho \subset Q$ and

$$\int \rho(y)dy = 1. \tag{3.5.12}$$

Set

$$\psi(x) = \int \rho(y - x)\phi(y)dy.$$

By Theorem 3.2.10 ψ is a C^∞ function. Moreover, if A_δ is the set of points in $\mathbf{R}^d$ whose distance in the sup norm from A is $\leq \delta$ then for $x \notin A_\delta$ and $y \in A$, $|x - y| > \delta$ and hence $\rho(y - x) = 0$. Thus for $x \notin A_\delta$

$$\int \rho(y - x)\phi(y)dy = \int_A \rho(y - x)\phi(y)dy = 0,$$

so ψ is supported on the compact set A_δ. Moreover, since $\delta < \frac{d}{2}$, $\operatorname{supp}\psi$ is contained in V. Finally note that by (3.5.12) and Exercise 3.4.iv

$$\int \rho(y - x)\, dy = \int \rho(y)\, dy = 1 \tag{3.5.13}$$

and hence

$$\phi(x) = \int \phi(x)\rho(y - x)dy$$

so

$$\phi(x) - \psi(x) = \int (\phi(x) - \phi(y))\rho(y - x)dy$$

and

$$|\phi(x) - \psi(x)| \leq \int |\phi(x) - \phi(y)|\rho(y-x)dy.$$

But $\rho(y-x) = 0$ for $|x-y| \geq \delta$; and $|\phi(x) - \phi(y)| < \varepsilon$ for $|x-y| \leq \delta$, so the integrand on the right is less than

$$\varepsilon \int \rho(y-x)\,dy,$$

and hence by equation (3.5.13) we have

$$|\phi(x) - \psi(x)| \leq \varepsilon. \qquad \square$$

To prove the identity (3.5.3), let $\gamma\colon \mathbf{R}^n \to \mathbf{R}$ be a C^∞ cut-off function which is one on a neighborhood V_1 of the support of ϕ is non-negative, and is compactly supported with $\operatorname{supp}\gamma \subset V$, and let

$$c = \int \gamma(y)\,dy.$$

By Theorem 3.5.11 there exists, for every $\varepsilon > 0$, a C^∞ function ψ, with support on V_1 satisfying

$$|\phi - \psi| \leq \frac{\varepsilon}{2c}. \tag{3.5.14}$$

Thus

$$\begin{aligned}\left|\int_V (\phi - \psi)(y)dy\right| &\leq \int_V |\phi - \psi|(y)dy \\ &\leq \int_V \gamma|\phi - \psi|(xy)dy \\ &\leq \frac{\varepsilon}{2c}\int \gamma(y)dy \leq \frac{\varepsilon}{2}\end{aligned}$$

so

$$\left|\int_V \phi(y)\,dy - \int_V \psi(y)\,dy\right| \leq \frac{\varepsilon}{2}. \tag{3.5.15}$$

Similarly, the expression

$$\left|\int_U (\phi - \psi) \circ f(x)|\det Df(x)|\,dx\right|$$

is less than or equal to the integral

$$\int_U (\gamma \circ f)(x)|(\phi - \psi) \circ f(x)|\,|\det Df(x)|\,dx$$

and by (3.5.14), $|(\phi - \psi) \circ f(x)| \leq \frac{\varepsilon}{2c}$, so this integral is less than or equal to

$$\frac{\varepsilon}{2c}\int (\gamma \circ f)(x)|\det Df(x)|\,dx$$

and hence by (3.5.3) is less than or equal to $\frac{\varepsilon}{2}$. Thus

$$\left| \int_U (\phi \circ f)(x) \,|\det Df(x)| dx - \int_U \psi \circ f(x) |\det Df(x)| \, dx \right| \le \frac{\varepsilon}{2}. \tag{3.5.16}$$

Combining (3.5.15), (3.5.16) and the identity

$$\int_V \psi(y)\,dy = \int \psi \circ f(x) |\det Df(x)|\, dx$$

we get, for all $\varepsilon > 0$,

$$\left| \int_V \phi(y)\,dy - \int_U (\phi \circ f)(x) |\det Df(x)|\, dx \right| \le \varepsilon$$

and hence

$$\int \phi(y)\,dy = \int (\phi \circ f)(x) |\det Df(x)|\, dx.$$

Exercises for §3.5

Exercise 3.5.i. Let $h \colon V \to \mathbf{R}$ be a non-negative continuous function. Show that if the improper integral

$$\int_V h(y)dy$$

is well-defined, then the improper integral

$$\int_U (h \circ f)(x) |\det Df(x)| dx$$

is well-defined and these two integrals are equal.

Hint: If $(\phi_i)_{i\ge 1}$ is a partition of unity on V then $\psi_i = \phi_i \circ f$ is a partition of unity on U and

$$\int \phi_i h dy = \int \psi_i (h \circ f(x)) |\det Df(x)| dx.$$

Now sum both sides of this identity over i.

Exercise 3.5.ii. Show that the result above is true without the assumption that h is non-negative.

Hint: $h = h_+ - h_-$, where $h_+ = \max(h, 0)$ and $h_- = \max(-h, 0)$.

Exercise 3.5.iii. Show that in equation (3.4.3) one can allow the function ϕ to be a *continuous* compactly supported function rather than a C^∞ compactly supported function.

Exercise 3.5.iv. Let $\mathbf{H}^n$ be the half-space (3.2.12) and U and V open subsets of $\mathbf{R}^n$. Suppose $f \colon U \to V$ is an orientation-preserving diffeomorphism mapping $U \cap \mathbf{H}^n$ onto $V \cap \mathbf{H}^n$. Show that for $\omega \in \Omega_c^n(V)$

$$\int_{U \cap \mathbf{H}^n} f^\star \omega = \int_{V \cap \mathbf{H}^n} \omega. \tag{3.5.17}$$

Hint: Interpret the left- and right-hand sides of this formula as improper integrals over $U \cap \text{int}(\mathbf{H}^n)$ and $V \cap \text{int}(\mathbf{H}^n)$.

Exercise 3.5.v. The boundary of $\mathbf{H}^n$ is the set

$$\partial \mathbf{H}^n := \{ (0, x_2, \dots, x_n) \,|\, (x_2, \dots, x_n) \in \mathbf{R}^{n-1} \}$$

so the map

$$\iota \colon \mathbf{R}^{n-1} \to \mathbf{H}^n , \quad (x_2, \dots, x_n) \mapsto (0, x_2, \dots, x_n)$$

in Exercise 3.2.viii maps $\mathbf{R}^{n-1}$ bijectively onto $\partial \mathbf{H}^n$.

(1) Show that the map $f \colon U \to V$ in Exercise 3.5.iv maps $U \cap \partial \mathbf{H}^n$ onto $V \cap \partial \mathbf{H}^n$.
(2) Let $U' = \iota^{-1}(U)$ and $V' = \iota^{-1}(V)$. Conclude from (1) that the restriction of f to $U \cap \partial \mathbf{H}^n$ gives one a diffeomorphism

$$g \colon U' \to V'$$

satisfying:

$$\iota \circ g = f \circ \iota.$$

(3) Let μ be in $\Omega_c^{n-1}(V)$. Conclude from equations (3.2.14) and (3.5.17):

$$\int_{U'} g^\star \iota^\star \mu = \int_{V'} \iota^\star \mu$$

and in particular show that the diffeomorphism $g \colon U' \to V'$ is orientation preserving.

3.6. Techniques for computing the degree of a mapping

Let U and V be open subsets of $\mathbf{R}^n$ and $f \colon U \to V$ a proper C^∞ mapping. In this section we will show how to compute the degree of f and, in particular, show that it is always an integer. From this fact we will be able to conclude that the degree of f is a topological invariant of f: if we deform f smoothly, its degree doesn't change.

Definition 3.6.1. A point $x \in U$ is a ***critical point*** of f if the derivative

$$Df(x) \colon \mathbf{R}^n \to \mathbf{R}^n$$

fails to be bijective, i.e., if $\det(Df(x)) = 0$.

We will denote the set of critical points of f by C_f. It is clear from the definition that this set is a closed subset of U and hence, by Exercise 3.4.iii, $f(C_f)$ is a closed subset of V. We will call this image the set of ***critical values*** of f and the complement of this image the set of ***regular values*** of f. Notice that $V \smallsetminus f(U)$ is contained in $V \smallsetminus f(C_f)$, so if a point $q \in V$ is not in the image of f, it is a regular value of f "by default", i.e., it contains no points of U in the preimage and hence, *a fortiori*, contains no critical points in its preimage. Notice also that C_f can be quite large. For instance, if $c \in V$ and $f \colon U \to V$ is the constant map which maps all of U onto c, then $C_f = U$. However, in this example, $f(C_f) = \{c\}$, so the set of regular values of f is $V \smallsetminus \{c\}$, and hence (in this example) is an open dense subset of V. We will show that this is true in general.

Theorem 3.6.2 (Sard). *If U and V are open subsets of $\mathbf{R}^n$ and $f : U \to V$ a proper C^∞ map, the set of regular values of f is an open dense subset of V.*

We will defer the proof of this to § 3.7 and in this section explore some of its implications. Picking a regular value q of f we will prove the following theorem.

Theorem 3.6.3. *The set $f^{-1}(q)$ is a finite set. Moreover, if $f^{-1}(q) = \{p_1, \dots, p_n\}$ there exist connected open neighborhoods U_i of p_i in Y and an open neighborhood W of q in V such that:*

(1) *for $i \neq j$ the sets U_i and U_j are disjoint;*
(2) *$f^{-1}(W) = U_1 \cup \cdots \cup U_n$;*
(3) *f maps U_i diffeomorphically onto W.*

Proof. If $p \in f^{-1}(q)$, then, since q is a regular value, $p \notin C_f$; so

$$Df(p) : \mathbf{R}^n \to \mathbf{R}^n$$

is bijective. Hence by the inverse function theorem, f maps a neighborhood, U_p of p diffeomorphically onto a neighborhood of q. The open sets

$$\{U_p \mid p \in f^{-1}(q)\}$$

are a covering of $f^{-1}(q)$; and, since f is proper, $f^{-1}(q)$ is compact. Thus we can extract a finite subcovering

$$\{U_{p_1}, \dots, U_{p_N}\}$$

and since p_i is the only point in U_{p_i} which maps onto q, we have that $f^{-1}(q) = \{p_1, \dots, p_N\}$.

Without loss of generality we can assume that the U_{p_i}'s are disjoint from each other; e.g., if not, we can replace them by smaller neighborhoods of the p_i's which have this property. By Theorem 3.4.7 there exists a connected open neighborhood W of q in V for which

$$f^{-1}(W) \subset U_{p_1} \cup \cdots \cup U_{p_N}.$$

To conclude the proof let $U_i \coloneqq f^{-1}(W) \cap U_{p_i}$. □

The main result of this section is a recipe for computing the degree of f by counting the number of p_i's above, keeping track of orientation.

Theorem 3.6.4. *For each $p_i \in f^{-1}(q)$ let $\sigma_{p_i} = +1$ if $f : U_i \to W$ is orientation preserving and -1 if $f : U_i \to W$ is orientation reversing. Then*

$$\deg(f) = \sum_{i=1}^{N} \sigma_{p_i}. \tag{3.6.5}$$

Proof. Let ω be a compactly supported n-form on W whose integral is one. Then

$$\deg(f) = \int_U f^\star \omega = \sum_{i=1}^{N} \int_{U_i} f^\star \omega.$$

Since $f: U_i \to W$ is a diffeomorphism

$$\int_{U_i} f^\star \omega = \pm \int_W \omega = \begin{cases} 1, & f \text{ is orientation preserving,} \\ -1, & f \text{ is not orientation preserving.} \end{cases}$$

Thus $\deg(f)$ is equal to the sum (3.6.5). □

As we pointed out above, a point $q \in V$ can qualify as a regular value of f "by default", i.e., by not being in the image of f. In this case the recipe (3.6.5) for computing the degree gives "by default" the answer zero. Let's corroborate this directly.

Theorem 3.6.6. *If $f: U \to V$ is not surjective, then* $\deg(f) = 0$.

Proof. By Exercise 3.4.iii, $V \smallsetminus f(U)$ is open; so if it is non-empty, there exists a compactly supported n-form ω with support in $V \smallsetminus f(U)$ and with integral equal to one. Since $\omega = 0$ on the image of f, $f^\star\omega = 0$; so

$$0 = \int_U f^\star \omega = \deg(f) \int_V \omega = \deg(f).$$ □

Remark 3.6.7. In applications the contrapositive of Theorem 3.6.6 is much more useful than the theorem itself.

Theorem 3.6.8. *If* $\deg(f) \neq 0$, *then f maps U surjectively onto V.*

In other words if $\deg(f) \neq 0$ the equation

$$f(x) = y$$

has a solution, $x \in U$ for *every* $y \in V$.

We will now show that the degree of f is a topological invariant of f: if we deform f by a "homotopy" we do not change its degree. To make this assertion precise, let's recall what we mean by a homotopy between a pair of C^∞ maps. Let U be an open subset of $\mathbf{R}^m$, V an open subset of $\mathbf{R}^n$, A an open subinterval of $\mathbf{R}$ containing 0 and 1, and $f_1, f_2 : U \to V$ a pair of C^∞ maps. Then a C^∞ map $F: U \times A \to V$ is a ***homotopy*** between f_0 and f_1 if $F(x,0) = f_0(x)$ and $F(x,1) = f_1(x)$. (See Definition 2.6.14.) Suppose now that f_0 and f_1 are proper.

Definition 3.6.9. A homotopy F between f_0 and f_1 is a ***proper homotopy*** if the map

$$F^\sharp : U \times A \to V \times A$$

defined by $(x,t) \mapsto (F(x,t),t)$ is proper.

Note that if F is a proper homotopy between f_0 and f_1, then for every t between 0 and 1, the map

$$f_t : U \to V\,, \quad f_t(x) := F(x,t)$$

is proper.

Now let U and V be open subsets of $\mathbf{R}^n$.

Theorem 3.6.10. *If f_0 and f_1 are properly homotopic, then* $\deg(f_0) = \deg(f_1)$.

Proof. Let

$$\omega = \phi(y)dy_1 \wedge \cdots \wedge dy_n$$

be a compactly supported n-form on V whose integral over V is 1. Then the degree of f_t is equal to

$$\int_U \phi(F_1(x,t),\ldots,F_n(x,t)) \det D_x F(x,t) dx. \tag{3.6.11}$$

The integrand in (3.6.11) is continuous and for $0 \leq t \leq 1$ is supported on a compact subset of $U \times [0,1]$, hence (3.6.11) is continuous as a function of t. However, as we've just proved, $\deg(f_t)$ is *integer* valued so this function is a constant. □

(For an alternative proof of this result see Exercise 3.6.ix below.)

Applications

We'll conclude this account of degree theory by describing a couple applications.

Application 3.6.12 (The Brouwer fixed point theorem). Let B^n be the closed unit ball in $\mathbf{R}^n$:

$$B^n := \{x \in \mathbf{R}^n \mid \|x\| \leq 1\}.$$

Theorem 3.6.13. *If $f: B^n \to B^n$ is a continuous mapping, then f has a fixed point, i.e., maps some point, $x_0 \in B^n$ onto itself.*

The idea of the proof will be to assume that there isn't a fixed point and show that this leads to a contradiction. Suppose that for every point $x \in B^n$ we have $f(x) \neq x$. Consider the ray through $f(x)$ in the direction of x:

$$f(x) + s(x - f(x)), \quad s \in [0,\infty).$$

This ray intersects the boundary $S^{n-1} := \partial B^n$ in a unique point $\gamma(x)$ (see Figure 3.6.1 below); and one of the exercises at the end of this section will be to show that the mapping $\gamma: B^n \to S^{n-1}$ given by $x \mapsto \gamma(x)$, is a continuous mapping. Also it is clear from Figure 3.6.1 that $\gamma(x) = x$ if $x \in S^{n-1}$, so we can extend γ to a continuous mapping of $\mathbf{R}^n$ into $\mathbf{R}^n$ by letting γ be the identity for $\|x\| \geq 1$. Note that this extended mapping has the property

$$\|\gamma(x)\| \geq 1$$

for all $x \in \mathbf{R}^n$ and

$$\gamma(x) = x \tag{3.6.14}$$

for all $\|x\| \geq 1$. To get a contradiction we'll show that γ can be approximated by a C^∞ map which has similar properties. For this we will need the following corollary of Theorem 3.5.11.

Lemma 3.6.15. *Let U be an open subset of $\mathbf{R}^n$, C a compact subset of U and $\phi: U \to \mathbf{R}$ a continuous function which is C^∞ on the complement of C. Then for*

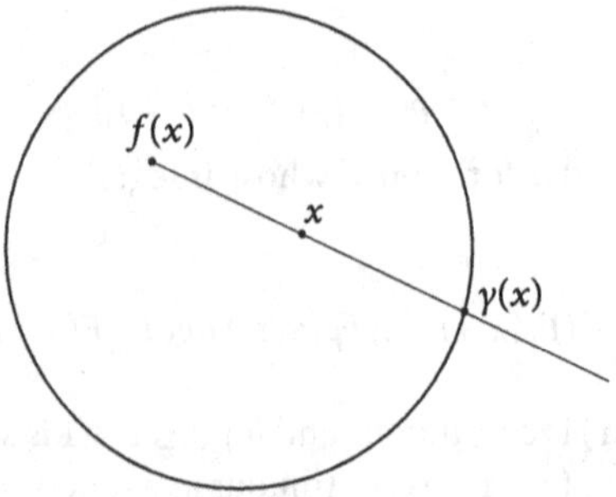

Figure 3.6.1. Brouwer fixed point theorem.

every $\varepsilon > 0$, *there exists a* C^∞ *function* $\psi\colon U \to \mathbf{R}$, *such that* $\phi - \psi$ *has compact support and* $|\phi - \psi| < \varepsilon$.

Proof. Let ρ be a bump function which is in $C_0^\infty(U)$ and is equal to 1 on a neighborhood of C. By Theorem 3.5.11 there exists a function $\psi_0 \in C_0^\infty(U)$ such that $|\rho\phi - \psi_0| < \varepsilon$. To complete the proof, let $\psi := (1-\rho)\phi + \psi_0$, and note that

$$\begin{aligned}\phi - \psi &= (1-\rho)\phi + \rho\phi - (1-\rho)\phi - \psi_0 \\ &= \rho\phi - \psi_0.\end{aligned}$$

□

By applying Lemma 3.6.15 to each of the coordinates of the map γ, one obtains a C^∞ map $g\colon \mathbf{R}^n \to \mathbf{R}^n$ such that

$$\|g - \gamma\| < \varepsilon < 1$$

and such that $g = \gamma$ on the complement of a compact set. However, by (3.6.14), this means that g is equal to the identity on the complement of a compact set and hence (see Exercise 3.6.ix) that g is proper and $\deg(g) = 1$. On the other hand by (3.6.19) and (3.6.14) we have $\|g(x)\| > 1 - \varepsilon$ for all $x \in \mathbf{R}^n$, so $0 \notin \operatorname{im}(g)$ and hence by Theorem 3.6.4 we have $\deg(g) = 0$, which provides the desired contradiction.

Application 3.6.16 (The fundamental theorem of algebra). Let

$$p(z) = z^n + a_{n-1}z^{n-1} + \cdots + a_1 z + a_0$$

be a polynomial of degree n with complex coefficients. If we identify the complex plane

$$\mathbf{C} := \{\, z = x + iy \mid x, y \in \mathbf{R} \,\}$$

with $\mathbf{R}^2$ via the map $\mathbf{R}^2 \to \mathbf{C}$ given by $(x, y) \mapsto z = x + iy$, we can think of p as defining a mapping

$$p\colon \mathbf{R}^2 \to \mathbf{R}^2\,,\ z \mapsto p(z).$$

We will prove the following theorem.

Theorem 3.6.17. *The mapping* $p\colon \mathbf{R}^2 \to \mathbf{R}^2$ *is proper and* $\deg(p) = n$.

Proof. For $t \in \mathbf{R}$ let

$$p_t(z) := (1-t)z^n + tp(z) = z^n + t\sum_{i=0}^{n-1} a_i z^i.$$

We will show that the mapping

$$g\colon \mathbf{R}^2 \times \mathbf{R} \to \mathbf{R}^2, \quad (z,t) \mapsto p_t(z)$$

is a proper homotopy. Let

$$C = \sup_{0\le i\le n-1} |a_i|.$$

Then for $|z| \ge 1$ we have

$$\begin{aligned} |a_0 + \cdots + a_{n-1}z^{n-1}| &\le |a_0| + |a_1||z| + \cdots + |a_{n-1}||z|^{n-1} \\ &\le Cn|z|^{n-1}, \end{aligned}$$

and hence, for $|t| \le a$ and $|z| \ge 2aCn$,

$$\begin{aligned} |p_t(z)| &\ge |z|^n - aCn|z|^{n-1} \\ &\ge aCn|z|^{n-1}. \end{aligned}$$

If $A \subset \mathbf{C}$ is compact, then for some $R > 0$, A is contained in the disk defined by $|w| \le R$, and hence the set

$$\{ z \in \mathbf{C} \mid (t, p_t(z)) \in [-a,a] \times A \}$$

is contained in the compact set

$$\{ z \in \mathbf{C} \mid aC|z|^{n-1} \le R \}.$$

This shows that g is a proper homotopy.

Thus for each $t \in \mathbf{R}$, the map $p_t\colon \mathbf{C} \to \mathbf{C}$ is proper and

$$\deg(p_t) = \deg(p_1) = \deg(p) = \deg(p_0).$$

However, $p_0\colon \mathbf{C} \to \mathbf{C}$ is just the mapping $z \mapsto z^n$ and an elementary computation (see Exercises 3.6.v and 3.6.vi below) shows that the degree of this mapping is n. □

In particular for $n > 0$ the degree of p is non-zero; so by Theorem 3.6.4 we conclude that $p\colon \mathbf{C} \to \mathbf{C}$ is surjective and hence has zero in its image.

Theorem 3.6.18 (Fundamental theorem of algebra). *Every positive-degree polynomial*

$$p(z) = z^n + a_{n-1}z^{n-1} + \cdots + a_0$$

with complex coefficients has a complex root: $p(z_0) = 0$ *for some* $z_0 \in \mathbf{C}$.

Exercises for §3.6

Exercise 3.6.i. Let W be a subset of $\mathbf{R}^n$ and let $a(x)$, $b(x)$ and $c(x)$ be real-valued functions on W of class C^r. Suppose that for every $x \in W$ the quadratic polynomial

$$a(x)s^2 + b(x)s + c(x)$$

has two distinct real roots, $s_+(x)$ and $s_-(x)$, with $s_+(x) > s_-(x)$. Prove that s_+ and s_- are functions of class C^r.

Hint: What *are* the roots of the quadratic polynomial: $as^2 + bs + c$?

Exercise 3.6.ii. Show that the function $\gamma(x)$ defined in Figure 3.6.1 is a continuous surjection $B^n \to S^{n-1}$.

Hint: $\gamma(x)$ lies on the ray,

$$f(x) + s(x - f(x)), \quad s \in [0, \infty)$$

and satisfies $\|\gamma(x)\| = 1$. Thus

$$\gamma(x) = f(x) + s_0(x - f(x)),$$

where s_0 is a non-negative root of the quadratic polynomial

$$\|f(x) + s(x - f(x))\|^2 - 1.$$

Argue from Figure 3.6.1 that this polynomial has to have two distinct real roots.

Exercise 3.6.iii. Show that the Brouwer fixed point theorem isn't true if one replaces the closed unit ball by the open unit ball.

Hint: Let U be the open unit ball (i.e., the interior of B^n). Show that the map

$$h\colon U \to \mathbf{R}^n, \quad h(x) := \frac{x}{1 - \|x\|^2}$$

is a diffeomorphism of U onto $\mathbf{R}^n$, and show that there are lots of mappings of $\mathbf{R}^n$ onto $\mathbf{R}^n$ which do not have fixed points.

Exercise 3.6.iv. Show that the fixed point in the Brouwer theorem doesn't have to be an interior point of B^n, i.e., show that it can lie on the boundary.

Exercise 3.6.v. If we identify $\mathbf{C}$ with $\mathbf{R}^2$ via the mapping $(x, y) \mapsto x + iy$, we can think of a $\mathbf{C}$-linear mapping of $\mathbf{C}$ into itself, i.e., a mapping of the form

$$z \mapsto cz$$

for a fixed $c \in \mathbf{C}$, as an $\mathbf{R}$-linear mapping of $\mathbf{R}^2$ into itself. Show that the determinant of this mapping is $|c|^2$.

Exercise 3.6.vi.

(1) Let $f : \mathbf{C} \to \mathbf{C}$ be the mapping $f(z) := z^n$. Show that $Df(z)$ is the linear map

$$Df(z) = nz^{n-1}$$

given by multiplication by nz^{n-1}.

Hint: Argue from first principles. Show that for $h \in \mathbf{C} = \mathbf{R}^2$

$$\frac{(z+h)^n - z^n - nz^{n-1}h}{|h|}$$

tends to zero as $|h| \to 0$.

(2) Conclude from Exercise 3.6.v that

$$\det(Df(z)) = n^2|z|^{2n-2}.$$

(3) Show that at every point $z \in \mathbf{C} \smallsetminus \{0\}$, f is orientation preserving.

(4) Show that every point, $w \in \mathbf{C} \smallsetminus \{0\}$ is a regular value of f and that

$$f^{-1}(w) = \{z_1, \dots, z_n\}$$

with $\sigma_{z_i} = +1$.

(5) Conclude that the degree of f is n.

Exercise 3.6.vii. Prove that the map f from Exercise 3.6.vi has degree n by deducing this directly from the definition of degree.

Hints:

- Show that in polar coordinates, f is the map $(r, \theta) \mapsto (r^n, n\theta)$.
- Let ω be the 2-form $\omega := g(x^2+y^2)dx \wedge dy$, where $g(t)$ is a compactly supported C^∞ function of t. Show that in polar coordinates $\omega = g(r^2)r dr \wedge d\theta$, and compute the degree of f by computing the integrals of ω and $f^\star\omega$ in polar coordinates and comparing them.

Exercise 3.6.viii. Let U be an open subset of $\mathbf{R}^n$, V an open subset of $\mathbf{R}^m$, A an open subinterval of $\mathbf{R}$ containing 0 and 1, $f_0, f_1 : U \to V$ a pair of C^∞ mappings, and $F : U \times A \to V$ a homotopy between f_0 and f_1.

(1) In Exercise 2.4.iv you proved that if $\mu \in \Omega^k(V)$ and $d\mu = 0$, then

$$f_0^\star \mu - f_1^\star \mu = d\nu \tag{3.6.19}$$

where ν is the $(k-1)$-form $Q\alpha$ in equation (2.6.17). Show (by careful inspection of the definition of $Q\alpha$) that if F is a *proper* homotopy and $\mu \in \Omega_c^k(V)$ then $\nu \in \Omega_c^{k-1}(U)$.

(2) Suppose in particular that U and V are open subsets of $\mathbf{R}^n$ and μ is in $\Omega_c^n(V)$. Deduce from equation (3.6.19) that

$$\int f_0^\star \mu = \int f_1^\star \mu$$

and deduce directly from the definition of degree that degree is a proper homotopy invariant.

Exercise 3.6.ix. Let U be an open connected subset of $\mathbf{R}^n$ and $f : U \to U$ a proper C^∞ map. Prove that if f is equal to the identity on the complement of a compact set C, then f is proper and $\deg(f) = 1$.

Hints:

- Show that for every subset $A \subset U$, we have $f^{-1}(A) \subset A \cup C$, and conclude from this that f is proper.
- Use the recipe (1.6.2) to compute $\deg(f)$ with $q \in U \smallsetminus f(C)$.

Exercise 3.6.x. Let $(a_{i,j})$ be an $n \times n$ matrix and $A\colon \mathbf{R}^n \to \mathbf{R}^n$ the linear mapping associated with this matrix. *Frobenius' Theorem* asserts: If the $a_{i,j}$ are non-negative then A has a non-negative eigenvalue. In other words there exist a $v \in \mathbf{R}^n$ and a $\lambda \in \mathbf{R}$, $\lambda \geq 0$, such that $Av = \lambda v$. Deduce this linear algebra result from the Brouwer fixed point theorem.

Hints:

- We can assume that A is bijective, otherwise 0 is an eigenvalue. Let S^{n-1} be the $(n-1)$-sphere, defined by $|x| = 1$, and $f\colon S^{n-1} \to S^{n-1}$ the map,

$$f(x) = \frac{Ax}{\|Ax\|}.$$

  Show that f maps the set

$$Q := \{ (x_1, \ldots, x_n) \in S^{n-1} \mid x_1, \ldots, x_n \geq 0 \}$$

  into itself.
- It is easy to prove that Q is homeomorphic to the unit ball B^{n-1}, i.e., that there exists a continuous map $g\colon Q \to B^{n-1}$ which is invertible and has a continuous inverse. Without bothering to prove this fact deduce from it Frobenius' theorem.

3.7. Appendix: Sard's theorem

The version of Sard's theorem stated in § 3.5 is a corollary of the following more general result.

Theorem 3.7.1. *Let U be an open subset of $\mathbf{R}^n$ and $f\colon U \to \mathbf{R}^n$ a C^∞ map. Then $\mathbf{R}^n \smallsetminus f(C_f)$ is dense in $\mathbf{R}^n$.*

Before undertaking to prove this we will make a few general comments about this result.

Remark 3.7.2. If $(U_m)_{m\geq 1}$ are open dense subsets of $\mathbf{R}^n$, the intersection $\bigcap_{m\geq 1} U_m$ is dense in $\mathbf{R}^n$. (This follows from the Baire category theorem; see, for instance, [7, Chapter 6 Theorem 34; 11, Theorem 48.2] or Exercise 3.7.iv.)

Remark 3.7.3. If $(A_n)_{n\geq 1}$ is a covering of U by compact sets, $O_n := \mathbf{R}^n \smallsetminus f(C_f \cap A_n)$ is open, so if we can prove that it is dense then by Remark 3.7.2 we will have proved Sard's theorem. Hence since we can always cover U by a countable collection of closed cubes, it suffices to prove the following: for every closed cube $A \subset U$, the subspace $\mathbf{R}^n \smallsetminus f(C_f \cap A)$ is dense in $\mathbf{R}^n$.

Remark 3.7.4. Let $g\colon W \to U$ be a diffeomorphism and let $h = f \circ g$. Then

$$f(C_f) = h(C_h) \tag{3.7.5}$$

so Sard's theorem for h implies Sard's theorem for f.

We will first prove Sard's theorem for the set of ***super-critical*** points of f, the set:

$$C_f^\sharp := \{p \in U \mid Df(p) = 0\}.$$

Proposition 3.7.6. *Let $A \subset U$ be a closed cube. Then the open set $\mathbf{R}^n \smallsetminus f(A \cap C_f^\sharp)$ is a dense subset of $\mathbf{R}^n$.*

We'll deduce this from the lemma below.

Lemma 3.7.7. *Given $\varepsilon > 0$ one can cover $f(A \cap C_f^\sharp)$ by a finite number of cubes of total volume less than ε.*

Proof. Let the length of each of the sides of A be ℓ. Given $\delta > 0$ one can subdivide A into N^n cubes, each of volume $(\ell/N)^n$, such that if x and y are points of any one of these subcubes

$$\left| \frac{\partial f_i}{\partial x_j}(x) - \frac{\partial f_i}{\partial x_j}(y) \right| < \delta. \tag{3.7.8}$$

Let $A_1, \dots, A_m$ be the cubes in this collection which intersect $C_f^\sharp$. Then for $z_0 \in A_i \cap C_f^\sharp$, $\frac{\partial f_i}{\partial x_j}(z_0) = 0$, so for $z \in A_i$ we have

$$\left| \frac{\partial f_i}{\partial x_j}(z) \right| < \delta \tag{3.7.9}$$

by equation (3.7.8). If x and y are points of A_i then by the mean value theorem there exists a point z on the line segment joining x to y such that

$$f_i(x) - f_i(y) = \sum_{j=1}^n \frac{\partial f_i}{\partial x_j}(z)(x_j - y_j)$$

and hence by (3.7.9)

$$|f_i(x) - f_i(y)| \le \delta \sum_{j=1}^n |x_j - y_j| \le n\delta \frac{\ell}{N}.$$

Thus $f(C_f \cap A_i)$ is contained in a cube B_i of volume $\left(n\frac{\delta\ell}{N}\right)^n$, and $f(C_f \cap A)$ is contained in a union of cubes B_i of total volume less than

$$N^n n^n \frac{\delta^n \ell^n}{N^n} = n^n \delta^n \ell^n$$

so if we choose δ such that $n^n \delta^n \ell^n < \varepsilon$, we're done. □

Proof of Proposition 3.7.6. To prove Proposition 3.7.6 we have to show that for every point $p \in \mathbf{R}^n$ and neighborhood, W of p, the set $W \smallsetminus f(C_f^\sharp \cap A)$ is non-empty. Suppose

$$W \subset f(C_f^\sharp \cap A). \tag{3.7.10}$$

Without loss of generality we can assume W is a cube of volume ε, but the lemma tells us that $f(C_f^\sharp \cap A)$ can be covered by a finite number of cubes whose total volume is *less* than ε, and hence by (3.7.10) W can be covered by a finite number of cubes of total volume less than ε, so its volume is less than ε. This contradiction proves that the inclusion (3.7.10) cannot hold. □

Now we prove Theorem 3.7.1.

Proof of Theorem 3.7.1. Let $U_{i,j}$ be the subset of U where $\frac{\partial f_i}{\partial x_j} \neq 0$. Then

$$U = C_f^\sharp \cup \bigcup_{1 \leq i,j \leq n} U_{i,j},$$

so to prove the theorem it suffices to show that $\mathbf{R}^n \smallsetminus f(U_{i,j} \cap C_f)$ is dense in $\mathbf{R}^n$, i.e., it suffices to prove the theorem with U replaced by $U_{i,j}$. Let $\sigma_i \colon \mathbf{R}^n \xrightarrow{\sim} \mathbf{R}^n$ be the involution which interchanges x_1 and x_i and leaves the remaining x_k's fixed. Letting $f_{\text{new}} = \sigma_i f_{\text{old}} \sigma_j$ and $U_{\text{new}} = \sigma_j U_{\text{old}}$, we have, for $f = f_{\text{new}}$ and $U = U_{\text{new}}$

$$\frac{\partial f_1}{\partial x_1}(p) \neq 0 \tag{3.7.11}$$

for all $p \in U$ so we're reduced to proving Theorem 3.7.1 for maps $f \colon U \to \mathbf{R}^n$ having the property (3.7.11). Let $g \colon U \to \mathbf{R}^n$ be defined by

$$g(x_1, \ldots, x_n) = (f_1(x), x_2, \ldots, x_n). \tag{3.7.12}$$

Then

$$g^\star x_1 = f^\star x_1 = f_1(x_1, \ldots, x_n)$$

and

$$\det(Dg) = \frac{\partial f_1}{\partial x_1} \neq 0.$$

Thus, by the inverse function theorem, g is locally a diffeomorphism at every point $p \in U$. This means that if A is a compact subset of U we can cover A by a finite number of open subsets $U_i \subset U$ such that g maps U_i diffeomorphically onto an open subset W_i in $\mathbf{R}^n$. To conclude the proof of the theorem we'll show that $\mathbf{R}^n \smallsetminus f(C_f \cap U_i \cap A)$ is a dense subset of $\mathbf{R}^n$. Let $h \colon W_i \to \mathbf{R}^n$ be the map $h = f \circ g^{-1}$. To prove this assertion it suffices by Remark 3.7.4 to prove that the set $\mathbf{R}^n \smallsetminus h(C_h)$ is dense in $\mathbf{R}^n$. This we will do by induction on n. First note that for $n = 1$, we have $C_f = C_f^\sharp$, so we've already proved Theorem 3.7.1 in dimension one. Now note that by (3.7.12) we have $h^\star x_1 = x_1$, i.e., h is a mapping of the form

$$h(x_1, \dots, x_n) = (x_1, h_2(x), \dots, h_n(x)).$$

Thus if we let

$$W_c := \{(x_2, \dots, x_n) \in \mathbf{R}^{n-1} \,|\, (c, x_2, \dots, x_n) \in W_i\}$$

and let $h_c \colon W_c \to \mathbf{R}^{n-1}$ be the map

$$h_c(x_2, \dots, x_n) = (h_2(c, x_2, \dots, x_n), \dots, h_n(c, x_2, \dots, x_n)).$$

Then

$$\det(Dh_c)(x_2, \dots, x_n) = \det(Dh)(c, x_2, \dots, x_n),$$

and hence

$$(c, x) \in W_i \cap C_h \iff x \in C_{h_c}. \tag{3.7.13}$$

Now let $p_0 = (c, x_0)$ be a point in $\mathbf{R}^n$. We have to show that every neighborhood V of p_0 contains a point $p \in \mathbf{R}^n \smallsetminus h(C_h)$. Let $V_c \subset \mathbf{R}^{n-1}$ be the set of points x for which $(c, x) \in V$. By induction V_c contains a point $x \in \mathbf{R}^{n-1} \smallsetminus h_c(C_{h_c})$ and hence $p = (c, x)$ is in V by definition and in $\mathbf{R}^n \smallsetminus h(C_h)$ by (3.7.13). □

Exercises for §3.7

Exercise 3.7.i. What are the set of critical points and the image of the set of critical points for the following maps $\mathbf{R} \to \mathbf{R}$?

(1) The map $f_1(x) = (x^2 - 1)^2$.
(2) The map $f_2(x) = \sin(x) + x$.
(3) The map

$$f_3(x) = \begin{cases} 0, & x \le 0, \\ e^{-\frac{1}{x}}, & x > 0. \end{cases}$$

Exercise 3.7.ii (Sard's theorem for affine maps). Let $f \colon \mathbf{R}^n \to \mathbf{R}^n$ be an *affine map*, i.e., a map of the form

$$f(x) = A(x) + x_0,$$

where $A \colon \mathbf{R}^n \to \mathbf{R}^n$ is a linear map and $x_0 \in \mathbf{R}^n$. Prove Sard's theorem for f.

Exercise 3.7.iii. Let $\rho \colon \mathbf{R} \to \mathbf{R}$ be a C^∞ function which is supported in the interval $(-1/2, 1/2)$ and has a maximum at the origin. Let $(r_i)_{i \ge 1}$ be an enumeration of the rational numbers, and let $f \colon \mathbf{R} \to \mathbf{R}$ be the map

$$f(x) = \sum_{i=1}^{\infty} r_i \rho(x - i)\,.$$

Show that f is a C^∞ map and show that the image of C_f is dense in $\mathbf{R}$.

The moral of this example: Sard's theorem says that the complement of C_f is dense in $\mathbf{R}$, but C_f can be dense as well.

Exercise 3.7.iv. Prove the assertion made in Remark 3.7.4.

Hint: You need to show that for every point $p \in \mathbf{R}^n$ and every neighborhood V of p, $V \cap \bigcap_{n\geq 1} U_n$ is non-empty. Construct, by induction, a family of closed balls $(B_k)_{k\geq 1}$ such that

- $B_k \subset V$,
- $B_{k+1} \subset B_k$,
- $B_k \subset \bigcap_{n\leq k} U_n$,
- The radius of B_k is less than $\frac{1}{k}$,

and show that $\bigcap_{k\geq 1} B_k \neq \varnothing$.

Exercise 3.7.v. Verify equation (3.7.5).

Chapter 4

Manifolds and Forms on Manifolds

4.1. Manifolds

Our agenda in this chapter is to extend to manifolds the results of Chapters 2 and 3 and to formulate and prove manifold versions of two of the fundamental theorems of integral calculus: Stokes' theorem and the divergence theorem. In this section we'll define what we mean by the term *manifold*, however, before we do so, a word of encouragement. Having had a course in multivariable calculus, you are already familiar with manifolds, at least in their one- and two-dimensional emanations, as curves and surfaces in $\mathbf{R}^3$, i.e., a manifold is basically just an n-dimensional surface in some high-dimensional Euclidean space. To make this definition precise let X be a subset of $\mathbf{R}^N$, Y a subset of $\mathbf{R}^n$ and $f\colon X \to Y$ a continuous map. We recall the following definition.

Definition 4.1.1. The map f is a C^∞ ***map*** if for every $p \in X$, there exist a neighborhood U_p of p in $\mathbf{R}^N$ and a C^∞ map $g_p\colon U_p \to \mathbf{R}^n$ which coincides with f on $U_p \cap X$.

We also recall the following theorem.

Theorem 4.1.2. *If $f\colon X \to Y$ is a C^∞ map, there exist a neighborhood U of X in $\mathbf{R}^N$ and a C^∞ map $g\colon U \to \mathbf{R}^n$ such that g coincides with f on X.*

(A proof of this can be found in Appendix A.)

We will say that f is a ***diffeomorphism*** if f is a bijection and f and f^{-1} are both C^∞ maps. In particular if Y is an open subset of $\mathbf{R}^n$, X is an example of an object which we will call a ***manifold***. More generally, we have the following definition.

Definition 4.1.3. Let N and n be non-negative integers with $n \leq N$. A subset $X \subset \mathbf{R}^N$ is an ***n-manifold*** if for every $p \in X$ there exist a neighborhood V of p in $\mathbf{R}^N$, an open subset U in $\mathbf{R}^n$, and a diffeomorphism $\phi\colon U \xrightarrow{\sim} X \cap V$.

Thus X is an n-manifold if, locally near every point p, X "looks like" an open subset of $\mathbf{R}^n$.

Examples 4.1.4.

(1) *Graphs of functions*: Let U be an open subset of $\mathbf{R}^n$ and $f : U \to \mathbf{R}$ a C^∞ function. Its graph

$$\Gamma_f = \{(x, f(x)) \in \mathbf{R}^{n+1} \mid x \in U\}$$

is an n-manifold in $\mathbf{R}^{n+1}$. In fact the map

$$\phi : U \to \mathbf{R}^{n+1}, \quad x \mapsto (x, f(x))$$

is a diffeomorphism of U onto Γ_f. (It's clear that ϕ is a C^∞ map, and it is a diffeomorphism since its inverse is the map $\pi : \Gamma_f \to U$ given by $\pi(x, t) := x$, which is also clearly C^∞.)

(2) *Graphs of mappings*: More generally if $f : U \to \mathbf{R}^k$ is a C^∞ map, its graph Γ_f is an n-manifold in $\mathbf{R}^{n+k}$.

(3) *Vector spaces*: Let V be an n-dimensional vector subspace of $\mathbf{R}^N$, and $(e_1, \dots, e_n)$ a basis of V. Then the linear map

$$\phi : \mathbf{R}^n \to V, \quad (x_1, \dots, x_n) \mapsto \sum_{i=1}^{n} x_i e_i \tag{4.1.5}$$

is a diffeomorphism of $\mathbf{R}^n$ onto V. Hence every n-dimensional vector subspace of $\mathbf{R}^N$ is automatically an n-dimensional submanifold of $\mathbf{R}^N$. Note, by the way, that if V is *any* n-dimensional vector space, not necessarily a subspace of $\mathbf{R}^N$, the map (4.1.5) gives us an identification of V with $\mathbf{R}^n$. This means that we can speak of subsets of V as being k-dimensional submanifolds if, via this identification, they get mapped onto k-dimensional submanifolds of $\mathbf{R}^n$. (This is a trivial, but useful, observation since a lot of interesting manifolds occur "in nature" as subsets of some abstract vector space rather than explicitly as subsets of some $\mathbf{R}^n$. An example is the manifold, $O(n)$, of orthogonal $n \times n$ matrices. This manifold occurs in nature as a submanifold of the vector space of $n \times n$ matrices.)

(4) *Affine subspaces of* $\mathbf{R}^n$: These are manifolds of the form $p + V$, where V is a vector subspace of $\mathbf{R}^N$ and p is a specified point in $\mathbf{R}^N$. In other words, they are diffeomorphic copies of vector subspaces with respect to the diffeomorphism

$$\tau_p : \mathbf{R}^N \xrightarrow{\sim} \mathbf{R}^N, \quad x \mapsto p + x.$$

If X is an arbitrary submanifold of $\mathbf{R}^N$, its *tangent space* at a point $p \in X$ is an example of a manifold of this type. (We'll have more to say about tangent spaces in §4.2.)

(5) *Product manifolds*: For $i = 1, 2$, let X_i be an n_i-dimensional submanifold of $\mathbf{R}^{N_i}$. Then the Cartesian product of X_1 and X_2

$$X_1 \times X_2 = \{(x_1, x_2) \mid x_i \in X_i\}$$

is an $(n_1 + n_2)$-dimensional submanifold of $\mathbf{R}^{N_1+N_2} = \mathbf{R}^{N_1} \times \mathbf{R}^{N_2}$.

We will leave for you to verify this fact as an exercise.

Hint: For $p_i \in X_i$, $i = 1, 2$, there exist a neighborhood, V_i, of p_i in $\mathbf{R}^{N_i}$, an open set, U_i in $\mathbf{R}^{n_i}$, and a diffeomorphism $\phi\colon U_i \to X_i \cap V_i$. Let $U = U_1 \times U_2$, $V = V_1 \times V_2$ and $X = X_1 \times X_2$, and let $\phi\colon U \to X \cap V$ be the product diffeomorphism, $(\phi(q_1), \phi_2(q_2))$.

(6) *The unit n-sphere*: This is the set of unit vectors in $\mathbf{R}^{n+1}$:

$$S^n = \{x \in \mathbf{R}^{n+1} \mid x_1^2 + \cdots + x_{n+1}^2 = 1\}.$$

To show that S^n is an n-manifold, let V be the open subset of $\mathbf{R}^{n+1}$ on which x_{n+1} is positive. If U is the open unit ball in $\mathbf{R}^n$ and $f : U \to \mathbf{R}$ is the function, $f(x) = (1 - (x_1^2 + \cdots + x_n^2))^{1/2}$, then $S^n \cap V$ is just the graph Γ_f of f. Hence we have a diffeomorphism

$$\phi\colon U \to S^n \cap V.$$

More generally, if $p = (x_1, \ldots, x_{n+1})$ is any point on the unit sphere, then x_i is non-zero for some i. If x_i is positive, then letting σ be the transposition, $i \leftrightarrow n+1$ and $f_\sigma\colon \mathbf{R}^{n+1} \to \mathbf{R}^{n+1}$, the map

$$f_\sigma(x_1, \ldots, x_n) = (x_{\sigma(1)}, \ldots, x_{\sigma(n)})$$

one gets a diffeomorphism $f_\sigma \circ \phi$ of U onto a neighborhood of p in S^n and if x_i is negative one gets such a diffeomorphism by replacing f_σ by $-f_\sigma$. In either case we have shown that for every point p in S^n, there is a neighborhood of p in S^n which is diffeomorphic to U.

(7) *The 2-torus*: In calculus books this is usually described as the surface of rotation in $\mathbf{R}^3$ obtained by taking the unit circle centered at the point, $(2, 0)$, in the (x_1, x_3) plane and rotating it about the x_3-axis. However, a slightly nicer description of it is as the product manifold $S^1 \times S^1$ in $\mathbf{R}^4$. (As an exercise, reconcile these two descriptions.)

We'll now turn to an alternative way of looking at manifolds: as *solutions of systems of equations*. Let U be an open subset of $\mathbf{R}^N$ and $f : U \to \mathbf{R}^k$ a C^∞ map.

Definition 4.1.6. A point $a \in \mathbf{R}^k$ is a ***regular value*** of f if for every point $p \in f^{-1}(a)$, the map f is a submersion at p.

Note that for f to be a submersion at p, the differential $Df(p)\colon \mathbf{R}^N \to \mathbf{R}^k$ has to be surjective, and hence k has to be less than or equal to N. Therefore this notion of regular value is interesting only if $N \geq k$.

Theorem 4.1.7. *Let $n := N - k$. If a is a regular value of f, the set $X := f^{-1}(a)$ is an n-manifold.*

Proof. Replacing f by $\tau_{-a} \circ f$ we can assume without loss of generality that $a = 0$. Let $p \in f^{-1}(0)$. Since f is a submersion at p, the canonical submersion theorem (see Theorem B.16) tells us that there exist a neighborhood O of 0 in $\mathbf{R}^N$, a neighborhood U_0 of p in U and a diffeomorphism $g\colon O \xrightarrow{\sim} U_0$ such that

$$f \circ g = \pi, \tag{4.1.8}$$

where π is the projection map

$$\mathbf{R}^N = \mathbf{R}^k \times \mathbf{R}^n \to \mathbf{R}^k, \quad (x, y) \mapsto x.$$

Hence $\pi^{-1}(0) = \{0\} \times \mathbf{R}^n = \mathbf{R}^n$ and by equation (4.1.5), g maps $\mathcal{O} \cap \pi^{-1}(0)$ diffeomorphically onto $U_0 \cap f^{-1}(0)$. However, $\mathcal{O} \cap \pi^{-1}(0)$ is a neighborhood of 0 in $\mathbf{R}^n$ and $U_0 \cap f^{-1}(0)$ is a neighborhood of p in X, and, as remarked, these two neighborhoods are diffeomorphic. □

Some examples are as follows.

Examples 4.1.9.

(1) *The n-sphere*: Let

$$f : \mathbf{R}^{n+1} \to \mathbf{R}$$

be the map,

$$(x_1, \ldots, x_{n+1}) \mapsto x_1^2 + \cdots + x_{n+1}^2 - 1.$$

Then

$$Df(x) = 2(x_1, \ldots, x_{n+1})$$

so, if $x \neq 0$, then f is a submersion at x. In particular, f is a submersion at all points x on the n-sphere

$$S^n = f^{-1}(0)$$

so the n-sphere is an n-dimensional submanifold of $\mathbf{R}^{n+1}$.

(2) *Graphs:* Let $g : \mathbf{R}^n \to \mathbf{R}^k$ be a C^∞ map and, as in Example 4.1.4 (2), let

$$\Gamma_g := \{(x, y) \in \mathbf{R}^n \times \mathbf{R}^k \mid y = g(x)\}.$$

We claim that Γ_g is an n-dimensional submanifold of $\mathbf{R}^{n+k} = \mathbf{R}^n \times \mathbf{R}^k$.

Proof. Let

$$f : \mathbf{R}^n \times \mathbf{R}^k \to \mathbf{R}^k$$

be the map $f(x, y) := y - g(x)$. Then

$$Df(x, y) = (-Dg(x) \quad \mathrm{id}_k),$$

where id_k is the identity map of $\mathbf{R}^k$ onto itself. This map is always of rank k. Hence $\Gamma_f = f^{-1}(0)$ is an n-dimensional submanifold of $\mathbf{R}^{n+k}$. □

(3) Let $\mathcal{M}_n$ be the set of all $n \times n$ matrices and let $\mathcal{S}_n$ be the set of all symmetric $n \times n$ matrices, i.e., the set

$$\mathcal{S}_n = \{A \in \mathcal{M}_n \mid A = A^\intercal\}.$$

The map

$$A = (a_{i,j}) \mapsto (a_{1,1}, a_{1,2}, \ldots, a_{1,n}, a_{2,1}, \ldots, a_{2,n}, \ldots)$$

gives us an identification

$$\mathcal{M}_n \xrightarrow{\sim} \mathbf{R}^{n^2}$$

and the map

$$A = (a_{i,j}) \mapsto (a_{1,1}, \ldots, a_{1,n}, a_{2,2}, \ldots, a_{2,n}, a_{3,3}, \ldots, a_{3,n}, \ldots)$$

gives us an identification

$$\mathcal{S}_n \xrightarrow{\sim} \mathbf{R}^{\frac{n(n+1)}{2}}.$$

(Note that if A is a symmetric matrix, then $a_{i,j} = a_{j,i}$, so this map avoids redundancies.) Let

$$O(n) = \{A \in \mathcal{M}_n \mid A^\intercal A = \mathrm{id}_n\}.$$

This is the set of ***orthogonal*** $n \times n$ matrices, and we will leave for you as an exercise to show that it's an $n(n-1)/2$-manifold.

Hint: Let $f : \mathcal{M}_n \to \mathcal{S}_n$ be the map $f(A) = A^\intercal A - \mathrm{id}_n$. Then

$$O(n) = f^{-1}(0).$$

These examples show that lots of interesting manifolds arise as zero sets of submersions $f : U \to \mathbf{R}^k$. This is, in fact, not just an accident. We will show that locally *every* manifold arises this way. More explicitly let $X \subset \mathbf{R}^N$ be an n-manifold, p a point of X, U a neighborhood of 0 in $\mathbf{R}^n$, V a neighborhood of p in $\mathbf{R}^N$ and $\phi\colon (U,0) \xrightarrow{\sim} (V \cap X, p)$ a diffeomorphism. We will for the moment think of ϕ as a C^∞ map $\phi\colon U \to \mathbf{R}^N$ whose image happens to lie in X.

Lemma 4.1.10. *The linear map $D\phi(0) : \mathbf{R}^n \to \mathbf{R}^N$ is injective.*

Proof. Since $\phi^{-1} : V \cap X \xrightarrow{\sim} U$ is a diffeomorphism, shrinking V if necessary, we can assume that there exists a C^∞ map $\psi\colon V \to U$ which coincides with ϕ^{-1}on $V \cap X$. Since ϕ maps U onto $V \cap X$, $\psi \circ \phi = \phi^{-1} \circ \phi$ is the identity map on U. Therefore,

$$D(\psi \circ \phi)(0) = (D\psi)(p)D\phi(0) = \mathrm{id}_n$$

by the chain rule, and hence if $D\phi(0)v = 0$, it follows from this identity that $v = 0$. ☐

Lemma 4.1.10 says that ϕ is an immersion at 0, so by the canonical immersion theorem (see Theorem B.18) there exist a neighborhood U_0 of 0 in U, a neighborhood V_p of p in V, and a diffeomorphism

$$g\colon (V_p, p) \xrightarrow{\sim} (U_0 \times \mathbf{R}^{N-n}, 0) \tag{4.1.11}$$

such that $g \circ \phi = \iota$, where ι is, as in Appendix B, the canonical immersion

$$\iota\colon U_0 \to U_0 \times \mathbf{R}^{N-n}, \quad x \mapsto (x, 0). \tag{4.1.12}$$

By (4.1.11) g maps $\phi(U_0)$ diffeomorphically onto $\iota(U_0)$. However, by (4.1.8) and (4.1.11) $\iota(U_0)$ is defined by the equations, $x_i = 0$, $i = n+1, \dots, N$. Hence if $g = (g_1, \dots, g_N)$ the set $\phi(U_0) = V_p \cap X$ is defined by the equations

$$g_i = 0, \quad i = n+1, \dots, N. \tag{4.1.13}$$

Let $\ell = N - n$, let

$$\pi\colon \mathbf{R}^N = \mathbf{R}^n \times \mathbf{R}^\ell \to \mathbf{R}^\ell$$

be the canonical submersion,

$$\pi(x_1, \dots, x_N) = (x_{n+1}, \dots, x_N)$$

and let $f = \pi \circ g$. Since g is a diffeomorphism, f is a submersion and (4.1.12) can be interpreted as saying that

$$V_p \cap X = f^{-1}(0). \tag{4.1.14}$$

Thus to summarize we have proved the following theorem.

Theorem 4.1.15. *Let X be an n-dimensional submanifold of $\mathbf{R}^N$ and let $\ell := N - n$. Then for every $p \in X$ there exist a neighborhood V_p of p in $\mathbf{R}^N$ and a submersion*

$$f : (V_p, p) \to (\mathbf{R}^\ell, 0)$$

such that $X \cap V_p$ is defined by equation (4.1.13).

A nice way of thinking about Theorem 4.1.15 is in terms of the coordinates of the mapping f. More specifically if $f = (f_1, \dots, f_k)$ we can think of $f^{-1}(a)$ as being the set of solutions of the system of equations

$$f_i(x) = a_i, \quad i = 1, \dots, k$$

and the condition that a be a regular value of f can be interpreted as saying that for every solution p of this system of equations the vectors

$$(df_i)_p = \sum_{j=1}^{n} \frac{\partial f_i}{\partial x_j}(0) dx_j$$

in $T_p^\star \mathbf{R}^n$ are linearly independent, i.e., the system (4.1.14) is an "independent system of defining equations" for X.

Exercises for §4.1

Exercise 4.1.i. Show that the set of solutions of the system of equations

$$\begin{cases} x_1^2 + \cdots + x_n^2 = 1, \\ x_1 + \cdots + x_n = 0 \end{cases}$$

is an $(n-2)$-dimensional submanifold of $\mathbf{R}^n$.

Exercise 4.1.ii. Let $S^{n-1} \subset \mathbf{R}^n$ be the $(n-1)$-sphere and let

$$X_a = \{x \in S^{n-1} \mid x_1 + \cdots + x_n = a\}.$$

For what values of a is X_a an $(n-2)$-dimensional submanifold of S^{n-1}?

Exercise 4.1.iii. Show that if X_i is an n_i-dimensional submanifold of $\mathbf{R}^{N_i}$, for $i = 1, 2$, then

$$X_1 \times X_2 \subset \mathbf{R}^{N_1} \times \mathbf{R}^{N_2}$$

is an $(n_1 + n_2)$-dimensional submanifold of $\mathbf{R}^{N_1} \times \mathbf{R}^{N_2}$.

Exercise 4.1.iv. Show that the set

$$X = \{ (x, v) \in S^{n-1} \times \mathbf{R}^n \mid x \cdot v = 0 \}$$

is a $(2n-2)$-dimensional submanifold of $\mathbf{R}^n \times \mathbf{R}^n$. (Here $x \cdot v := \sum_{i=1}^{n} x_i v_i$ is the dot product.)

Exercise 4.1.v. Let $g \colon \mathbf{R}^n \to \mathbf{R}^k$ be a C^∞ map and let $X = \Gamma_g$ be the graph of g. Prove directly that X is an n-manifold by proving that the map

$$\gamma_g \colon \mathbf{R}^n \to X, \qquad x \mapsto (x, g(x))$$

is a diffeomorphism.

Exercise 4.1.vi. Prove that the orthogonal group $O(n)$ is an $n(n-1)/2$-manifold.

Hints:

- Let $f : \mathcal{M}_n \to \mathcal{S}_n$ be the map
$$f(A) = A^{\mathsf{T}} A - \mathrm{id}_n.$$
Show that $O(n) = f^{-1}(0)$.
- Show that
$$f(A + \varepsilon B) = A^{\mathsf{T}} A + \varepsilon(A^{\mathsf{T}} B + B^{\mathsf{T}} A) + \varepsilon^2 B^{\mathsf{T}} B - \mathrm{id}_n.$$
- Conclude that the derivative of f at A is the map given by

$$B \mapsto A^{\mathsf{T}} B + B^{\mathsf{T}} A. \tag{4.1.16}$$

- Let A be in $O(n)$. Show that if C is in $\mathcal{S}_n$ and $B = AC/2$ then the map (4.1.16) maps B onto C.
- Conclude that the derivative of f is surjective at A.
- Conclude that 0 is a regular value of the mapping f.

The next five exercises, which are somewhat more demanding than the exercises above, are an introduction to *Grassmannian geometry*.

Exercise 4.1.vii.

(1) Let $e_1, \ldots, e_n$ be the standard basis of $\mathbf{R}^n$ and let $W = \mathrm{span}(e_{k+1}, \ldots, e_n)$. Prove that if V is a k-dimensional subspace of $\mathbf{R}^n$ and

$$V \cap W = 0, \tag{4.1.17}$$

then one can find a *unique* basis of V of the form

$$v_i = e_i + \sum_{j=1}^{\ell} b_{i,j} e_{k+j}, \quad i = 1, \ldots, k, \tag{4.1.18}$$

where $\ell = n - k$.

(2) Let G_k be the set of k-dimensional subspaces of $\mathbf{R}^n$ having the property (4.1.17) and let $\mathcal{M}_{k,\ell}$ be the vector space of $k \times \ell$ matrices. Show that one gets from the identities (4.1.18) a bijective map:

$$\gamma \colon \mathcal{M}_{k,\ell} \to G_k. \tag{4.1.19}$$

Exercise 4.1.viii. Let S_n be the vector space of linear mappings of $\mathbf{R}^n$ into itself which are self-adjoint, i.e., have the property $A = A^\top$.

(1) Given a k-dimensional subspace V of $\mathbf{R}^n$, let $\pi_V : \mathbf{R}^n \to \mathbf{R}^n$ be the orthogonal projection of $\mathbf{R}^n$ onto V. Show that π_V is in S_n and is of rank k, and show that $(\pi_V)^2 = \pi_V$.

(2) Conversely suppose A is an element of S_n which is of rank k and has the property $A^2 = A$. Show that if V is the image of A in $\mathbf{R}^n$, then $A = \pi_V$.

Definition 4.1.20. We call an $A \in S_n$ of the form $A = \pi_V$ above a ***rank k projection operator***.

Exercise 4.1.ix. Composing the map

$$\rho \colon G_k \to S_n, \quad V \mapsto \pi_V$$

with the map (4.1.19) we get a map

$$\phi \colon \mathcal{M}_{k,\ell} \to S_n, \quad \phi = \rho \cdot \gamma.$$

Prove that ϕ is C^∞.

Hints:

- By Gram–Schmidt one can convert (4.1.18) into an orthonormal basis

$$e_{1,B}, \ldots, e_{k,B}$$

of V. Show that the $e_{i,B}$'s are C^∞ functions of the matrix $B = (b_{i,j})$.

- Show that π_V is the linear mapping

$$v \mapsto \sum_{i=1}^{k} (v \cdot e_{i,B}) e_{i,B}.$$

Exercise 4.1.x. Let $V_0 = \operatorname{span}(e_1, \ldots, e_k)$ and let $\widetilde{G}_k = \rho(G_k)$. Show that ϕ maps a neighborhood of V_0 in $\mathcal{M}_{k,\ell}$ diffeomorphically onto a neighborhood of π_{V_0} in $\widetilde{G}_k$.

Hints: π_V is in $\widetilde{G}_k$ if and only if V satisfies (4.1.17). For $1 \le i \le k$ let

$$w_i = \pi_V(e_i) = \sum_{j=1}^{k} a_{i,j} e_j + \sum_{r=1}^{\ell} c_{i,r} e_{k+r}.$$

- Show that if the matrix $A = (a_{i,j})$ is invertible, π_V is in $\widetilde{G}_k$.
- Let $O \subset \widetilde{G}_k$ be the set of all π_V's for which A is invertible. Show that $\phi^{-1} : O \to \mathcal{M}_{k,\ell}$ is the map

$$\phi^{-1}(\pi_V) = B = A^{-1}C,$$

where $C = (c_{i,j})$.

Exercise 4.1.xi. Let $G(k,n) \subset S_n$ be the set of rank k projection operators. Prove that $G(k,n)$ is a $k\ell$-dimensional submanifold of the Euclidean space $S_n \cong \mathbf{R}^{\frac{n(n+1)}{2}}$.

Hints:

- Show that if V is any k-dimensional subspace of $\mathbf{R}^n$ there exists a linear mapping $A \in O(n)$ mapping V_0 to V.
- Show that $\pi_V = A\pi_{V_0}A^{-1}$.
- Let $K_A : S_n \to S_n$ be the linear mapping,

$$K_A(B) = ABA^{-1}.$$

Show that

$$K_A \circ \phi \colon \mathcal{M}_{k,\ell} \to S_n$$

maps a neighborhood of V_0 in $\mathcal{M}_{k,\ell}$ diffeomorphically onto a neighborhood of π_V in $G(k,n)$.

Remark 4.1.21. Let $\mathrm{Gr}(k,n)$ be the set of all k-dimensional subspaces of $\mathbf{R}^n$. The identification of $\mathrm{Gr}(k,n)$ with $G(k,n)$ given by $V \mapsto \pi_V$ allows us to restate the result above in the form.

Theorem 4.1.22. *The Grassmannian* $\mathrm{Gr}(k,n)$ *of k-dimensional subspaces of* $\mathbf{R}^n$ *is a* $k\ell$*-dimensional submanifold of* $S_n \cong \mathbf{R}^{\frac{n(n+1)}{2}}$.

Exercise 4.1.xii. Show that $\mathrm{Gr}(k,n)$ is a compact submanifold of S_n.
Hint: Show that it's closed and bounded.

4.2. Tangent spaces

We recall that a subset X of $\mathbf{R}^N$ is an n-dimensional manifold if for every $p \in X$ there exist an open set $U \subset \mathbf{R}^n$, a neighborhood V of p in $\mathbf{R}^N$, and a C^∞ diffeomorphism $\phi \colon U \xrightarrow{\sim} X \cap V$.

Definition 4.2.1. We will call ϕ a ***parameterization*** of X at p.

Our goal in this section is to define the notion of the *tangent space* T_pX to X at p and describe some of its properties. Before giving our official definition we'll discuss some simple examples.

Example 4.2.2. Let $f : \mathbf{R} \to \mathbf{R}$ be a C^∞ function and let $X = \Gamma_f$ be the graph of f.
Then in Figure 4.2.1 the tangent line ℓ to X at $p_0 = (x_0, y_0)$ is defined by the equation

$$y - y_0 = a(x - x_0),$$

where $a = f'(x_0)$. In other words if p is a point on ℓ then $p = p_0 + \lambda v_0$ where $v_0 = (1,a)$ and $\lambda \in \mathbf{R}$. We would, however, like the tangent space to X at p_0 to be a subspace of the tangent space to $\mathbf{R}^2$ at p_0, i.e., to be a subspace of the space $T_{p_0}\mathbf{R}^2 = \{p_0\} \times \mathbf{R}^2$, and this we'll achieve by defining

$$T_{p_0}X := \{(p_0, \lambda v_0) \mid \lambda \in \mathbf{R}\}.$$

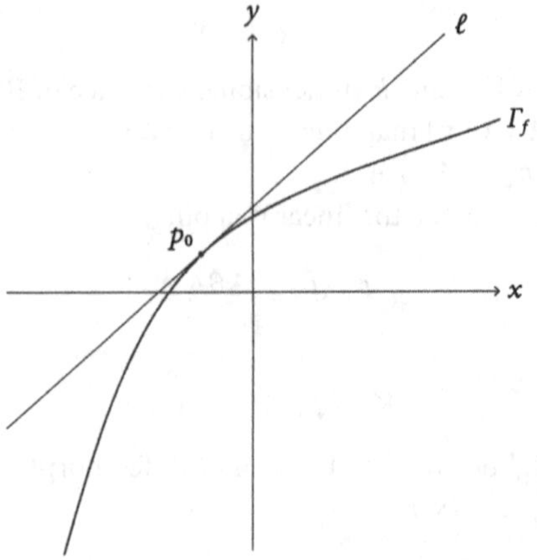

Figure 4.2.1. The tangent line ℓ to the graph of f at p_0.

Example 4.2.3. Let S^2 be the unit 2-sphere in $\mathbf{R}^3$. The tangent plane to S^2 at p_0 is usually defined to be the plane

$$\{\, p_0 + v \,|\, v \in \mathbf{R}^3 \text{ and } v \perp p_0 \,\}.$$

However, this tangent plane is easily converted into a subspace of $T_p\mathbf{R}^3$ via the map $p_0 + v \mapsto (p_0, v)$ and the image of this map

$$\{\, (p_0, v) \,|\, v \in \mathbf{R}^3 \text{ and } v \perp p_0 \,\}$$

will be our definition of $T_{p_0}S^2$.

Let's now turn to the general definition. As above let X be an n-dimensional submanifold of $\mathbf{R}^N$, p a point of X, V a neighborhood of p in $\mathbf{R}^N$, U an open set in $\mathbf{R}^n$ and

$$\phi\colon (U, q) \xrightarrow{\sim} (X \cap V, p)$$

a parameterization of X. We can think of ϕ as a C^∞ map

$$\phi\colon (U, q) \to (V, p)$$

whose image happens to lie in $X \cap V$ and we proved in §4.1 that its derivative at q

$$(d\phi)_q\colon T_q\mathbf{R}^n \to T_p\mathbf{R}^N$$

is injective.

Definition 4.2.4. The *tangent space* T_pX to X at p is the image of the linear map

$$(d\phi)_q\colon T_q\mathbf{R}^n \to T_p\mathbf{R}^N.$$

In other words, $w \in T_p\mathbf{R}^N$ is in T_pX if and only if $w = d\phi_q(v)$ for some $v \in T_q\mathbf{R}^n$. More succinctly,

$$T_pX := (d\phi_q)(T_q\mathbf{R}^n). \tag{4.2.5}$$

(Since $d\phi_q$ is injective, T_pX is an n-dimensional vector subspace of $T_p\mathbf{R}^N$.)

One problem with this definition is that it appears to depend on the choice of ϕ. To get around this problem, we'll give an alternative definition of T_pX. In §4.1 we showed that there exist a neighborhood V of p in $\mathbf{R}^N$ (which we can without loss of generality take to be the same as V above) and a C^∞ map

$$f\colon (V,p) \to (\mathbf{R}^k, 0), \quad \text{where } k = N - n, \tag{4.2.6}$$

such that $X \cap V = f^{-1}(0)$ and such that f is a submersion at all points of $X \cap V$, and in particular at p. Thus

$$df_p\colon T_p\mathbf{R}^N \to T_0\mathbf{R}^k$$

is surjective, and hence the kernel of df_p has dimension n. Our alternative definition of T_pX is

$$T_pX \coloneqq \ker(df_p). \tag{4.2.7}$$

Lemma 4.2.8. *The spaces (4.2.5) and (4.2.7) are both n-dimensional subspaces of $T_p\mathbf{R}^N$, and we claim that these spaces are the same.*

(Notice that the definition (4.2.7) of T_pX does not depend on ϕ, so if we can show that these spaces are the same, the definitions (4.2.5) and (4.2.7) will depend *neither* on ϕ *nor* on f.)

Proof. Since $\phi(U)$ is contained in $X \cap V$ and $X \cap V$ is contained in $f^{-1}(0)$, $f \circ \phi = 0$, so by the chain rule

$$df_p \circ d\phi_q = d(f \circ \phi)_q = 0. \tag{4.2.9}$$

Hence if $v \in T_p\mathbf{R}^n$ and $w = d\phi_q(v)$, $df_p(w) = 0$. This shows that the space (4.2.5) is contained in the space (4.2.7). However, these two spaces are n-dimensional so they coincide. □

From the proof above one can extract a slightly stronger result.

Theorem 4.2.10. *Let W be an open subset of $\mathbf{R}^\ell$ and $h\colon (W,q) \to (\mathbf{R}^N, p)$ a C^∞ map. Suppose $h(W)$ is contained in X. Then the image of the map*

$$dh_q\colon T_q\mathbf{R}^\ell \to T_p\mathbf{R}^N$$

is contained in T_pX.

Proof. Let f be the map (4.2.6). We can assume without loss of generality that $h(W)$ is contained in V, and so, by assumption, $h(W) \subset X \cap V$. Therefore, as above, $f \circ h = 0$, and hence $dh_q(T_q\mathbf{R}^\ell)$ is contained in the kernel of df_p. □

This result will enable us to define the ***derivative*** of a mapping between manifolds. Explicitly: Let X be a submanifold of $\mathbf{R}^N$, Y a submanifold of $\mathbf{R}^m$, and

$g: (X, p) \to (Y, y_0)$ a C^∞ map. By Definition 4.1.1 there exist a neighborhood O of X in $\mathbf{R}^N$ and a C^∞ map $\tilde{g}: O \to \mathbf{R}^m$ extending g. We will define

$$(dg)_p: T_pX \to T_{y_0}Y \tag{4.2.11}$$

to be the restriction of the map

$$(d\tilde{g})_p: T_p\mathbf{R}^N \to T_{y_0}\mathbf{R}^m$$

to T_pX. There are two obvious problems with this definition:

(1) Is the space $(d\tilde{g})_p(T_pX)$ contained in $T_{y_0}Y$?
(2) Does the definition depend on $\tilde{g}$?

To show that the answer to (1) is yes and the answer to (2) is no, let

$$\phi: (U, x_0) \xrightarrow{\approx} (X \cap V, p)$$

be a parameterization of X, and let $h = \tilde{g} \circ \phi$. Since $\phi(U) \subset X$, $h(U) \subset Y$ and hence by Theorem 4.2.10

$$dh_{x_0}(T_{x_0}\mathbf{R}^n) \subset T_{y_0}Y.$$

But by the chain rule

$$dh_{x_0} = d\tilde{g}_p \circ d\phi_{x_0},$$

so by (4.2.5)

$$(d\tilde{g}_p)(T_pX) \subset T_pY$$

and

$$(d\tilde{g}_p)(T_pX) = (dh)_{x_0}(T_{x_0}\mathbf{R}^n)$$

Thus the answer to (1) is yes, and since $h = \tilde{g} \circ \phi = g \circ \phi$, the answer to (2) is no.

From equations (4.2.9) and (4.2.11) one easily deduces the following.

Theorem 4.2.12 (Chain rule for mappings between manifolds). *Let Z be a submanifold of $\mathbf{R}^\ell$ and $\psi: (Y, y_0) \to (Z, z_0)$ a C^∞ map. Then $d\psi_{y_0} \circ dg_p = d(\psi \circ g)_p$.*

We will next prove manifold versions of the inverse function theorem and the canonical immersion and submersion theorems.

Theorem 4.2.13 (Inverse function theorem for manifolds). *Let X and Y be n-manifolds and $f : X \to Y$ a C^∞ map. Suppose that at $p \in X$ the map*

$$df_p: T_pX \to T_qY, \quad q \coloneqq f(p),$$

is bijective. Then f maps a neighborhood U of p in X diffeomorphically onto a neighborhood V of q in Y.

Proof. Let U and V be open neighborhoods of p in X and q in Y and let $\phi_0: (U_0, 0) \to (U, p)$ and $\psi_0: (V_0, 0) \to (V, q)$ be parameterizations of these neighborhoods. Shrinking U_0 and U we can assume that $f(U) \subset V$. Let

$$g: (U_0, p_0) \to (V_0, q_0)$$

be the map $\psi_0^{-1} \circ f \circ \phi_0$. Then $\psi_0 \circ g = f \circ \phi$, so by the chain rule

$$(d\psi_0)_{q_0} \circ (dg)_{p_0} = (df)_p \circ (d\phi_0)_{p_0}.$$

Since $(d\psi_0)_{q_0}$ and $(d\phi_0)_{p_0}$ are bijective it's clear from this identity that if df_p is bijective the same is true for $(dg)_{p_0}$. Hence by the inverse function theorem for open subsets of $\mathbf{R}^n$, g maps a neighborhood of p_0 in U_0 diffeomorphically onto a neighborhood of q_0 in V_0. Shrinking U_0 and V_0 we assume that these neighborhoods are U_0 and V_0 and hence that g is a diffeomorphism. Thus since $f : U \to V$ is the map $\psi_0 \circ g \circ \phi_0^{-1}$, it is a diffeomorphism as well. □

Theorem 4.2.14 (The canonical submersion theorem for manifolds). *Let X and Y be manifolds of dimension n and m, where $m < n$, and let $f : X \to Y$ be a C^∞ map. Suppose that at $p \in X$ the map*

$$df_p \colon T_pX \to T_qY, \quad q := f(p),$$

is surjective. Then there exist an open neighborhood U of p in X, and open neighborhood, V of $f(U)$ in Y and parameterizations $\phi_0 \colon (U_0, 0) \to (U, p)$ and $\psi_0 \colon (V_0, 0) \to (V, q)$ such that in the square

$$\begin{array}{ccc} U_0 & \xrightarrow{\psi_0^{-1} f \phi_0} & V_0 \\ {\scriptstyle \phi_0}\downarrow & & \downarrow{\scriptstyle \psi_0} \\ U & \xrightarrow[f]{} & V \end{array}$$

the map $\psi_0^{-1} \circ f \circ \phi_0$ is the canonical submersion π.

Proof. Let U and V be open neighborhoods of p and q and $\phi_0 \colon (U_0, 0) \to (U, p)$ and $\psi_0 \colon (V_0, 0) \to (V, q)$ be parameterizations of these neighborhoods. Composing ϕ_0 and ψ_0 with the translations we can assume that p_0 is the origin in $\mathbf{R}^n$ and q_0 the origin in $\mathbf{R}^m$, and shrinking U we can assume $f(U) \subset V$. As above let $g \colon (U_0, 0) \to (V_0, 0)$ be the map $\psi_0^{-1} \circ f \circ \phi_0$. By the chain rule

$$(d\psi_0)_0 \circ (dg)_0 = df_p \circ (d\phi_0)_0,$$

therefore, since $(d\psi_0)_0$ and $(d\phi_0)_0$ are bijective it follows that $(dg)_0$ is surjective. Hence, by Theorem 4.1.2, we can find an open neighborhood U of the origin in $\mathbf{R}^n$ and a diffeomorphism $\phi_1 \colon (U_1, 0) \xrightarrow{\sim} (U_0, 0)$ such that $g \circ \phi_1$ is the canonical submersion. Now replace U_0 by U_1 and ϕ_0 by $\phi_0 \circ \phi_1$. □

Theorem 4.2.15 (The canonical immersion theorem for manifolds). *Let X and Y be manifolds of dimension n and m, where $n < m$, and $f \colon X \to Y$ a C^∞ map. Suppose that at $p \in X$ the map*

$$df_p \colon T_pX \to T_qY\,, \quad q = f(p)$$

is injective. Then there exist an open neighborhood U of p in X, an open neighborhood V of $f(U)$ in Y and parameterizations $\phi_0 \colon (U_0, 0) \to (U, p)$ and $\psi_0 \colon (V_0, 0) \to$

(V, q) such that in the square

$$\begin{array}{ccc} U_0 & \xrightarrow{\psi_0^{-1} f \phi_0} & V_0 \\ {\scriptstyle \phi_0}\downarrow & & \downarrow{\scriptstyle \psi_0} \\ U & \xrightarrow[f]{} & V \end{array}$$

the map $\psi_0^{-1} \circ f \circ \phi_0$ is the canonical immersion ι.

Proof. The proof is identical with the proof of Theorem 4.2.14 except for the last step. In the last step one converts g into the canonical immersion via a map $\psi_1 : (V_1, 0) \to (V_0, 0)$ with the property $g \circ \psi_1 = \iota$ and then replaces ψ_0 by $\psi_0 \circ \psi_1$. □

Exercises for §4.2

Exercise 4.2.i. What is the tangent space to the quadric

$$Q = \{ (x_1, \dots, x_n) \in \mathbf{R}^n \mid x_n = x_1^2 + \cdots + x_{n-1}^2 \}$$

at the point $(1, 0, \dots, 0, 1)$?

Exercise 4.2.ii. Show that the tangent space to the $(n-1)$-sphere S^{n-1} at p is the space of vectors $(p, v) \in T_p\mathbf{R}^n$ satisfying $p \cdot v = 0$.

Exercise 4.2.iii. Let $f : \mathbf{R}^n \to \mathbf{R}^k$ be a C^∞ map and let $X = \Gamma_f$. What is the tangent space to X at $(a, f(a))$?

Exercise 4.2.iv. Let $\sigma : S^{n-1} \to S^{n-1}$ be the antipodal map $\sigma(x) = -x$. What is the derivative of σ at $p \in S^{n-1}$?

Exercise 4.2.v. Let $X_i \subset \mathbf{R}^{N_i}$, $i = 1, 2$, be an n_i-manifold and let $p_i \in X_i$. Define X to be the Cartesian product

$$X_1 \times X_2 \subset \mathbf{R}^{N_1} \times \mathbf{R}^{N_2}$$

and let $p = (p_1, p_2)$. Show that $T_pX \cong T_{p_1}X_1 \oplus T_{p_2}X_2$.

Exercise 4.2.vi. Let $X \subset \mathbf{R}^N$ be an n-manifold and for $i = 1, 2$, let $\phi_i : U_i \to X \cap V_i$ be two parameterizations. From these parameterizations one gets an overlap diagram

$$\begin{array}{ccc} \phi_1^{-1}(X \cap V_1 \cap V_2) & \xrightarrow{\quad \phi_2^{-1} \circ \phi_1 \quad} & \phi_2^{-1}(X \cap V_1 \cap V_2) \\ & {\scriptstyle \phi_1}\searrow \quad \swarrow{\scriptstyle \phi_2} & \\ & X \cap V_1 \cap V_2 & \end{array} \tag{4.2.16}$$

(1) Let $p \in X \cap V_1 \cap V_2$ and let $q_i = \phi_i^{-1}(p)$. Derive from the overlap diagram (4.2.16) an overlap diagram of linear maps

$$\begin{array}{ccc} T_{q_1}\mathbf{R}^n & \xrightarrow{d(\phi_2^{-1}\circ\phi_1)_{q_1}} & T_{q_2}\mathbf{R}^n \\ & d(\phi_1)_{q_1} \searrow \quad \swarrow d(\phi_2)_{q_2} & \\ & T_p\mathbf{R}^N & \end{array}$$

(2) Use overlap diagrams to give another proof that T_pX is intrinsically defined.

4.3. Vector fields and differential forms on manifolds

A *vector field* on an open subset $U \subset \mathbf{R}^n$ is a function $\boldsymbol{v}$ which assigns to each $p \in U$ an element $\boldsymbol{v}(p)$ of T_pU, and a *k-form* is a function ω which assigns to each $p \in U$ an element ω_p of $\Lambda^k(T_p^\star U)$. These definitions have obvious generalizations to manifolds.

Definition 4.3.1. Let X be a manifold. A ***vector field*** on X is a function $\boldsymbol{v}$ which assigns to each $p \in X$ an element $\boldsymbol{v}(p)$ of T_pX. A ***k-form*** is a function ω which assigns to each $p \in X$ an element ω_p of $\Lambda^k(T_p^\star X)$.

We'll begin our study of vector fields and k-forms on manifolds by showing that, like their counterparts on open subsets of $\mathbf{R}^n$, they have nice pullback and pushforward properties with respect to mappings. Let X and Y be manifolds and $f: X \to Y$ a C^∞ mapping.

Definition 4.3.2. Given a vector field $\boldsymbol{v}$ on X and a vector field $\boldsymbol{w}$ on Y, we'll say that $\boldsymbol{v}$ and $\boldsymbol{w}$ are ***f-related*** if for all $p \in X$ and $q = f(p)$ we have

$$(df)_p\boldsymbol{v}(p) = \boldsymbol{w}(q). \tag{4.3.3}$$

In particular, if f is a diffeomorphism, and we're given a vector field $\boldsymbol{v}$ on X we can define a vector field $\boldsymbol{w}$ on Y by requiring that for every point $q \in Y$, the identity (4.3.3) holds at the point $p = f^{-1}(q)$. In this case we'll call $\boldsymbol{w}$ the ***pushforward*** of $\boldsymbol{v}$ by f and denote it by $f_\star\boldsymbol{v}$. Similarly, given a vector field $\boldsymbol{w}$ on Y we can define a vector field $\boldsymbol{v}$ on X by applying the same construction to the inverse diffeomorphism $f^{-1}: Y \to X$. We will call the vector field $(f^{-1})_\star\boldsymbol{w}$ the ***pullback*** of $\boldsymbol{w}$ by f (and also denote it by $f^\star\boldsymbol{w}$).

For differential forms the situation is even nicer. Just as in §2.6 we can define the pullback operation on forms for *any* C^∞ map $f: X \to Y$. Specifically: Let ω be a k-form on Y. For every $p \in X$, and $q = f(p)$ the linear map

$$df_p: T_pX \to T_qY$$

induces by (1.8.2) a pullback map

$$(df_p)^\star: \Lambda^k(T_q^\star Y) \to \Lambda^k(T_p^\star X)$$

and, as in §2.6, we'll define the pullback $f^\star\omega$ of ω to X by defining it at p by the identity

$$(f^\star\omega)(p) = (df_p)^\star\omega(q).$$

The following results about these operations are proved in exactly the same way as in §2.6.

Proposition 4.3.4. *Let X, Y, and Z be manifolds and $f: X \to Y$ and $g: Y \to Z$ C^∞ maps. Then if ω is a k-form on Z*

$$f^\star(g^\star\omega) = (g \circ f)^\star\omega.$$

If $\boldsymbol{v}$ is a vector field on X and f and g are diffeomorphisms, then

$$(g \circ f)_\star v = g_\star(f_\star v)\,. \tag{4.3.5}$$

Our first application of these identities will be to define what one means by a "C^∞ vector field" and a "C^∞ k-form".

Definition 4.3.6. Let X be an n-manifold and U an open subset of X. The set U is a ***parameterizable open set*** if there exist an open set U_0 in $\mathbf{R}^n$ and a diffeomorphism $\phi_0: U_0 \to U$.

In other words, U is parameterizable if there exists a parameterization having U as its image. (Note that X being a manifold means that every point is contained in a parameterizable open set.)

Now let $U \subset X$ be a parameterizable open set and $\phi_0: U_0 \to U$ a parameterization of U.

Definition 4.3.7. A k-form ω on U is C^∞ or ***smooth*** if $\phi_0^\star\omega$ is C^∞.

This definition appears to depend on the choice of the parameterization ϕ, but we claim it does not. To see this let $\phi_1: U_1 \to U$ be another parameterization of U and let

$$\psi: U_0 \to U_1$$

be the composite map $\phi_1^{-1} \circ \phi_0$. Then $\phi_0 = \phi_1 \circ \psi$ and hence by Proposition 4.3.4

$$\phi_0^\star\omega = \psi^\star\phi_1^\star\omega,$$

so by (2.5.11) $\phi_0^\star\omega$ is C^∞ if $\phi_1^\star\omega$ is C^∞. The same argument applied to ψ^{-1} shows that $\phi_1^\star\omega$ is C^∞ if $\phi_0^\star\omega$ is C^∞.

The notion of "C^∞" for vector fields is defined similarly.

Definition 4.3.8. A vector field $\boldsymbol{v}$ on U is ***smooth***, or simply C^∞, if $\phi_0^\star\boldsymbol{v}$ is C^∞.

By Proposition 4.3.4 $\phi_0^\star\boldsymbol{v} = \psi^\star\phi_1^\star\boldsymbol{v}$, so, as above, this definition is independent of the choice of parameterization.

We now globalize these definitions.

Definition 4.3.9. A k-form ω on X is C^∞ if, for every point $p \in X$, ω is C^∞ on a neighborhood of p. Similarly, a vector field $\boldsymbol{v}$ on X is C^∞ if, for every point $p \in X$, $\boldsymbol{v}$ is C^∞ on a neighborhood of p.

We will also use the identities (4.3.5) and (4.3.18) to prove the following two results.

Proposition 4.3.10. *Let X and Y be manifolds and $f\colon X \to Y$ a C^∞ map. Then if ω is a C^∞ k-form on Y, the pullback $f^\star\omega$ is a C^∞ k-form on X.*

Proof. For $p \in X$ and $q = f(p)$ let $\phi_0\colon U_0 \to U$ and $\psi_0\colon V_0 \to V$ be parameterizations with $p \in U$ and $q \in V$. Shrinking U if necessary we can assume that $f(U) \subset V$. Let $g\colon U_0 \to V_0$ be the map $g = \psi_0^{-1} \circ f \circ \phi_0$. Then $\psi_0 \circ g = f \circ \phi_0$, so $g^\star\psi_0^\star\omega = \phi_0^\star f^\star\omega$. Since ω is C^∞, we see that $\psi_0^\star\omega$ is C^∞, so by equation (2.6.11) we have $g^\star\psi_0^\star\omega$ is C^∞, and hence $\phi_0^\star f^\star\omega$ is C^∞. Thus by definition $f^\star\omega$ is C^∞ on U. □

By exactly the same argument one proves.

Proposition 4.3.11. *If $\boldsymbol{w}$ is a C^∞ vector field on Y and f is a diffeomorphism, then $f^\star\boldsymbol{w}$ is a C^∞ vector field on X.*

Notation 4.3.12.

(1) We'll denote the space of C^∞ k-forms on X by $\Omega^k(X)$.

(2) For $\omega \in \Omega^k(X)$ we'll define the ***support*** of ω by

$$\operatorname{supp}(\omega) := \overline{\{\, p \in X \mid \omega(p) \neq 0 \,\}}$$

and we'll denote by $\Omega^k_c(X) \subset \Omega^k(X)$ the space of ***compactly supported k-forms.***

(3) For a vector field $\boldsymbol{v}$ on X we'll define the support of $\boldsymbol{v}$ to be the set

$$\operatorname{supp}(\boldsymbol{v}) := \overline{\{\, p \in X \mid \boldsymbol{v}(p) \neq 0 \,\}}.$$

We will now review some of the results about vector fields and differential forms that we proved in Chapter 2 and show that they have analogues for manifolds.

Integral curves

Let $I \subset \mathbf{R}$ be an open interval and $\gamma\colon I \to X$ a C^∞ curve. For $t_0 \in I$ we will call $\vec{u} = (t_0, 1) \in T_{t_0}\mathbf{R}$ the ***unit vector*** in $T_{t_0}\mathbf{R}$ and if $p = \gamma(t_0)$ we will call the vector

$$d\gamma_{t_0}(\vec{u}) \in T_pX$$

the ***tangent vector*** to γ at p. If $\boldsymbol{v}$ is a vector field on X we will say that γ is an ***integral curve*** of $\boldsymbol{v}$ if for all $t_0 \in I$

$$\boldsymbol{v}(\gamma(t_0)) = d\gamma_{t_0}(\vec{u}).$$

Proposition 4.3.13. *Let X and Y be manifolds and $f\colon X \to Y$ a C^∞ map. If $\boldsymbol{v}$ and $\boldsymbol{w}$ are vector fields on X and Y which are f-related, then integral curves of $\boldsymbol{v}$ get mapped by f onto integral curves of $\boldsymbol{w}$.*

Proof. If the curve $\gamma\colon I \to X$ is an integral curve of $\boldsymbol{v}$ we have to show that $f \circ \gamma\colon I \to Y$ is an integral curve of $\boldsymbol{w}$. If $\gamma(t) = p$ and $q = f(p)$ then by the chain rule

$$\begin{aligned}\boldsymbol{w}(q) &= df_p(\boldsymbol{v}(p)) = df_p(d\gamma_t(\vec{u}))\\ &= d(f \circ \gamma)_t(\vec{u}).\end{aligned}$$

□

From this result it follows that the local existence, uniqueness and "smooth dependence on initial data" results about vector fields that we described in § 2.1 are true for vector fields on manifolds. More explicitly, let U be a parameterizable open subset of X and $\phi: U_0 \to U$ a parameterization. Since U_0 is an open subset of $\mathbf{R}^n$ these results are true for the vector field $\boldsymbol{w} = \phi_0^* \boldsymbol{v}$ and hence since $\boldsymbol{w}$ and $\boldsymbol{v}$ are ϕ_0-related they are true for $\boldsymbol{v}$. In particular, we have the following propositions.

Proposition 4.3.14 (Local existence). *For every $p \in U$ there exists an integral curve $\gamma(t)$ defined for $-\varepsilon < t < \varepsilon$, of $\boldsymbol{v}$ with $\gamma(0) = p$.*

Proposition 4.3.15 (Local uniqueness). *For $i = 1, 2$, let $\gamma_i: I_i \to U$ be integral curves of $\boldsymbol{v}$ and let $I = I_1 \cap I_2$. Suppose $\gamma_2(t) = \gamma_1(t)$ for some $t \in I$. Then there exists a unique integral curve, $\gamma: I_1 \cup I_2 \to U$ with $\gamma = \gamma_1$ on I_1 and $\gamma = \gamma_2$ on I_2.*

Proposition 4.3.16 (Smooth dependence on initial data). *For every $p \in U$, there exist a neighborhood O of p in U, an interval $(-\varepsilon, \varepsilon)$ and a C^∞ map $h: O \times (-\varepsilon, \varepsilon) \to U$ such that for every $p \in O$ the curve*

$$\gamma_p(t) = h(p, t), \quad -\varepsilon < t < \varepsilon,$$

is an integral curve of $\boldsymbol{v}$ with $\gamma_p(0) = p$.

As in Chapter 2 we will say that $\boldsymbol{v}$ is ***complete*** if, for every $p \in X$ there exists an integral curve $\gamma(t)$ defined for $-\infty < t < \infty$, with $\gamma(0) = p$. In Chapter 2 we showed that one simple criterion for a vector field to be complete is that it be compactly supported. We will prove that the same is true for manifolds.

Theorem 4.3.17. *If X is compact or, more generally, if $\boldsymbol{v}$ is compactly supported, $\boldsymbol{v}$ is complete.*

Proof. It's not hard to prove this by the same argument that we used to prove this theorem for vector fields on $\mathbf{R}^n$, but we'll give a simpler proof that derives this directly from the $\mathbf{R}^n$ result. Suppose X is a submanifold of $\mathbf{R}^N$. Then for $p \in X$,

$$T_p X \subset T_p \mathbf{R}^N = \{(p, v) \mid v \in \mathbf{R}^N\},$$

so $\boldsymbol{v}(p)$ can be regarded as a pair $(p, v(p))$, where $v(p)$ is in $\mathbf{R}^N$. Let

$$f_v: X \to \mathbf{R}^N \tag{4.3.18}$$

be the map $f_v(p) = v(p)$. It is easy to check that $\boldsymbol{v}$ is C^∞ if and only if f_v is C^∞. (See Exercise 4.3.x.) Hence (see Appendix B) there exist a neighborhood O of X and a map $g: O \to \mathbf{R}^N$ extending f_v. Thus the vector field $\boldsymbol{w}$ on O defined by $\boldsymbol{w}(q) = (q, g(q))$ extends the vector field $\boldsymbol{v}$ to O. In other words if $\iota: X \hookrightarrow O$ is the inclusion map then $\boldsymbol{v}$ and $\boldsymbol{w}$ are ι-related. Thus by Proposition 4.3.13 the integral curves of $\boldsymbol{v}$ are just integral curves of $\boldsymbol{w}$ that are contained in X.

Suppose now that $\boldsymbol{v}$ is compactly supported. Then there exists a function $\rho \in C_0^\infty(O)$ which is 1 on the support of $\boldsymbol{v}$, so, replacing $\boldsymbol{w}$ by $\rho\boldsymbol{w}$, we can assume that $\boldsymbol{w}$ is compactly supported. Thus $\boldsymbol{w}$ is complete. Let $\gamma(t)$, defined for $-\infty < t < \infty$, be an integral curve of $\boldsymbol{w}$. We will prove that if $\gamma(0) \in X$, then this curve is an integral curve of $\boldsymbol{v}$. We first observe the following lemma.

Lemma 4.3.19. *The set of points $t \in \mathbf{R}$ for which $\gamma(t) \in X$ is both open and closed.*

Proof. If $p \notin \operatorname{supp}(\boldsymbol{v})$ then $\boldsymbol{w}(p) = 0$ so if $\gamma(t) = p$, $\gamma(t)$ is the constant curve $\gamma = p$, and there's nothing to prove. Thus we are reduced to showing that the set

$$\{ t \in \mathbf{R} \mid \gamma(t) \in \operatorname{supp}(\boldsymbol{v}) \} \tag{4.3.20}$$

is both open and closed. Since $\operatorname{supp}(\boldsymbol{v})$ is compact this set is clearly closed. To show that it's open suppose $\gamma(t_0) \in \operatorname{supp}(\boldsymbol{v})$. By local existence there exist an interval $(-\varepsilon + t_0, \varepsilon + t_0)$ and an integral curve $\gamma_1(t)$ of $\boldsymbol{v}$ defined on this interval and taking the value $\gamma_1(t_0) = \gamma(t_0)$ at p. However since $\boldsymbol{v}$ and $\boldsymbol{w}$ are ι-related γ_1 is also an integral curve of $\boldsymbol{w}$ and so it has to coincide with γ on the interval $(-\varepsilon + t_0, \varepsilon + t_0)$. In particular, for t on this interval $\gamma(t) \in \operatorname{supp}(\boldsymbol{v})$ so the set (4.3.20) is open. □

To conclude the proof of Theorem 4.3.17 we note that since $\mathbf{R}$ is connected it follows that if $\gamma(t_0) \in X$ for some $t_0 \in \mathbf{R}$ then $\gamma(t) \in X$ for *all* $t \in \mathbf{R}$, and hence γ is an integral curve of $\boldsymbol{v}$. Thus in particular *every* integral curve of $\boldsymbol{v}$ exists for all time, so $\boldsymbol{v}$ is complete. □

Since $\boldsymbol{w}$ is complete it generates a one-parameter group of diffeomorphisms $g_t \colon O \xrightarrow{\sim} O$, defined for $-\infty < t < \infty$, having the property that the curve

$$g_t(p) = \gamma_p(t), \quad -\infty < t < \infty$$

is the unique integral curve of $\boldsymbol{w}$ with initial point $\gamma_p(0) = p$. But if $p \in X$ this curve is an integral curve of $\boldsymbol{v}$, so the restriction

$$f_t = g_t|_X$$

is a one-parameter group of diffeomorphisms of X with the property that for $p \in X$ the curve

$$f_t(p) = \gamma_p(t), \quad -\infty < t < \infty$$

is the unique integral curve of $\boldsymbol{v}$ with initial point $\gamma_p(0) = p$.

The exterior differentiation operation

Let ω be a C^∞ k-form on X and $U \subset X$ a parameterizable open set. Given a parameterization $\phi_0 \colon U_0 \to U$ we define the exterior derivative $d\omega$ of ω on X by the formula

$$d\omega = (\phi_0^{-1})^\star d\phi_0^\star \omega. \tag{4.3.21}$$

(Notice that since U_0 is an open subset of $\mathbf{R}^n$ and $\phi_0^\star \omega$ a k-form on U_0, the "d" on the right is well-defined.) We claim that this definition does not depend on the choice of parameterization. To see this let $\phi_1 \colon U_1 \to U$ be another parameterization of

U and let $\psi: U_0 \to U_1$ be the diffeomorphism $\phi_1^{-1} \circ \phi_0$. Then $\phi_0 = \phi_1 \circ \psi$ and hence

$$d\phi_0^\star \omega = d\psi^\star \phi_1^\star \omega = \psi^\star d\phi_1^\star \omega$$
$$= \phi_0^\star (\phi_1^{-1})^\star d\phi_1^\star \omega$$

so

$$(\phi_0^{-1})^\star d\phi_0^\star \omega = (\phi_1^{-1})^\star d\phi_1^\star \omega$$

as claimed. We can, therefore, define the exterior derivative $d\omega$ globally by defining it to be equal to (4.3.21) on every parameterizable open set.

It's easy to see from the definition (4.3.21) that this exterior differentiation operation inherits from the exterior differentiation operation on open subsets of $\mathbf{R}^n$ the properties (2.4.3) and (2.4.4) and that for 0-forms, i.e., C^∞ functions $f: X \to \mathbf{R}$, df is the "intrinsic" df defined in §2.1, i.e., for $p \in X$, df_p is the derivative of f

$$df_p : T_pX \to \mathbf{R}$$

viewed as an element of $\Lambda^1(T_p^\star X)$. Let's check that it also has the property (2.6.12).

Theorem 4.3.22. *Let X and Y be manifolds and $f: X \to Y$ a C^∞ map. Then for $\omega \in \Omega^k(Y)$ we have*

$$f^\star(d\omega) = d(f^\star \omega).$$

Proof. For every $p \in X$ we'll check that this equality holds in a neighborhood of p. Let $q = f(p)$ and let U and V be parameterizable neighborhoods of p and q. Shrinking U if necessary we can assume $f(U) \subset V$. Given parameterizations

$$\phi: U_0 \to U$$

and

$$\psi: V_0 \to V,$$

we get by composition a map

$$g: U_0 \to V_0\,, \quad g = \psi^{-1} \circ f \circ \phi$$

with the property $\psi \circ g = f \circ \phi$. Thus

$$\begin{aligned}
\phi^\star d(f^\star \omega) &= d\phi^\star f^\star \omega && \text{(by definition of } d) \\
&= d(f \circ \phi)^\star \omega \\
&= d(\psi \circ g)^\star \omega \\
&= dg^\star(\psi^\star \omega) \\
&= g^\star d \circ^\star \omega && \text{(by (2.6.12))} \\
&= g^\star \psi^\star d\omega && \text{(by definition of } d) \\
&= \phi^\star f^\star d\omega.
\end{aligned}$$

Hence $df^\star \omega = f^\star d\omega$. □

The interior product and Lie derivative operation

Given a k-form $\omega \in \Omega^k(X)$ and a C^∞ vector field $\boldsymbol{v}$ we will define the interior product

$$\iota_v \omega \in \Omega^{k-1}(X), \tag{4.3.23}$$

as in §2.5, by setting

$$(\iota_v \omega)_p = \iota_{v(p)} \omega_p$$

and the Lie derivative $L_v \omega \in \Omega^k(X)$ by setting

$$L_v \omega := \iota_v d\omega + d\iota_v \omega. \tag{4.3.24}$$

It's easily checked that these operations satisfy Properties 2.5.3 and 2.5.10 (since, just as in §2.5, these identities are deduced from the definitions (4.3.23) and (4.3.3) by purely formal manipulations). Moreover, if $\boldsymbol{v}$ is complete and

$$f_t : X \to X\,, \quad -\infty < t < \infty$$

is the one-parameter group of diffeomorphisms of X generated by $\boldsymbol{v}$ the Lie derivative operation can be defined by the alternative recipe

$$L_v \omega = \frac{d}{dt} f_t^\star \omega \bigg|_{t=0} \tag{4.3.25}$$

as in (2.6.22). (Just as in §2.6 one proves this by showing that the operation (4.3.25) satisfies Properties 2.5.10 and hence that it agrees with the operation (4.3.24) provided the two operations agree on 0-forms.)

Exercises for §4.3

Exercise 4.3.i. Let $X \subset \mathbf{R}^3$ be the paraboloid defined by $x_3 = x_1^2 + x_2^2$ and let $\boldsymbol{w}$ be the vector field

$$\boldsymbol{w} := x_1 \frac{\partial}{\partial x_1} + x_2 \frac{\partial}{\partial x_2} + 2x_3 \frac{\partial}{\partial x_3}.$$

(1) Show that $\boldsymbol{w}$ is tangent to X and hence defines by restriction a vector field $\boldsymbol{v}$ on X.
(2) What are the integral curves of $\boldsymbol{v}$?

Exercise 4.3.ii. Let S^2 be the unit 2-sphere, $x_1^2 + x_2^2 + x_3^2 = 1$, in $\mathbf{R}^3$ and let $\boldsymbol{w}$ be the vector field

$$\boldsymbol{w} := x_1 \frac{\partial}{\partial x_2} - x_2 \frac{\partial}{\partial x_1}.$$

(1) Show that $\boldsymbol{w}$ is tangent to S^2, and hence by restriction defines a vector field $\boldsymbol{v}$ on S^2.
(2) What are the integral curves of $\boldsymbol{v}$?

Exercise 4.3.iii. As in Exercise 4.3.ii let S^2 be the unit 2-sphere in $\mathbf{R}^3$ and let $\boldsymbol{w}$ be the vector field

$$\boldsymbol{w} := \frac{\partial}{\partial x_3} - x_3 \left(x_1 \frac{\partial}{\partial x_1} + x_2 \frac{\partial}{\partial x_2} + x_3 \frac{\partial}{\partial x_3} \right).$$

(1) Show that $\boldsymbol{w}$ is tangent to S^2 and hence by restriction defines a vector field $\boldsymbol{v}$ on S^2.
(2) What do its integral curves look like?

Exercise 4.3.iv. Let S^1 be the unit circle, $x_1^2 + x_2^2 = 1$, in $\mathbf{R}^2$ and let $X = S^1 \times S^1$ in $\mathbf{R}^4$ with defining equations

$$f_1 = x_1^2 + x_2^2 - 1 = 0,$$
$$f_2 = x_3^2 + x_4^2 - 1 = 0.$$

(1) Show that the vector field

$$\boldsymbol{w} := x_1 \frac{\partial}{\partial x_2} - x_2 \frac{\partial}{\partial x_1} + \lambda \left(x_4 \frac{\partial}{\partial x_3} - x_3 \frac{\partial}{\partial x_4} \right),$$

$\lambda \in \mathbf{R}$, is tangent to X and hence defines by restriction a vector field $\boldsymbol{v}$ on X.
(2) What are the integral curves of $\boldsymbol{v}$?
(3) Show that $L_{\boldsymbol{w}} f_i = 0$.

Exercise 4.3.v. For the vector field $\boldsymbol{v}$ in Exercise 4.3.iv, describe the one-parameter group of diffeomorphisms it generates.

Exercise 4.3.vi. Let X and $\boldsymbol{v}$ be as in Exercise 4.3.i and let $f : \mathbf{R}^2 \to X$ be the map $f(x_1, x_2) = (x_1, x_2, x_1^2 + x_2^2)$. Show that if $\boldsymbol{u}$ is the vector field,

$$\boldsymbol{u} := x_1 \frac{\partial}{\partial x_1} + x_2 \frac{\partial}{\partial x_2},$$

then $f_\star \boldsymbol{u} = \boldsymbol{v}$.

Exercise 4.3.vii. Let X be a submanifold of $\mathbf{R}^N$ and U an open subset of $\mathbf{R}^N$ containing X, and let $\boldsymbol{v}$ and $\boldsymbol{w}$ be the vector fields on X and U. Denoting by ι the inclusion map of X into U, show that $\boldsymbol{v}$ and $\boldsymbol{w}$ are ι-related if and only if $\boldsymbol{w}$ is tangent to X and its restriction to X is $\boldsymbol{v}$.

Exercise 4.3.viii. An elementary result in number theory asserts the following theorem.

Theorem 4.3.26. *A number $\lambda \in \mathbf{R}$ is irrational if and only if the set*

$$\{ m + \lambda n \mid m, n \in \mathbf{Z} \}$$

is a dense subset of $\mathbf{R}$.

Let $\boldsymbol{v}$ be the vector field in Exercise 4.3.iv. Using Theorem 4.3.26 prove that if λ is irrational then for every integral curve $\gamma(t)$, defined for $-\infty < t < \infty$, of $\boldsymbol{v}$ the set of points on this curve is a dense subset of X.

Exercise 4.3.ix. Let X be an n-dimensional submanifold of $\mathbf{R}^N$. Prove that a vector field $\boldsymbol{v}$ on X is C^∞ if and only if the map (4.3.18) is C^∞.

Hint: Let U be a parameterizable open subset of X and $\phi\colon U_0 \to U$ a parameterization of U. Composing ϕ with the inclusion map $\iota\colon X \to \mathbf{R}^N$ one gets a map $\iota \circ \phi\colon U \to \mathbf{R}^N$. Show that if

$$\phi^\star \boldsymbol{v} = \sum v_i \frac{\partial}{\partial x_j}$$

then

$$\phi^\star f_i = \sum \frac{\partial \phi_i}{\partial x_j} v_j,$$

where $f_1, \dots, f_N$ are the coordinates of the map $f_{\boldsymbol{v}}$ and $\phi_1, \dots, \phi_N$ the coordinates of $\iota \circ \phi$.

Exercise 4.3.x. Let $\boldsymbol{v}$ be a vector field on X and let $\phi\colon X \to \mathbf{R}$ be a C^∞ function. Show that if the function

$$L_{\boldsymbol{v}}\phi = \iota_{\boldsymbol{v}} d\phi$$

is zero, then ϕ is constant along integral curves of $\boldsymbol{v}$.

Exercise 4.3.xi. Suppose that $\phi\colon X \to \mathbf{R}$ is proper. Show that if $L_{\boldsymbol{v}}\phi = 0$, $\boldsymbol{v}$ is complete.

Hint: For $p \in X$ let $a = \phi(p)$. By assumption, $\phi^{-1}(a)$ is compact. Let $\rho \in C_0^\infty(X)$ be a "bump" function which is one on $\phi^{-1}(a)$ and let $\boldsymbol{w}$ be the vector field $\rho\boldsymbol{v}$. By Theorem 4.3.17, $\boldsymbol{w}$ is complete and since

$$L_{\boldsymbol{w}}\phi = \iota_{\rho\boldsymbol{v}} d\phi = \rho\iota_{\boldsymbol{v}} d\phi = 0$$

ϕ is constant along integral curves of $\boldsymbol{w}$. Let $\gamma(t)$, $-\infty < t < \infty$, be the integral curve of $\boldsymbol{w}$ with initial point $\gamma(0) = p$. Show that γ is an integral curve of $\boldsymbol{v}$.

4.4. Orientations

The last part of Chapter 4 will be devoted to the "integral calculus" of forms on manifolds. In particular we will prove manifold versions of two basic theorems of integral calculus on $\mathbf{R}^n$, Stokes' theorem and the divergence theorem, and also develop a manifold version of degree theory. However, to extend the integral calculus to manifolds without getting involved in horrendously technical "orientation" issues we will confine ourselves to a special class of manifolds: orientable manifolds. The goal of this section will be to explain what this term means.

Definition 4.4.1. Let X be an n-manifold. An ***orientation*** of X is a rule for assigning to each $p \in X$ an orientation of T_pX.

Thus by Definition 1.9.3 one can think of an orientation as a "labeling" rule which, for every $p \in X$, labels one of the two components of the set $\Lambda^n(T_p^\star X) \setminus \{0\}$ by $\Lambda^n(T_p^\star X)_+$, which we'll henceforth call the "plus" part of $\Lambda^n(T_p^\star X)$, and the other component by $\Lambda^n(T_p^\star X)_-$, which we'll henceforth call the "minus" part of $\Lambda^n(T_p^\star X)$.

Definition 4.4.2. An orientation of a manifold X is ***smooth***, or simply C^∞, if for every $p \in X$, there exist a neighborhood U of p and a non-vanishing n-form $\omega \in \Omega^n(U)$ with the property

$$\omega_q \in \Lambda^n(T_q^\star X)_+$$

for every $q \in U$.

Remark 4.4.3. If we're given an orientation of X we can define another orientation by assigning to each $p \in X$ the opposite orientation to the orientation we already assigned, i.e., by switching the labels on $\Lambda^n(T_p^\star)_+$ and $\Lambda^n(T_p^\star)_-$. We will call this the ***reversed orientation*** of X. We will leave for you to check as an exercise that if X is connected and equipped with a smooth orientation, the only smooth orientations of X are the given orientation and its reversed orientation.

Hint: Given any smooth orientation of X the set of points where it agrees with the given orientation is open, and the set of points where it does not is also open. Therefore one of these two sets has to be empty.

Note that if $\omega \in \Omega^n(X)$ is a non-vanishing n-form one gets from ω a smooth orientation of X by requiring that the "labeling rule" above satisfy

$$\omega_p \in \Lambda^n(T_p^\star X)_+$$

for every $p \in X$. If ω has this property we will call ω a ***volume form***. It's clear from this definition that if ω_1 and ω_2 are volume forms on X then $\omega_2 = f_{2,1}\omega_1$ where $f_{2,1}$ is an everywhere positive C^∞ function.

Example 4.4.4. Let U be an open subset $\mathbf{R}^n$. We will usually assign to U its ***standard orientation***, by which we will mean the orientation defined by the n-form $dx_1 \wedge \cdots \wedge dx_n$.

Example 4.4.5. Let $f : \mathbf{R}^N \to \mathbf{R}^k$ be a C^∞ map. If zero is a regular value of f, the set $X := f^{-1}(0)$ is a submanifold of $\mathbf{R}^N$ of dimension $n := N-k$ (by Theorem 4.2.13). Moreover, for $p \in X$, T_pX is the kernel of the surjective map

$$df_p \colon T_p\mathbf{R}^N \to T_0\mathbf{R}^k$$

so we get from df_p a bijective linear map

$$T_p\mathbf{R}^N/T_pX \to T_0\mathbf{R}^k. \tag{4.4.6}$$

As explained in Example 4.4.4, $T_p\mathbf{R}^N$ and $T_0\mathbf{R}^k$ have "standard" orientations, hence if we require that the map (4.4.6) be orientation preserving, this gives $T_p\mathbf{R}^N/T_pX$ an orientation and, by Theorem 1.9.9, gives T_pX an orientation. It's intuitively clear that since df_p varies smoothly with respect to p this orientation does as well; however, this fact requires a proof, and we'll supply a sketch of such a proof in the exercises.

Example 4.4.7. A special case of Example 4.4.5 is the n-sphere

$$S^n = \{(x_1, \ldots, x_{n+1}) \in \mathbf{R}^{n+1} \mid x_1^2 + \cdots + x_{n+1}^2 = 1\},$$

which acquires an orientation from its defining map $f: \mathbf{R}^{n+1} \to \mathbf{R}$ given by

$$f(x) := x_1^2 + \cdots + x_{n+1}^2 - 1.$$

Example 4.4.8. Let X be an oriented submanifold of $\mathbf{R}^N$. For every $p \in X$, T_pX sits inside $T_p\mathbf{R}^N$ as a vector subspace, hence, via the identification $T_p\mathbf{R}^N \cong \mathbf{R}^N$, one can think of T_pX as a vector subspace of $\mathbf{R}^N$. In particular from the standard Euclidean inner product on $\mathbf{R}^N$ one gets, by restricting this inner product to vectors in T_pX, an inner product

$$B_p : T_pX \times T_pX \to \mathbf{R}$$

on T_pX. Let σ_p be the volume element in $\Lambda^n(T_p^\star X)$ associated with B_p (see Exercise 1.9.xi) and let $\sigma = \sigma_X$ be the non-vanishing n-form on X defined by the assignment

$$p \mapsto \sigma_p.$$

In the exercises at the end of this section we'll sketch a proof of the following.

Theorem 4.4.9. *The form σ_X is C^∞ and hence, in particular, is a volume form. (We will call this form the* ***Riemannian volume form****.)*

Example 4.4.10 (Möbius strip). The Möbius strip is a surface in $\mathbf{R}^3$ which is *not* orientable. It is obtained from the rectangle

$$R := [0,1] \times (-1,1) = \{\, (x,y) \in \mathbf{R}^2 \mid 0 \le x \le 1 \quad \text{and} \quad -1 < y < 1 \,\}$$

by gluing the ends together in the wrong way, i.e., by gluing $(1, y)$ to $(0, -y)$. It is easy to see that the Möbius strip cannot be oriented by taking the standard orientation at $p = (1, 0)$ and moving it along the line $(t, 0)$, $0 \le t \le 1$ to the point $(0, 0)$ (which is *also* the point p after we have glued the ends of the rectangle together).

We'll next investigate the "compatibility" question for diffeomorphisms between oriented manifolds. Let X and Y be n-manifolds and $f: X \xrightarrow{\sim} Y$ a diffeomorphism. Suppose both of these manifolds are equipped with orientations. We will say that f is ***orientation preserving*** if for all $p \in X$ and $q = f(p)$, the linear map

$$df_p : T_pX \to T_qY$$

is orientation preserving. It's clear that if ω is a volume form on Y then f is orientation preserving if and only if $f^\star\omega$ is a volume form on X, and from (1.9.5) and the chain rule one easily deduces the following theorem.

Theorem 4.4.11. *If Z is an oriented n-manifold and $g: Y \to Z$ a diffeomorphism, then if both f and g are orientation preserving, so is $g \circ f$.*

If $f: X \xrightarrow{\sim} Y$ is a diffeomorphism then the set of points $p \in X$ at which the linear map

$$df_p : T_pX \to T_qY\,, \quad q = f(p)\,,$$

is orientation preserving is open, and the set of points at which is orientation reversing is open as well. Hence if X is connected, df_p has to be orientation preserving at all points or orientation reversing at all points. In the latter case we'll say that f is ***orientation reversing***.

If U is a parameterizable open subset of X and $\phi : U_0 \to U$ a parameterization of U we'll say that this parameterization is an ***oriented parameterization*** if ϕ is orientation preserving with respect to the standard orientation of U_0 and the given orientation on U. Notice that if this parameterization isn't oriented we can convert it into one that is by replacing every connected component V_0 of U_0 on which ϕ isn't orientation preserving by the open set

$$V_0^\sharp = \{(x_1, \dots, x_n) \in \mathbf{R}^n \mid (x_1, \dots, x_{n-1}, -x_n) \in V_0\} \tag{4.4.12}$$

and replacing ϕ by the map

$$\psi(x_1, \dots, x_n) = \phi(x_1, \dots, x_{n-1}, -x_n). \tag{4.4.13}$$

If $\phi_i : U_i \to U$, $i = 0, 1$, are oriented parameterizations of U and $\psi : U_0 \to U_1$ is the diffeomorphism $\phi_1^{-1} \circ \phi_0$, then by the theorem above ψ is orientation preserving or in other words

$$\det\left(\frac{\partial \psi_i}{\partial x_j}\right) > 0 \tag{4.4.14}$$

at every point on U_0.

We'll conclude this section by discussing some orientation issues which will come up when we discuss Stokes' theorem and the divergence theorem in § 4.6. First a definition.

Definition 4.4.15. An open subset D of X is a ***smooth domain*** if

(1) the boundary ∂D is an $(n-1)$-dimensional submanifold of X, and
(2) the boundary of D coincides with the boundary of the closure of D.

Examples 4.4.16.

(1) The n-ball, $x_1^2 + \dots + x_n^2 < 1$, whose boundary is the sphere, $x_1^2 + \dots + x_n^2 = 1$.
(2) The n-dimensional annulus,

$$1 < x_1^2 + \dots + x_n^2 < 2$$

whose boundary consists of the union of the two spheres

$$\begin{cases} x_1^2 + \dots + x_n^2 = 1, \\ x_1^2 + \dots + x_n^2 = 2. \end{cases}$$

(3) Let S^{n-1} be the unit sphere, $x_1^2 + \dots + x_2^2 = 1$ and let $D = \mathbf{R}^n \setminus S^{n-1}$. Then the boundary of D is S^{n-1} but D is not a smooth domain since the boundary of its closure is empty.
(4) The simplest example of a smooth domain is the half-space

$$\mathbf{H}^n = \{(x_1, \dots, x_n) \in \mathbf{R}^n \mid x_1 < 0\}$$

whose boundary

$$\partial\mathbf{H}^n = \{(x_1, \dots, x_n) \in \mathbf{R}^n \mid x_1 = 0\}$$

we can identify with $\mathbf{R}^{n-1}$ via the map $\mathbf{R}^{n-1} \to \partial\mathbf{H}^n$ given by

$$(x_2, \dots, x_n) \mapsto (0, x_2, \dots, x_n).$$

We will show that every bounded domain looks locally like this example.

Theorem 4.4.17. *Let D be a smooth domain and p a boundary point of D. Then there exist a neighborhood U of p in X, an open set U_0 in $\mathbf{R}^n$, and a diffeomorphism $\psi: U_0 \xrightarrow{\sim} U$ such that ψ maps $U_0 \cap \mathbf{H}^n$ onto $U \cap D$.*

Proof. Let $Z := \partial D$. First we prove the following lemma.

Lemma 4.4.18. *For every $p \in Z$ there exist an open set U in X containing p and a parameterization*

$$\psi: U_0 \to U$$

of U with the property

$$\psi(U_0 \cap \partial\mathbf{H}^n) = U \cap Z. \tag{4.4.19}$$

Proof. Since X is locally diffeomorphic at p to an open subset of $\mathbf{R}^n$, it suffices to prove this assertion for $X = \mathbf{R}^n$. However, if Z is an $(n-1)$-dimensional submanifold of $\mathbf{R}^n$ then by Theorem 4.2.15 there exist, for every $p \in Z$, a neighborhood U of p in $\mathbf{R}^n$ and a function $\phi \in C^\infty(U)$ with the properties

$$x \in U \cap Z \iff \phi(x) = 0 \tag{4.4.20}$$

and

$$d\phi_p \neq 0\,. \tag{4.4.21}$$

Without loss of generality we can assume by equation (4.4.21) that

$$\frac{\partial \phi}{\partial x_1}(p) \neq 0.$$

Hence if $\rho: U \to \mathbf{R}^n$ is the map

$$\rho(x_1, \dots, x_n) := (\phi(x), x_2, \dots, x_n), \tag{4.4.22}$$

$(d\rho)_p$ is bijective, and hence ρ is locally a diffeomorphism at p. Shrinking U we can assume that ρ is a diffeomorphism of U onto an open set U_0. By (4.4.20) and (4.4.22) ρ maps $U \cap Z$ onto $U_0 \cap \partial\mathbf{H}^n$ hence if we take ψ to be ρ^{-1}, it will have the property (4.4.19). □

We will now prove Theorem 4.4.9. Without loss of generality we can assume that the open set U_0 in Lemma 4.4.18 is an open ball with center at $q \in \partial\mathbf{H}^n$ and that the diffeomorphism ψ maps q to p. Thus for $\psi^{-1}(U \cap D)$ there are three possibilities:

(i) $\psi^{-1}(U \cap D) = (\mathbf{R}^n - \partial\mathbf{H}^n) \cap U_0$.
(ii) $\psi^{-1}(U \cap D) = (\mathbf{R}^n - \overline{\mathbf{H}}^n) \cap U_0$.
(iii) $\psi^{-1}(U \cap D) = \mathbf{H}^n \cap U_0$.

However, (i) is excluded by the second hypothesis in Definition 4.4.15 and if (ii) occurs we can rectify the situation by composing ϕ with the map $(x_1, \ldots, x_n) \mapsto (-x_1, x_2, \ldots, x_n)$. □

Definition 4.4.23. We will call an open set U with the properties above a *D-**adapted parameterizable open set***.

We will now show that if X is oriented and $D \subset X$ is a smooth domain then the boundary $Z \coloneqq \partial D$ of D acquires from X a natural orientation. To see this we first observe the following.

Lemma 4.4.24. *The diffeomorphism $\psi: U_0 \xrightarrow{\sim} U$ in Theorem 4.4.17 can be chosen to be orientation preserving.*

Proof. If it is not, then by replacing ψ with the diffeomorphism

$$\psi^\sharp(x_1, \ldots, x_n) \coloneqq \psi(x_1, \ldots, x_{n-1}, -x_n),$$

we get a D-adapted parameterization of U which *is* orientation preserving. (See (4.4.12)–(4.4.13).) □

Let $V_0 = U_0 \cap \mathbf{R}^{n-1}$ be the boundary of $U_0 \cap \mathbf{H}^n$. The restriction of ψ to V_0 is a diffeomorphism of V_0 onto $U \cap Z$, and we will orient $U \cap Z$ by requiring that this map be an oriented parameterization. To show that this is an "intrinsic" definition, i.e., does not depend on the choice of ψ. We'll prove the following theorem.

Theorem 4.4.25. *If $\psi_i: U_i \to U$, $i = 0, 1$, are oriented parameterizations of U with the property*

$$\psi_i: U_i \cap \mathbf{H}^n \to U \cap D$$

the restrictions of ψ_i to $U_i \cap \mathbf{R}^{n-1}$ induce compatible orientations on $U \cap X$.

Proof. To prove this we have to prove that the map $\phi_1^{-1} \circ \phi_0$ restricted to $U \cap \partial\mathbf{H}^n$ is an orientation-preserving diffeomorphism of $U_0 \cap \mathbf{R}^{n-1}$ onto $U_1 \cap \mathbf{R}^{n-1}$. Thus we have to prove the following proposition.

Proposition 4.4.26. *Let U_0 and U_1 be open subsets of $\mathbf{R}^n$ and $f: U_0 \to U_1$ an orientation-preserving diffeomorphism which maps $U_0 \cap \mathbf{H}^n$ onto $U_1 \cap \mathbf{H}^n$. Then the restriction g of f to the boundary $U_0 \cap \mathbf{R}^{n-1}$ of $U_0 \cap \mathbf{H}^n$ is an orientation-preserving diffeomorphism*

$$g: U_0 \cap \mathbf{R}^{n-1} \xrightarrow{\sim} U_1 \cap \mathbf{R}^{n-1}.$$

Let $f(x) = (f_1(x), \ldots, f_n(x))$. By assumption $f_1(x_1, \ldots, x_n)$ is less than zero if x_1 is less than zero and equal to zero if x_1 is equal to zero, hence

$$\frac{\partial f_1}{\partial x_1}(0, x_2, \ldots, x_n) \geq 0 \tag{4.4.27}$$

and

$$\frac{\partial f_1}{\partial x_i}(0, x_2, \ldots, x_n) = 0$$

for $i > 1$. Moreover, since g is the restriction of f to the set $x_1 = 0$

$$\frac{\partial f_i}{\partial x_j}(0, x_2, \ldots, x_n) = \frac{\partial g_i}{\partial x_j}(x_2, \ldots, x_n)$$

for $i, j \geq 2$. Thus on the set defined by $x_1 = 0$ we have

$$\det\left(\frac{\partial f_i}{\partial x_j}\right) = \frac{\partial f_1}{\partial x_1} \det\left(\frac{\partial g_i}{\partial x_j}\right). \tag{4.4.28}$$

Since f is orientation preserving the left-hand side of equation (4.4.28) is positive at all points $(0, x_2, \ldots, x_n) \in U_0 \cap \mathbf{R}^{n-1}$ hence by (4.4.27) the same is true for $\frac{\partial f_1}{\partial x_1}$ and $\det\left(\frac{\partial g_i}{\partial x_j}\right)$. Thus g is orientation preserving. □

Remark 4.4.29. For an alternative proof of this result see Exercises 3.2.viii and 3.6.iv.

We will now orient the boundary of D by requiring that for every D-adapted parameterizable open set U the orientation of Z coincides with the orientation of $U \cap Z$ that we described above. We will conclude this discussion of orientations by proving a global version of Proposition 4.4.26.

Proposition 4.4.30. *For $i = 1, 2$, let X_i be an oriented manifold, $D_i \subset X_i$ a smooth domain and $Z_i := \partial D_i$ its boundary. Then if f is an orientation-preserving diffeomorphism of (X_1, D_1) onto (X_2, D_2) the restriction g of f to Z_1 is an orientation-preserving diffeomorphism of Z_1 onto Z_2.*

Proof. Let U be an open subset of X_1 and $\phi: U_0 \to U$ an oriented D_1-compatible parameterization of U. Then if $V = f(U)$ the map $f\phi: U_0 \to V$ is an oriented D_2-compatible parameterization of V and hence $g: U \cap Z_1 \to V \cap Z_2$ is orientation preserving. □

Exercises for §4.4

Exercise 4.4.i. Let V be an oriented n-dimensional vector space, B an inner product on V and $e_1, \ldots, e_n \in V$ an oriented orthonormal basis. Given vectors $v_1, \ldots, v_n \in V$, show that if

$$b_{i,j} = B(v_i, v_j) \tag{4.4.31}$$

and

$$v_i = \sum_{j=1}^{n} a_{j,i} e_j,$$

the matrices $\mathbf{A} = (a_{i,j})$ and $\mathbf{B} = (b_{i,j})$ satisfy the identity:

$$\mathbf{B} = \mathbf{A}^{\top}\mathbf{A}$$

and conclude that $\det(\mathbf{B}) = \det(\mathbf{A})^2$. (In particular conclude that $\det(\mathbf{B}) > 0$.)

Exercise 4.4.ii. Let V and W be oriented n-dimensional vector spaces. Suppose that each of these spaces is equipped with an inner product, and let $e_1, \ldots, e_n \in V$ and $f_1, \ldots, f_n \in W$ be oriented orthonormal bases. Show that if $A \colon W \to V$ is an orientation-preserving linear mapping and $Af_i = v_i$ then

$$A^\star \operatorname{vol}_V = \det((b_{i,j}))^{\frac{1}{2}} \operatorname{vol}_W,$$

where $\operatorname{vol}_V = e_1^\star \wedge \cdots \wedge e_n^\star$, $\operatorname{vol}_W = f_1^\star \wedge \cdots \wedge f_n^\star$ and $(b_{i,j})$ is the matrix (4.4.31).

Exercise 4.4.iii. Let X be an oriented n-dimensional submanifold of $\mathbf{R}^n$, U an open subset of X, U_0 an open subset of $\mathbf{R}^n$, and $\phi \colon U_0 \to U$ an oriented parameterization. Let $\phi_1, \ldots, \phi_N$ be the coordinates of the map

$$U_0 \to U \hookrightarrow \mathbf{R}^N,$$

the second map being the inclusion map. Show that if σ is the Riemannian volume form on X then

$$\phi^\star \sigma = (\det(\phi_{i,j}))^{\frac{1}{2}}\, dx_1 \wedge \cdots \wedge dx_n,$$

where

$$\phi_{i,j} = \sum_{k=1}^{N} \frac{\partial \phi_k}{\partial x_i} \frac{\partial \phi_k}{\partial x_j} \quad \text{for } 1 \le i, j \le n. \tag{4.4.32}$$

Conclude that σ is a smooth n-form and hence that it *is* a volume form.

Hint: For $p \in U_0$ and $q = \phi(p)$ apply Exercise 4.4.ii with $V = T_qX$, $W = T_p\mathbf{R}^n$, $A = (d\phi)_p$ and $v_i = (d\phi)_p \left(\frac{\partial}{\partial x_i}\right)_p$.

Exercise 4.4.iv. Given a C^∞ function $f \colon \mathbf{R} \to \mathbf{R}$, its graph

$$X \coloneqq \Gamma_f = \{(x, f(x)) \in \mathbf{R} \times \mathbf{R} \,|\, x \in \mathbf{R}\}$$

is a submanifold of $\mathbf{R}^2$ and

$$\phi \colon \mathbf{R} \to X, \quad x \mapsto (x, f(x))$$

is a diffeomorphism. Orient X by requiring that ϕ be orientation preserving and show that if σ is the Riemannian volume form on X then

$$\phi^\star \sigma = \left(1 + \left(\frac{df}{dx}\right)^2\right)^{\frac{1}{2}} dx.$$

Hint: Exercise 4.4.iii.

Exercise 4.4.v. Given a C^∞ function $f \colon \mathbf{R}^n \to \mathbf{R}$, the graph Γ_f of f is a submanifold of $\mathbf{R}^{n+1}$ and

$$\phi \colon \mathbf{R}^n \to X, \quad x \mapsto (x, f(x)) \tag{4.4.33}$$

is a diffeomorphism. Orient X by requiring that ϕ is orientation preserving and show that if σ is the Riemannian volume form on X then

$$\phi^\star \sigma = \left(1 + \sum_{i=1}^{n} \left(\frac{\partial f}{\partial x_i}\right)^2\right)^{\frac{1}{2}} dx_1 \wedge \cdots \wedge dx_n.$$

Hints:

- Let $v = (c_1, \ldots, c_n) \in \mathbf{R}^n$. Show that if $C: \mathbf{R}^n \to \mathbf{R}$ is the linear mapping defined by the matrix $(c_i c_j)$ then $Cv = (\sum_{i=1}^n c_i^2)v$ and $Cw = 0$ if $w \cdot v = 0$.
- Conclude that the eigenvalues of C are $\lambda_1 = \sum c_i^2$ and $\lambda_2 = \cdots = \lambda_n = 0$.
- Show that the determinant of $I + C$ is $1 + \sum_{i=1}^n c_i^2$.
- Compute the determinant of the matrix (4.4.32) where ϕ is the mapping (4.4.33).

Exercise 4.4.vi. Let V be an oriented N-dimensional vector space and $\ell_1, \ldots, \ell_k \in V^\star$ be linearly independent vectors in $V^\star$. Define

$$L: V \to \mathbf{R}^k$$

to be the map $v \mapsto (\ell_1(v), \ldots, \ell_k(v))$.

(1) Show that L is surjective and that the kernel W of L is of dimension $n = N - k$.

(2) Show that one gets from this mapping a bijective linear mapping

$$V/W \to \mathbf{R}^k \tag{4.4.34}$$

and hence from the standard orientation on $\mathbf{R}^k$ an induced orientation on V/W and on W.

Hint: Exercise 1.2.viii and Theorem 1.9.9.

(3) Let ω be an element of $\Lambda^N(V^\star)$. Show that there exists a $\mu \in \Lambda^n(V^\star)$ with the property

$$\ell_1 \wedge \cdots \wedge \ell_k \wedge \mu = \omega. \tag{4.4.35}$$

Hint: Choose an oriented basis, $e_1, \ldots, e_N$ of V such that $\omega = e_1^\star \wedge \cdots \wedge e_N^\star$ and $\ell_i = e_i^\star$ for $i = 1, \ldots, k$, and let $\mu = e_{i+1}^\star \wedge \cdots \wedge e_N^\star$.

(4) Show that if ν is an element of $\Lambda^n(V^\star)$ with the property

$$\ell_1 \wedge \cdots \wedge \ell_k \wedge \nu = 0$$

then there exist elements $\nu_1, \ldots, \nu_k$, of $\Lambda^{n-1}(V^\star)$ such that

$$\nu = \sum_{i=1}^k \ell_i \wedge \nu_i.$$

Hint: Same hint as in part (3).

(5) Show that if μ_1 and μ_2 are elements of $\Lambda^n(V^\star)$ with the property (4.4.35) and $\iota: W \to V$ is the inclusion map then $\iota^\star\mu_1 = \iota^\star\mu_2$.

Hint: Let $\nu = \mu_1 - \mu_2$. Conclude from part (4) that $\iota^\star\nu = 0$.

(6) Conclude that if μ is an element of $\Lambda^n(V^\star)$ satisfying (4.4.35) the element, $\sigma = \iota^\star\mu$, of $\Lambda^n(W^\star)$ is *intrinsically* defined independent of the choice of μ.

(7) Show that σ lies in $\Lambda^n(W^\star)_+$.

Exercise 4.4.vii. Let U be an open subset of $\mathbf{R}^N$ and $f: U \to \mathbf{R}^k$ a C^∞ map. If zero is a regular value of f, the set $X = f^{-1}(0)$ is a manifold of dimension $n = N - k$. Show that this manifold has a natural smooth orientation.

Some suggestions:

- Let $f = (f_1, \ldots, f_k)$ and let
$$df_1 \wedge \cdots \wedge df_k = \sum f_I dx_I$$
summed over multi-indices which are strictly increasing. Show that for every $p \in X$, $f_I(p) \neq 0$ for some multi-index, $I = (i_1, \ldots, i_k)$, $1 \leq i_1 < \cdots < i_k \leq N$.
- Let $J = (j_1, \ldots, j_n)$, $1 \leq j_1 < \cdots < j_n \leq N$ be the complementary multi-index to I, i.e., $j_r \neq i_s$ for all r and s. Show that
$$df_1 \wedge \cdots \wedge df_k \wedge dx_J = \pm f_I dx_1 \wedge \cdots \wedge dx_N$$
and conclude that the n-form
$$\mu = \pm \frac{1}{f_I} dx_J$$
is a C^∞ n-form on a neighborhood of p in U and has the property:
$$df_1 \wedge \cdots \wedge df_k \wedge \mu = dx_1 \wedge \cdots \wedge dx_N.$$
- Let $\iota\colon X \to U$ be the inclusion map. Show that the assignment
$$p \mapsto (\iota^\star \mu)_p$$
defines an *intrinsic* nowhere vanishing n-form $\sigma \in \Omega^n(X)$ on X.
- Show that the orientation of X defined by σ coincides with the orientation that we described earlier in this section.

Exercise 4.4.viii. Let S^n be the n-sphere and $\iota\colon S^n \to \mathbf{R}^{n+1}$ the inclusion map. Show that if $\omega \in \Omega^n(\mathbf{R}^{n+1})$ is the n-form $\omega = \sum_{i=1}^{n+1} (-1)^{i-1} x_i dx_1 \wedge \cdots \wedge \widehat{dx_i} \wedge \cdots \wedge dx_{n+1}$, the n-form $\iota^\star \omega \in \Omega^n(S^n)$ is the Riemannian volume form.

Exercise 4.4.ix. Let S^{n+1} be the $(n+1)$-sphere and let
$$S^{n+1}_+ = \{(x_1, \ldots, x_{n+2}) \in S^{n+1} \mid x_1 < 0\}$$
be the lower hemisphere in S^{n+1}.

(1) Prove that S^{n+1}_+ is a smooth domain.
(2) Show that the boundary of S^{n+1}_+ is S^n.
(3) Show that the boundary orientation of S^n agrees with the orientation of S^n in Exercise 4.4.viii.

4.5. Integration of forms on manifolds

In this section we will show how to integrate differential forms over manifolds. In what follows X will be an oriented n-manifold and W an open subset of X, and our goal will be to make sense of the integral

$$\int_W \omega \tag{4.5.1}$$

where ω is a compactly supported n-form. We'll begin by showing how to define this integral when the support of ω is contained in a parameterizable open set U. Let U_0 be an open subset of $\mathbf{R}^n$ and $\phi_0 : U_0 \to U$ a parameterization. As we noted in §4.4 we can assume without loss of generality that this parameterization is oriented. Making this assumption, we'll define

$$\int_W \omega = \int_{W_0} \phi_0^\star \omega, \tag{4.5.2}$$

where $W_0 = \phi_0^{-1}(U \cap W)$. Notice that if $\phi^\star \omega = f dx_1 \wedge \cdots \wedge dx_n$, then, by assumption, f is in $C_0^\infty(U_0)$. Hence since

$$\int_{W_0} \phi_0^\star \omega = \int_{W_0} f dx_1 \wedge \cdots \wedge dx_n$$

and since f is a bounded continuous function and is compactly supported the Riemann integral on the right is well-defined. (See Appendix B.) Moreover, if $\phi_1 : U_1 \to U$ is another oriented parameterization of U and $\psi : U_0 \to U_1$ is the map $\psi = \phi_1^{-1} \circ \phi_0$ then $\phi_0 = \phi_1 \circ \psi$, so by Proposition 4.3.4

$$\phi_0^\star \omega = \psi^\star \phi_1^\star \omega.$$

Moreover, by (4.3.18) ψ is orientation preserving. Therefore since

$$W_1 = \psi(W_0) = \phi_1^{-1}(U \cap W)$$

Theorem 3.5.2 tells us that

$$\int_{W_1} \phi_1^\star \omega = \int_{W_0} \phi_0^\star \omega. \tag{4.5.3}$$

Thus the definition (4.5.2) is a legitimate definition. It does not depend on the parameterization that we use to define the integral on the right. From the usual additivity properties of the Riemann integral one gets analogous properties for the integral (4.5.2). Namely for $\omega_1, \omega_2 \in \Omega_c^n(U)$,

$$\int_W (\omega_1 + \omega_2) = \int_W \omega_1 + \int_W \omega_2 \tag{4.5.4}$$

and for $\omega \in \Omega_c^n(U)$ and $c \in \mathbf{R}$

$$\int_W c\omega = c \int_W \omega.$$

We will next show how to define the integral (4.5.1) for *any* compactly supported n-form. This we will do in more or less the same way that we defined improper Riemann integrals in Appendix B: by using partitions of unity. We'll begin by deriving from the partition of unity theorem in Appendix B a manifold version of this theorem.

Theorem 4.5.5. *Let*

$$\mathcal{U} = \{U_\alpha\}_{\alpha \in I} \tag{4.5.6}$$

a covering of X by open subsets. Then there exists a family of functions $\rho_i \in C_0^\infty(X)$, *for* $i \geq 1$, *with the following properties:*

(1) $\rho_i \geq 0$.
(2) *For every compact set* $C \subset X$, *there exists a positive integer N such that if* $i > N$ *we have* $\operatorname{supp}(\rho_i) \cap C = \emptyset$.
(3) $\sum_{i=1}^\infty \rho_i = 1$.
(4) *For every* $i \geq 1$ *there exists an index* $\alpha \in I$ *such that* $\operatorname{supp}(\rho_i) \subset U_\alpha$.

Remark 4.5.7. Conditions (1)–(3) say that the ρ_i's are a partition of unity and (4) says that this partition of unity is subordinate to the covering (4.5.5).

Proof. For each $p \in X$ and for some U_α containing a p choose an open set O_p in $\mathbf{R}^N$ with $p \in O_p$ and with

$$\overline{O_p \cap X} \subset U_\alpha. \tag{4.5.8}$$

Let $O := \bigcup_{p\in X} O_p$ and let $\tilde{\rho}_i \in C_0^\infty(O)$, for $i \geq 1$, be a partition of unity subordinate to the covering of O by the O_p's. By (4.5.8) the restriction ρ_i of $\tilde{\rho}_i$ to X has compact support and it is clear that the ρ_i's inherit from the $\tilde{\rho}_i$'s the properties (1)–(4). □

Now let the covering (4.5.6) be any covering of X by parameterizable open sets and let $\rho_i \in C_0^\infty(X)$, for $i \geq 1$ be a partition of unity subordinate to this covering. Given $\omega \in \Omega_c^n(X)$ we will define the integral of ω over W by the sum

$$\sum_{i=1}^\infty \int_W \rho_i \omega. \tag{4.5.9}$$

Note that since each ρ_i is supported in some U_α the individual summands in this sum are well-defined and since the support of ω is compact all but finitely many of these summands are zero by part (2) of Theorem 4.5.5. Hence the sum itself is well-defined. Let's show that this sum does not depend on the choice of $\mathcal{U}$ and the ρ_i's. Let $\mathcal{U}'$ be another covering of X by parameterizable open sets and $(\rho'_j)_{j\geq 1}$ a partition of unity subordinate to $\mathcal{U}'$. Then

$$\sum_{j=1}^\infty \int_W \rho'_j \omega = \sum_{j=1}^\infty \int_W \sum_{i=1}^\infty \rho'_j \rho_i \omega = \sum_{j=1}^\infty \left(\sum_{i=1}^\infty \int_W \rho'_j \rho_i \omega \right) \tag{4.5.10}$$

by equation (4.5.4). Interchanging the orders of summation and resuming with respect to the j's this sum becomes

$$\sum_{i=1}^\infty \int_W \sum_{j=1}^\infty \rho'_j \rho_i \omega$$

or

$$\sum_{i=1}^\infty \int_W \rho_i \omega.$$

Hence

$$\sum_{i=1}^{\infty} \int_W \rho_j' \omega = \sum_{i=1}^{\infty} \int_W \rho_i \omega,$$

so the two sums are the same.

From equations (4.5.4) and (4.5.9) one easily deduces the following proposition:

Proposition 4.5.11. *For* $\omega_1, \omega_2 \in \Omega_c^n(X)$,

$$\int_W \omega_1 + \omega_2 = \int_W \omega_1 + \int_W \omega_2$$

and for $\omega \in \Omega_c^n(X)$ *and* $c \in \mathbf{R}$

$$\int_W c\omega = c \int_W \omega.$$

The definition of the integral (4.5.1) depends on the choice of an orientation of X, but it's easy to see *how* it depends on this choice. We pointed out in §4.4 that if X is connected, there is just one way to orient it smoothly other than by its given orientation, namely by *reversing* the orientation of T_pX at each point p and it's clear from the definitions (4.5.2) and (4.5.9) that the effect of doing this is to change the sign of the integral, i.e., to change $\int_X \omega$ to $-\int_X \omega$.

In the definition of the integral (4.5.1) we have allowed W to be an arbitrary open subset of X but required ω to be compactly supported. This integral is also well-defined if we allow ω to be an arbitrary element of $\Omega^n(X)$ but require the closure of W in X to be compact. To see this, note that under this assumption the sum (4.5.8) is still a finite sum, so the definition of the integral still makes sense, and the double sum on the right side of (4.5.10) is still a finite sum so it's still true that the definition of the integral does not depend on the choice of partitions of unity. In particular if the closure of W in X is compact we will define the volume of W to be the integral,

$$\mathrm{vol}(W) = \int_W \sigma_{\mathrm{vol}},$$

where σ_{vol} is the Riemannian volume form and if X itself is compact we'll define its volume to be the integral

$$\mathrm{vol}(X) = \int_X \sigma_{\mathrm{vol}}.$$

We'll next prove a manifold version of the change of variables formula (3.5.3).

Theorem 4.5.12 (change of variables formula). *Let* X' *and* X *be oriented* n*-manifolds and* $f\colon X' \to X$ *an orientation-preserving diffeomorphism. If* W *is an open subset of* X *and* $W' := f^{-1}(W)$, *then*

$$\int_{W'} f^\star \omega = \int_W \omega \tag{4.5.13}$$

for all $\omega \in \Omega_c^n(X)$.

Proof. By (4.5.9) the integrand of the integral above is a finite sum of C^∞-forms, each of which is supported on a parameterizable open subset, so we can assume that ω itself as this property. Let V be a parameterizable open set containing the support of ω and let $\phi_0 \colon U \xrightarrow{\sim} V$ be an oriented parameterization of V. Since f is a diffeomorphism its inverse exists and is a diffeomorphism of X onto X_1. Let $V' \coloneqq f^{-1}(V)$ and $\phi_0' \coloneqq f^{-1} \circ \phi_0$. Then $\phi_0' \colon U \xrightarrow{\sim} V'$ is an oriented parameterization of V'. Moreover, $f \circ \phi_0' = \phi_0$ so if $W_0 = \phi_0^{-1}(W)$ we have

$$W_0 = (\phi_0')^{-1}(f^{-1}(W)) = (\phi_0')^{-1}(W')$$

and by the chain rule we have

$$\phi_0^\star \omega = (f \circ \phi_0')^\star \omega = (\phi_0')^\star f^\star \omega$$

hence

$$\int_W \omega = \int_{W_0} \phi_0^\star \omega = \int_{W_0} (\phi_0')^\star (f^\star \omega) = \int_{W'} f^\star \omega.$$ □

As an exercise, show that if $f \colon X' \to X$ is orientation reversing

(4.5.14) $$\int_{W'} f^\star \omega = -\int_W \omega.$$

We'll conclude this discussion of "integral calculus on manifolds" by proving a preliminary version of Stokes' theorem.

Theorem 4.5.15. *If μ is in $\Omega_c^{n-1}(X)$ then*

$$\int_X d\mu = 0.$$

Proof. Let $(\rho_i)_{i \geq 1}$ be a partition of unity with the property that each ρ_i is supported in a parameterizable open set $U_i = U$. Replacing μ by $\rho_i \mu$ it suffices to prove the theorem for $\mu \in \Omega_c^{n-1}(U)$. Let $\phi \colon U_0 \to U$ be an oriented parameterization of U. Then

$$\int_U d\mu = \int_{U_0} \phi^\star d\mu = \int_{U_0} d(\phi^\star \mu) = 0$$

by Theorem 3.3.1. □

Exercises for §4.5

Exercise 4.5.i. Let $f \colon \mathbf{R}^n \to \mathbf{R}$ be a C^∞ function. Orient the graph $X \coloneqq \Gamma_f$ of f by requiring that the diffeomorphism

$$\phi \colon \mathbf{R}^n \to X\,, \; x \mapsto (x, f(x))$$

be orientation preserving. Given a bounded open set U in $\mathbf{R}^n$ compute the Riemannian volume of the image

$$X_U = \phi(U)$$

of U in X as an integral over U.

Hint: Exercise 4.4.v.

Exercise 4.5.ii. Evaluate this integral for the open subset X_U of the paraboloid defined by $x_3 = x_1^2 + x_2^2$, where U is the disk $x_1^2 + x_2^2 < 2$.

Exercise 4.5.iii. In Exercise 4.5.i let $\iota: X \hookrightarrow \mathbf{R}^{n+1}$ be the inclusion map of X onto $\mathbf{R}^{n+1}$.

(1) If $\omega \in \Omega^n(\mathbf{R}^{n+1})$ is the n-form $x_{n+1}dx_1 \wedge \cdots \wedge dx_n$, what is the integral of $\iota^\star\omega$ over the set X_U? Express this integral as an integral over U.
(2) Same question for $\omega = x_{n+1}^2 dx_1 \wedge \cdots \wedge dx_n$.
(3) Same question for $\omega = dx_1 \wedge \cdots \wedge dx_n$.

Exercise 4.5.iv. Let $f: \mathbf{R}^n \to (0, +\infty)$ be a positive C^∞ function, U a bounded open subset of $\mathbf{R}^n$, and W the open set of $\mathbf{R}^{n+1}$ defined by the inequalities

$$0 < x_{n+1} < f(x_1, \ldots, x_n)$$

and the condition $(x_1, \ldots, x_n) \in U$.

(1) Express the integral of the $(n+1)$-form $\omega = x_{n+1}dx_1 \wedge \cdots \wedge dx_{n+1}$ over W as an integral over U.
(2) Same question for $\omega = x_{n+1}^2 dx_1 \wedge \cdots \wedge dx_{n+1}$.
(3) Same question for $\omega = dx_1 \wedge \cdots \wedge dx_n$.

Exercise 4.5.v. Integrate the "Riemannian area" form

$$x_1 dx_2 \wedge dx_3 + x_2 dx_3 \wedge dx_1 + x_3 dx_1 \wedge dx_2$$

over the unit 2-sphere S^2. (See Exercise 4.4.viii.)

Hint: An easier problem: Using polar coordinates integrate $\omega = x_3 dx_1 \wedge dx_2$ over the hemisphere defined by $x_3 = \sqrt{1 - x_1^2 - x_2^2}$, where $x_1^2 + x_2^2 < 1$.

Exercise 4.5.vi. Let α be the 1-form $\sum_{i=1}^n y_i dx_i$ in equation (2.8.2) and let $\gamma(t)$, $0 \leq t \leq 1$, be a trajectory of the Hamiltonian vector field (2.8.2). What is the integral of α over $\gamma(t)$?

4.6. Stokes' theorem and the divergence theorem

Let X be an oriented n-manifold and $D \subset X$ a smooth domain. We showed in §4.4 that ∂D acquires from D a natural orientation. Hence if $\iota: \partial D \to X$ is the inclusion map and μ is in $\Omega_c^{n-1}(X)$, the integral

$$\int_{\partial D} \iota^\star \mu$$

is well-defined. We will prove the following theorem:

Theorem 4.6.1 (Stokes' theorem). *For $\mu \in \Omega_c^{n-1}(X)$ we have*

$$\int_{\partial D} \iota^\star \mu = \int_D d\mu. \tag{4.6.2}$$

Proof. Let $(\rho_i)_{i\geq 1}$ be a partition of unity such that for each i, the support of ρ_i is contained in a parameterizable open set $U_i = U$ of one of the following three types:

(a) $U \subset \text{int}(D)$.
(b) $U \subset \text{ext}(D)$.
(c) There exist an open subset U_0 of $\mathbf{R}^n$ and an oriented D-adapted parameterization

$$\phi : U_0 \xrightarrow{\simeq} U.$$

Replacing μ by the finite sum $\sum_{i=1}^{\infty} \rho_i \mu$ it suffices to prove (4.6.2) for each $\rho_i \mu$ separately. In other words we can assume that the support of μ itself is contained in a parameterizable open set U of type (a), (b), or (c). But if U is of type (a)

$$\int_D d\mu = \int_U d\mu = \int_X d\mu$$

and $\iota^\star \mu = 0$. Hence the left-hand side of equation (4.6.2) is zero and, by Theorem 4.5.15, the right-hand side is as well. If U is of type (b) the situation is even simpler: $\iota^\star \mu$ is zero and the restriction of μ to D is zero, so both sides of equation (4.6.2) are automatically zero. Thus one is reduced to proving (4.6.2) when U is an open subset of type (c). In this case the restriction of the map (4.6.2) to $U_0 \cap \partial \mathbf{H}^n$ is an orientation-preserving diffeomorphism

$$\psi : U_0 \cap \partial \mathbf{H}^n \to U \cap Z \tag{4.6.3}$$

and

$$\iota_Z \circ \psi = \phi \circ \iota_{\mathbf{R}^{n-1}}, \tag{4.6.4}$$

where the maps $\iota = \iota_Z$ and

$$\iota_{\mathbf{R}^{n-1}} : \mathbf{R}^{n-1} \hookrightarrow \mathbf{R}^n$$

are the inclusion maps of Z into X and $\partial \mathbf{H}^n$ into $\mathbf{R}^n$. (Here we're identifying $\partial \mathbf{H}^n$ with $\mathbf{R}^{n-1}$.) Thus

$$\int_D d\mu = \int_{\mathbf{H}^n} \phi^\star d\mu = \int_{\mathbf{H}^n} d\phi^\star \mu$$

and by (4.6.4)

$$\begin{aligned} \int_Z \iota_Z^\star \mu &= \int_{\mathbf{R}^{n-1}} \psi^\star \iota_Z^\star \mu \\ &= \int_{\mathbf{R}^{n-1}} \iota_{\mathbf{R}^{n-1}}^\star \phi^\star \mu \\ &= \int_{\partial \mathbf{H}^n} \iota_{\mathbf{R}^{n-1}}^\star \phi^\star \mu. \end{aligned}$$

Thus it suffices to prove Stokes' theorem with μ replaced by $\phi^\star \mu$, or, in other words, to prove Stokes' theorem for $\mathbf{H}^n$; and this we will now do.

Theorem 4.6.5 (Stokes' theorem for $\mathbf{H}^n$). *Let*

$$\mu = \sum_{i=1}^{n} (-1)^{i-1} f_i dx_1 \wedge \cdots \wedge \widehat{dx_i} \wedge \cdots \wedge dx_n.$$

Then

$$d\mu = \sum_{i=1}^{n} \frac{\partial f_i}{\partial x_i} dx_1 \wedge \cdots \wedge dx_n$$

and

$$\int_{\mathbf{H}^n} d\mu = \sum_{i=1}^{n} \int_{\mathbf{H}^n} \frac{\partial f_i}{\partial x_i} dx_1 \cdots dx_n.$$

We will compute each of these summands as an iterated integral doing the integration with respect to dx_i first. For $i > 1$ the dx_i integration ranges over the interval $-\infty < x_i < \infty$ and hence since f_i is compactly supported

$$\int_{-\infty}^{\infty} \frac{\partial f_i}{\partial x_i} dx_i = f_i(x_1, \ldots, x_i, \ldots, x_n)\Big|_{x_i=-\infty}^{x_i=+\infty} = 0.$$

On the other hand the dx_1 integration ranges over the interval $-\infty < x_1 < 0$ and

$$\int_{-\infty}^{0} \frac{\partial f_1}{\partial x_1} dx_1 = f(0, x_2, \ldots, x_n).$$

Thus integrating with respect to the remaining variables we get

$$\int_{\mathbf{H}^n} d\mu = \int_{\mathbf{R}^{n-1}} f(0, x_2, \ldots, x_n)\, dx_2 \wedge \cdots \wedge dx_n. \tag{4.6.6}$$

On the other hand, since $\iota_{\mathbf{R}^{n-1}}^{\star} x_1 = 0$ and $\iota_{\mathbf{R}^{n-1}}^{\star} x_i = x_i$ for $i > 1$,

$$\iota_{\mathbf{R}^{n-1}}^{\star} \mu = f_1(0, x_2, \ldots, x_n)\, dx_2 \wedge \cdots \wedge dx_n$$

so

$$\int_{\mathbf{R}^{n-1}} \iota_{\mathbf{R}^{n-1}}^{\star} \mu = \int_{\mathbf{R}^{n-1}} f(0, x_2, \ldots, x_n)\, dx_2 \wedge \cdots \wedge dx_n. \tag{4.6.7}$$

Hence the two sides, (4.6.6) and (4.6.7), of Stokes' theorem are equal. □

One important variant of Stokes' theorem is the divergence theorem: Let ω be in $\Omega_c^n(X)$ and let $\boldsymbol{v}$ be a vector field on X. Then

$$L_{\boldsymbol{v}}\omega = \iota_{\boldsymbol{v}} d\omega + d\iota_{\boldsymbol{v}}\omega = d\iota_{\boldsymbol{v}}\omega,$$

hence, denoting by ι_Z the inclusion map of Z into X we get from Stokes' theorem, with $\mu = \iota_{\boldsymbol{v}}\omega$.

Theorem 4.6.8 (Divergence theorem for manifolds).

$$\int_D L_{\boldsymbol{v}}\omega = \int_Z \iota_Z^{\star}(\iota_{\boldsymbol{v}}\omega).$$

If D is an open domain in $\mathbf{R}^n$ this reduces to the usual divergence theorem of multi-variable calculus. Namely if $\omega = dx_1 \wedge \cdots \wedge dx_n$ and $\boldsymbol{v} = \sum_{i=1}^{n} v_i \frac{\partial}{\partial x_i}$ then by (2.5.14)

$$L_{\boldsymbol{v}} dx_1 \wedge \cdots \wedge dx_n = \operatorname{div}(\boldsymbol{v})\, dx_1 \wedge \cdots \wedge dx_n,$$

where

$$\operatorname{div}(\boldsymbol{v}) = \sum_{i=1}^{n} \frac{\partial v_i}{\partial x_i}.$$

Thus if $Z = \partial D$ and ι_Z is the inclusion $Z \hookrightarrow \mathbf{R}^n$,

$$\int_D \operatorname{div}(\boldsymbol{v})dx = \int_Z \iota_Z^\star(\iota_{\boldsymbol{v}} dx_1 \wedge \cdots \wedge dx_n). \tag{4.6.9}$$

The right-hand side of this identity can be interpreted as the "flux" of the vector field $\boldsymbol{v}$ through the boundary of D. To see this let $f: \mathbf{R}^n \to \mathbf{R}$ be a C^∞ defining function for D, i.e., a function with the properties

$$p \in D \iff f(p) < 0$$

and $df_p \neq 0$ if $p \in \partial D$. This second condition says that zero is a regular value of f and hence that $Z := \partial D$ is defined by the non-degenerate equation:

$$p \in Z \iff f(p) = 0. \tag{4.6.10}$$

Let $\boldsymbol{w}$ be the vector field

$$\boldsymbol{w} := \left(\sum_{i=1}^{n} \left(\frac{\partial f}{\partial x_i} \right)^2 \right)^{-1} \sum_{i=1}^{n} \frac{\partial f_i}{\partial x_i} \frac{\partial}{\partial x_i}.$$

In view of (4.6.10), this vector field is well-defined on a neighborhood U of Z and satisfies

$$\iota_{\boldsymbol{w}} df = 1.$$

Now note that since $df \wedge dx_1 \wedge \cdots \wedge dx_n = 0$

$$\begin{aligned} 0 &= \iota_{\boldsymbol{w}}(df \wedge dx_1 \wedge \cdots \wedge dx_n) \\ &= (\iota_{\boldsymbol{w}} df) dx_1 \wedge \cdots \wedge dx_n - df \wedge \iota_{\boldsymbol{w}} dx_1 \wedge \cdots \wedge dx_n \\ &= dx_1 \wedge \cdots \wedge dx_n - df \wedge \iota_{\boldsymbol{w}} dx_1 \wedge \cdots \wedge dx_n, \end{aligned}$$

hence letting ν be the $(n-1)$-form $\iota_{\boldsymbol{w}} dx_1 \wedge \cdots \wedge dx_n$ we get the identity

$$dx_1 \wedge \cdots \wedge dx_n = df \wedge \nu \tag{4.6.11}$$

and by applying the operation $\iota_{\boldsymbol{v}}$ to both sides of (4.6.11) the identity

$$\iota_{\boldsymbol{v}} dx_1 \wedge \cdots \wedge dx_n = (L_{\boldsymbol{v}} f)\nu - df \wedge \iota_{\boldsymbol{v}} \nu. \tag{4.6.12}$$

Let $\nu_Z = \iota_Z^\star \nu$ be the restriction of ν to Z. Since $\iota_Z^\star = 0$, $\iota_Z^\star df = 0$ and hence by (4.6.12)

$$\iota_Z^\star(\iota_{\boldsymbol{v}} dx_1 \wedge \cdots \wedge dx_n) = \iota_Z^\star(L_{\boldsymbol{v}} f)\nu_Z,$$

and the formula (4.6.9) now takes the form

$$\int_D \operatorname{div}(\boldsymbol{v})dx = \int_Z L_{\boldsymbol{v}} f \nu_Z, \tag{4.6.13}$$

where the term on the right is by definition the flux of v through Z. In calculus books this is written in a slightly different form. Letting

$$\sigma_Z = \left(\sum_{i=1}^{n}\left(\frac{\partial f}{\partial x_i}\right)^2\right)^{\frac{1}{2}} \nu_Z$$

and letting

$$\vec{n} := \left(\sum_{i=1}^{n}\left(\frac{\partial f}{\partial x_i}\right)^2\right)^{-\frac{1}{2}} \left(\frac{\partial f}{\partial x_1}, \dots, \frac{\partial f}{\partial x_n}\right)$$

and $\vec{v} := (v_1, \dots, v_n)$ we have

$$L_v \nu_Z = (\vec{n} \cdot \vec{v})\sigma_Z$$

and hence

$$\int_D \operatorname{div}(v)dx = \int_Z (\vec{n} \cdot \vec{v})\sigma_Z. \tag{4.6.14}$$

In three dimensions σ_Z is just the standard "infinitesimal element of area" on the surface Z and n_p the unit outward normal to Z at p, so this version of the divergence theorem is the version one finds in most calculus books.

As an application of Stokes' theorem, we'll give a very short alternative proof of the Brouwer fixed point theorem. As we explained in §3.6 the proof of this theorem basically comes down to proving the following theorem.

Theorem 4.6.15. *Let B^n be the closed unit ball in $\mathbf{R}^n$ and S^{n-1} its boundary. Then the identity map $\mathrm{id}_{S^{n-1}}$ on S^{n-1} cannot be extended to a C^∞ map $f: B^n \to S^{n-1}$.*

Proof. Suppose that f is such a map. Then for every $(n-1)$-form $\mu \in \Omega^{n-1}(S^{n-1})$,

$$\int_{B^n} d(f^\star \mu) = \int_{S^{n-1}} (\iota_{S^{n-1}})^\star f^\star \mu. \tag{4.6.16}$$

But $d(f^\star\mu) = f^\star(d\mu) = 0$ since μ is an $(n-1)$-form and S^{n-1} is an $(n-1)$-manifold, and since f is the identity map on S^{n-1}, $(\iota_{S_{n-1}})^\star f^\star \mu = (f \circ \iota_{S^{n-1}})^\star \mu = \mu$. Thus for *every* $\mu \in \Omega^{n-1}(S^{n-1})$, equation (4.6.16) says that the integral of μ over S^{n-1} is zero. Since there are lots of $(n-1)$-forms for which this is not true, this shows that a map f with the property above cannot exist. □

Exercises for §4.6

Exercise 4.6.i. Let B^n be the open unit ball in $\mathbf{R}^n$ and S^{n-1} the unit $(n-1)$-sphere. Show that $\operatorname{vol}(S^{n-1}) = n \operatorname{vol}(B^n)$.

Hint: Apply Stokes' theorem to the $(n-1)$-form

$$\mu := \sum_{i=1}^{n} (-1)^{i-1} x_i dx_1 \wedge \cdots \wedge \widehat{dx_i} \wedge \cdots \wedge dx_n$$

and note by Exercise 4.4.ix that μ is the Riemannian volume form of S^{n-1}.

Exercise 4.6.ii. Let $D \subset \mathbf{R}^n$ be a smooth domain. Show that there exist a neighborhood U of ∂D in $\mathbf{R}^n$ and a C^∞ defining function $g\colon U \to \mathbf{R}$ for D with the properties:

(1) $p \in U \cap D$ if and only if $g(p) < 0$,
(2) and $dg_p \neq 0$ if $p \in Z$.

Hint: Deduce from Theorem 4.4.9 that a local version of this result is true. Show that you can cover Z by a family $\mathcal{U} = \{U_\alpha\}_{\alpha\in I}$ of open subsets of $\mathbf{R}^n$ such that for each there exists a function $g_\alpha\colon U_\alpha \to \mathbf{R}$ with the properties (1) and (2). Now let $(\rho_i)_{i\geq 1}$ be a partition of unity subordinate to $\mathcal{U}$ and let $g = \sum_{i=1}^\infty \rho_i g_{\alpha_i}$ where $\mathrm{supp}(\rho_i) \subset U_{\alpha_i}$.

Exercise 4.6.iii. In Exercise 4.6.ii suppose Z is compact. Show that there exists a global defining function $f\colon \mathbf{R}^n \to \mathbf{R}$ for D with properties (1) and (2).

Hint: Let $\rho \in C_0^\infty(U)$ be a function such that $0 \leq \rho(x) \leq 1$ for all $x \in U$ and ρ is identically 1 on a neighborhood of Z, and replace g by the function

$$f(x) = \begin{cases} \rho(x)g(x) + (1-\rho(x)), & x \in \mathbf{R}^n \smallsetminus D,\\ g(x), & x \in \partial D,\\ \rho(x) - (1-\rho(x))g(x), & x \in \mathrm{int}(D). \end{cases}$$

Exercise 4.6.iv. Show that the form $L_v f \nu_Z$ in equation (4.6.13) does not depend on what choice we make of a defining function f for D.

Hints:

- Show that if g is another defining function, then, at $p \in Z$, $df_p = \lambda dg_p$, where λ is a positive constant.
- Show that if one replaces df_p by $(dg)_p$ the first term in the product $(L_v f)(p)(\nu_Z)_p$ changes by a factor λ and the second term by a factor $1/\lambda$.

Exercise 4.6.v. Show that the form ν_Z is *intrinsically defined* in the sense that if ν is *any* $(n-1)$-form satisfying equation (4.6.11), then $\nu_Z = \iota_Z^\star \nu$.

Hint: Exercise 4.5.vi.

Exercise 4.6.vi. Show that the form σ_Z in equation (4.6.14) is the Riemannian volume form on Z.

Exercise 4.6.vii. Show that the $(n-1)$-form

$$\mu = (x_1^2 + \cdots + x_n^2)^{-n} \sum_{r=1}^{n} (-1)^{r-1} x_r dx_1 \wedge \cdots \wedge \widehat{dx_r} \wedge \cdots dx_n$$

is closed and prove directly that Stokes' theorem holds for the annulus $a < x_1^2 + \cdots + x_n^2 < b$ by showing that the integral of μ over the sphere, $x_1^2 + \cdots + x_n^2 = a$, is equal to the integral over the sphere, $x_1^2 + \cdots + x_n^2 = b$.

Exercise 4.6.viii. Let $f\colon \mathbf{R}^{n-1} \to \mathbf{R}$ be an everywhere positive C^∞ function and let U be a bounded open subset of $\mathbf{R}^{n-1}$. Verify directly that Stokes' theorem is true if D is the domain defined by

$$0 < x_n < f(x_1, \dots, x_{n-1}) \,, \quad (x_1, \dots, x_{n-1}) \in U$$

and μ an $(n-1)$-form of the form

$$\phi(x_1, \dots, x_n) dx_1 \wedge \cdots \wedge dx_{n-1},$$

where ϕ is in $C_0^\infty(\mathbf{R}^n)$.

Exercise 4.6.ix. Let X be an oriented n-manifold and $\boldsymbol{v}$ a vector field on X which is complete. Verify that for $\omega \in \Omega_c^n(X)$

$$\int_X L_{\boldsymbol{v}} \omega = 0$$

in the following ways:

(1) Directly by using the divergence theorem.
(2) Indirectly by showing that

$$\int_X f_t^* \omega = \int_X \omega,$$

where $f_t \colon X \to X$, $-\infty < t < \infty$, is the one-parameter group of diffeomorphisms of X generated by $\boldsymbol{v}$.

Exercise 4.6.x. Let X be an oriented n-manifold and $D \subset X$ a smooth domain whose closure is compact. Show that if $Z := \partial D$ is the boundary of D and $g \colon Z \to Z$ a diffeomorphism g *cannot* be extended to a smooth map $f \colon \overline{D} \to Z$.

4.7. Degree theory on manifolds

In this section we'll show how to generalize to manifolds the results about the "degree" of a proper mapping that we discussed in Chapter 3. We'll begin by proving the manifold analogue of Theorem 3.3.1.

Theorem 4.7.1. *Let X be an oriented connected n-manifold and $\omega \in \Omega_c^n(X)$ a compactly supported n-form. Then the following conditions are equivalent:*

(1) $\int_X \omega = 0$.
(2) $\omega = d\mu$ *for some* $\mu \in \Omega_c^{n-1}(X)$.

Proof. We have already verified the assertion (2) ⇒(1) (see Theorem 4.5.15), so what is left to prove is the converse assertion. The proof of this is more or less identical with the proof of the "(1) ⇒(2)" part of Theorem 3.2.2:

Step 1. *Let U be a connected parameterizable open subset of X. If $\omega \in \Omega_c^n(U)$ has property* (1), *then $\omega = d\mu$ for some $\mu \in \Omega_c^{n-1}(U)$.*

Proof of Step 1. Let $\phi \colon U_0 \to U$ be an oriented parameterization of U. Then

$$\int_{U_0} \phi^* \omega = \int_X \omega = 0$$

and since U_0 is a connected open subset of $\mathbf{R}^n$, $\phi^*\omega = d\nu$ for some $\nu \in \Omega_c^{n-1}(U_0)$ by Theorem 3.3.1. Let $\mu = (\phi^{-1})^*\nu$. Then $d\mu = (\phi^{-1})^* d\nu = \omega$. □

Step 2. *Fix a base point $p_0 \in X$ and let p be any point of X. Then there exists a collection of connected parameterizable open sets $W_1, \ldots, W_N$ with $p_0 \in W_1$ and $p \in W_N$ such that for $1 \leq i \leq N-1$, the intersection $W_i \cap W_{i+1}$ is non-empty.*

Proof of Step 2. The set of points, $p \in X$, for which this assertion is true is open and the set for which it is not true is open. Moreover, this assertion is true for $p = p_0$. □

Step 3. *We deduce Theorem 4.7.1 from a slightly stronger result. Introduce an equivalence relation on $\Omega_c^n(X)$ by declaring that two n-forms $\omega_1, \omega_2 \in \Omega_c^n(X)$ are* ***equivalent*** *if $\omega_1 - \omega_2 \in d\Omega_x^{n-1}(X)$. Denote this equivalence relation by a $\omega_1 \sim \omega_2$.*

We will prove the following theorem.

Theorem 4.7.2. *For ω_1 and $\omega_2 \in \Omega_c^n(X)$ the following conditions are equivalent:*

(1) $\int_X \omega_1 = \int_X \omega_2$.
(2) $\omega_1 \sim \omega_2$.

Applying this result to a form $\omega \in \Omega_c^n(X)$ whose integral is zero, we conclude that $\omega \sim 0$, which means that $\omega = d\mu$ for some $\mu \in \Omega_c^{n-1}(X)$. Hence Theorem 4.7.2 implies Theorem 4.7.1. Conversely, if $\int_X \omega_1 = \int_X \omega_2$, then $\int_X(\omega_1 - \omega_2) = 0$, so $\omega_1 - \omega_2 = d\mu$ for some $\mu \in \Omega_c^n(X)$. Hence Theorem 4.7.1 implies Theorem 4.7.2.

Step 4. *By a partition of unity argument it suffices to prove Theorem 4.7.2 for $\omega_1 \in \Omega_c^n(U_1)$ and $\omega_2 \in \Omega_c^n(U_2)$ where U_1 and U_2 are connected parameterizable open sets. Moreover, if the integrals of ω_1 and ω_2 are zero then $\omega_i = d\mu_i$ for some $\mu_i \in \Omega_c^n(U_i)$ by Step 1, so in this case, the theorem is true. Suppose on the other hand that*

$$\int_X \omega_1 = \int_X \omega_2 = c \neq 0.$$

Then dividing by c, we can assume that the integrals of ω_1 and ω_2 are both equal to 1.

Step 5. *Let $W_1, \ldots, W_N$ be, as in Step 2, a sequence of connected parameterizable open sets with the property that the intersections, $W_1 \cap U_1$, $W_N \cap U_2$ and $W_i \cap W_{i+1}$, for $i = 1, \ldots, N-1$, are all non-empty. Select n-forms, $\alpha_0 \in \Omega_c^n(U_1 \cap W_1)$, $\alpha_N \in \Omega_c^n(W_N \cap U_2)$ and $\alpha_i \in \Omega_c^n(W_i \cap W_{i+1})$, for $i = 1, \ldots, N-1$, such that the integral of each α_i over X is equal to 1. By Step 1 we see that Theorem 4.7.1 is true for U_1, U_2 and the $W_1, \ldots, W_N$, hence Theorem 4.7.2 is true for U_1, U_2 and the $W_1, \ldots, W_N$, so*

$$\omega_1 \sim \alpha_0 \sim \alpha_1 \sim \cdots \sim \alpha_N \sim \omega_2$$

and thus $\omega_1 \sim \omega_2$. □

Just as in (3.4.2) we get as a corollary of the theorem above the following "definition–theorem" of the degree of a differentiable mapping.

Theorem 4.7.3. *Let X and Y be compact oriented n-manifolds and let Y be connected. Given a proper C^∞ mapping, $f : X \to Y$, there exists a topological invariant* $\deg(f)$ *with the defining property:*

$$\int_X f^\star \omega = \deg(f) \int_Y \omega. \tag{4.7.4}$$

Proof. As in the proof of Theorem 3.4.6 pick an n-form $\omega_0 \in \Omega_c^n(Y)$ whose integral over Y is one and define the degree of f to be the integral over X of $f^\star \omega_0$, i.e., set

$$\deg(f) := \int_X f^\star \omega_0. \tag{4.7.5}$$

Now let ω be any n-form in $\Omega_c^n(Y)$ and let

$$c := \int_Y \omega. \tag{4.7.6}$$

Then the integral of $\omega - c\omega_0$ over Y is zero so there exists an $(n-1)$-form $\mu \in \Omega_c^{n-1}(Y)$ for which $\omega - c\omega_0 = d\mu$. Hence $f^\star\omega = cf^\star\omega_0 + df^\star\mu$, so

$$\int_X f^\star \omega = c\int_X f^\star \omega_0 = \deg(f) \int_Y \omega$$

by equations (4.7.5) and (4.7.6). □

It's clear from the formula (4.7.4) that the degree of f is independent of the choice of ω_0. (Just apply this formula to any $\omega \in \Omega_c^n(Y)$ having integral over Y equal to one.) It's also clear from (4.7.4) that "degree" behaves well with respect to composition of mappings.

Theorem 4.7.7. *Let Z be an oriented, connected n-manifold and $g : Y \to Z$ a proper C^∞ map. Then*

$$\deg(g \circ f) = \deg(f)\deg(g). \tag{4.7.8}$$

Proof. Let $\omega \in \Omega_c^n(Z)$ be a compactly supported form with the property that $\int_Z \omega = 1$. Then

$$\begin{aligned}\deg(g \circ f) &= \int_X (g \circ f)^\star \omega = \int_X f^\star \circ g^\star \omega = \deg(f) \int_Y g^\star \omega \\ &= \deg(f)\deg(g).\end{aligned}$$

□

We will next show how to compute the degree of f by generalizing to manifolds the formula for $\deg(f)$ that we derived in §3.6.

Definition 4.7.9. A point $p \in X$ is a ***critical point*** of f if the map

$$df_p : T_pX \to T_{f(p)}Y \tag{4.7.10}$$

is not bijective.

We'll denote by C_f the set of all critical points of f, and we'll call a point $q \in Y$ a ***critical value*** of f if it is in the image $f(C_f)$ of C_f under f and a ***regular value*** if it's not. (Thus the set of regular values is the set $Y \setminus f(C_f)$.) If q is a regular value, then as we observed in §3.6, the map (4.7.10) is bijective for every $p \in f^{-1}(q)$ and hence by Theorem 4.2.13, f maps a neighborhood U_p of p diffeomorphically onto a neighborhood V_p of q. In particular, $U_p \cap f^{-1}(q) = p$. Since f is proper the set $f^{-1}(q)$ is compact, and since the sets U_p are a covering of $f^{-1}(q)$, this covering must be a finite covering. In particular, the set $f^{-1}(q)$ itself has to be a finite set. As in §2.7 we can shrink the U_p's so as to insure that they have the following properties:

(i) Each U_p is a parameterizable open set.
(ii) $U_p \cap U_{p'}$ is empty for $p \neq p'$.
(iii) $f(U_p) = f(U_{p'}) = V$ for all p and p'.
(iv) V is a parameterizable open set.
(v) $f^{-1}(V) = \bigcup_{p \in f^{-1}(q)} U_p$.

To exploit these properties let ω be an n-form in $\Omega_c^n(V)$ with integral equal to 1. Then by (v):

$$\deg(f) = \int_X f^\star \omega = \sum_p \int_{U_p} f^\star \omega.$$

But $f : U_p \to V$ is a diffeomorphism, hence by (4.5.13) and (4.5.14)

$$\int_{U_p} f^\star \omega = \int_V \omega$$

if $f : U_p \to V$ is orientation preserving and

$$\int_{U_p} f^\star \omega = -\int_V \omega$$

if $f : U_p \to V$ is orientation reversing. Thus we have proved the following.

Theorem 4.7.11. *The degree of f is equal to the sum*

$$\deg(f) = \sum_{p \in f^{-1}(q)} \sigma_p, \tag{4.7.12}$$

where $\sigma_p = +1$ if the map (4.7.10) is orientation preserving and $\sigma_p = -1$ if the map (4.7.10) is orientation reversing.

We will next show that Sard's theorem is true for maps between manifolds and hence that there exist lots of regular values. We first observe that if U is a parameterizable open subset of X and V a parameterizable open neighborhood of $f(U)$ in Y, then Sard's theorem is true for the map $f : U \to V$ since, up to diffeomorphism, U and V are just open subsets of $\mathbf{R}^n$. Now let q be any point in Y, let B be a compact neighborhood of q, and let V be a parameterizable open set containing B. Then if $A = f^{-1}(B)$ it follows from Theorem 3.4.7 that A can be covered by a finite collection of parameterizable open sets, $U_1, \dots, U_N$ such that $f(U_i) \subset V$. Hence since Sard's theorem is true for each of the maps $f : U_i \to V$ and $f^{-1}(B)$ is contained in the union of the U_i's we conclude that *the set of regular values of f intersects the interior of B in an open dense set.* Thus, since q is an arbitrary point of Y, we have proved the following theorem.

Theorem 4.7.13. *If X and Y are n-manifolds and $f : X \to Y$ is a proper C^∞ map the set of regular values of f is an open dense subset of Y.*

Since there exist lots of regular values the formula (4.7.12) gives us an effective way of computing the degree of f. We'll next justify our assertion that $\deg(f)$ is a topological invariant of f. To do so, let's generalize Definition 2.6.14 to manifolds.

Definition 4.7.14. Let X and Y be manifolds and $f_0, f_1 : X \to Y$ be C^∞ maps. A C^∞ map

$$F : X \times [0,1] \to Y \tag{4.7.15}$$

is a ***homotopy*** between f_0 and f_1 if for all $x \in X$ we have $F(x,0) = f_0(x)$ and $F(x,1) = f_1(x)$. Moreover, if f_0 and f_1 are proper maps, the homotopy, F, is a ***proper*** homotopy if F is proper as a C^∞ map, i.e., for every compact set C of Y, $F^{-1}(C)$ is compact.

Let's now prove the manifold analogue of Theorem 3.6.10.

Theorem 4.7.16. *Let X and Y be oriented n-manifolds and assume that Y is connected. If $f_0, f_2 : X \to Y$ are proper maps and the map (4.7.8) is a proper homotopy, then* $\deg(f_1) = \deg(f_2)$.

Proof. Let ω be an n-form in $\Omega^n_c(Y)$ whose integral over Y is equal to 1, and let $C := \mathrm{supp}(\omega)$ be the support of ω. Then if F is a proper homotopy between f_0 and f_1, the set, $F^{-1}(C)$, is compact and its projection on X

$$\{\, x \in X \mid (x,t) \in F^{-1}(C) \text{ for some } t \in [0,1] \,\} \tag{4.7.17}$$

is compact. Let $f_t : X \to Y$ be the map: $f_t(x) = F(x,t)$. By our assumptions on F, f_t is a proper C^∞ map. Moreover, for all t the n-form $f_t^\star \omega$ is a C^∞ function of t and is supported on the fixed compact set (4.7.17). Hence it's clear from the definitions of the integral of a form and f_t that the integral

$$\int_X f_t^\star \omega$$

is a C^∞ function of t. On the other hand this integral is by definition the degree of f_t and hence by Theorem 4.7.3, $\deg(f_t)$ is an integer, so it does not depend on t. In particular, $\deg(f_0) = \deg(f_1)$. □

Exercises for §4.7

Exercise 4.7.i. Let $f : \mathbf{R} \to \mathbf{R}$ be the map $x \mapsto x^n$. Show that $\deg(f) = 0$ if n is even and 1 if n is odd.

Exercise 4.7.ii. Let $f\colon \mathbf{R} \to \mathbf{R}$ be the polynomial function,

$$f(x) = x^n + a_1 x^{n-1} + \cdots + a_{n-1}x + a_n\,,$$

where the a_i's are in $\mathbf{R}$. Show that if n is even, $\deg(f) = 0$ and if n is odd, $\deg(f) = 1$.

Exercise 4.7.iii. Let S^1 be the unit circle in the complex plane and let $f\colon S^1 \to S^1$ be the map $e^{i\theta} \mapsto e^{iN\theta}$, where N is a positive integer. What is the degree of f?

Exercise 4.7.iv. Let S^{n-1} be the unit sphere in $\mathbf{R}^n$ and $\sigma\colon S^{n-1} \to S^{n-1}$ the antipodal map $x \mapsto -x$. What is the degree of σ?

Exercise 4.7.v. Let $A \in O(n)$ be an orthogonal $n \times n$ matrix and let

$$f_A\colon S^{n-1} \to S^{n-1}$$

be the map $x \mapsto Ax$. What is the degree of f_A?

Exercise 4.7.vi. A manifold Y is ***contractible*** if for some point $p_0 \in Y$, the identity map of Y onto itself is homotopic to the constant map $f_{p_0}\colon Y \to Y$ given by $f_{p_0}(y) := p_0$. Show that if Y is an oriented contractible n-manifold and X an oriented connected n-manifold then for every proper mapping $f\colon X \to Y$ we have $\deg(f) = 0$. In particular show that if $n > 0$ and Y is compact then Y cannot be contractible.

Hint: Let f be the identity map of Y onto itself.

Exercise 4.7.vii. Let X and Y be oriented connected n-manifolds and $f\colon X \to Y$ a proper C^∞ map. Show that if $\deg(f) \neq 0$, then f is surjective.

Exercise 4.7.viii. Using Sard's theorem prove that if X and Y are manifolds of dimension k and ℓ, with $k < \ell$ and $f\colon X \to Y$ is a proper C^∞ map, then the complement of the image of X in Y is open and dense.

Hint: Let $r := \ell - k$ and apply Sard's theorem to the map

$$g\colon X \times S^r \to Y,\quad g(x,a) := f(x).$$

Exercise 4.7.ix. Prove that the 2-sphere S^2 and the torus $S^1 \times S^1$ are not diffeomorphic.

4.8. Applications of degree theory

The purpose of this section will be to describe a few typical applications of degree theory to problems in analysis, geometry, and topology. The first of these applications will be yet another variant of the Brouwer fixed point theorem.

Application 4.8.1. Let X be an oriented $(n+1)$-dimensional manifold, $D \subset X$ a smooth domain and $Z := \partial D$ the boundary of D. Assume that the closure $\overline{D} = Z \cup D$ of D is compact (and in particular that X is compact).

Theorem 4.8.2. *Let Y be an oriented connected n-manifold and $f\colon Z \to Y$ a C^∞ map. Suppose there exists a C^∞ map $F\colon \overline{D} \to Y$ whose restriction to Z is f. Then* $\deg(f) = 0$.

Proof. Let μ be an element of $\Omega_c^n(Y)$. Then $d\mu = 0$, so $dF^\star\mu = F^\star d\mu = 0$. On the other hand if $\iota\colon Z \to X$ is the inclusion map,

$$\int_D dF^\star\mu = \int_Z \iota^\star F^\star\mu = \int_Z f^\star\mu = \deg(f)\int_Y \mu$$

by Stokes' theorem since $F \circ \iota = f$. Hence $\deg(f)$ has to be zero. □

Application 4.8.3 (A nonlinear eigenvalue problem). This application is a nonlinear generalization of a standard theorem in linear algebra. Let $A\colon \mathbf{R}^n \to \mathbf{R}^n$ be a linear map. If n is even, A may not have *real* eigenvalues. (For instance for the map

$$A\colon \mathbf{R}^2 \to \mathbf{R}^2, \quad (x,y) \mapsto (-y,x),$$

the eigenvalues of A are $\pm\sqrt{-1}$.) However, if n is odd it is a standard linear algebra fact that there exist a vector, $v \in \mathbf{R}^n \smallsetminus \{0\}$, and a $\lambda \in \mathbf{R}$ such that $Av = \lambda v$. Moreover replacing v by $\frac{v}{|v|}$ one can assume that $|v| = 1$. This result turns out to be a special case of a much more general result. Let S^{n-1} be the unit $(n-1)$-sphere in $\mathbf{R}^n$ and let $f\colon S^{n-1} \to \mathbf{R}^n$ be a C^∞ map.

Theorem 4.8.4. *There exist a vector $v \in S^{n-1}$ and a number $\lambda \in \mathbf{R}$ such that $f(v) = \lambda v$.*

Proof. The proof will be by contradiction. If the theorem isn't true the vectors v and $f(v)$, are linearly independent and hence the vector

(4.8.5) $$g(v) = f(v) - (f(v)\cdot v)v$$

is non-zero. Let

(4.8.6) $$h(v) = \frac{g(v)}{|g(v)|}.$$

By (4.8.5)–(4.8.6), $|v| = |h(v)| = 1$ and $v \cdot h(v) = 0$, i.e., v and $h(v)$ are both unit vectors and are perpendicular to each other. Let

(4.8.7) $$\gamma_t\colon S^{n-1} \to S^{n-1}, \quad 0 \le t \le 1$$

be the map

(4.8.8) $$\gamma_t(v) = \cos(\pi t)v + \sin(\pi t)h(v).$$

For $t = 0$ this map is the identity map and for $t = 1$, it is the antipodal map $\sigma(v) = v$, hence (4.8.7) asserts that the identity map and the antipodal map are homotopic and therefore that the degree of the antipodal map is one. On the other hand the antipodal map is the restriction to S^{n-1} of the map $(x_1,\dots,x_n) \mapsto (-x_1,\dots,-x_n)$ and the volume form ω on S^{n-1} is the restriction to S^{n-1} of the $(n-1)$-form

(4.8.9) $$\sum_{i=1}^{n}(-1)^{i-1}x_i dx_i \wedge \cdots \wedge \widehat{dx_i} \wedge \cdots \wedge dx_n.$$

If we replace x_i by $-x_i$ in equation (4.8.9) the sign of this form changes by $(-1)^n$ hence $\sigma^\star\omega = (-1)^n\omega$. Thus if n is odd, σ is an orientation-reversing diffeomorphism of S^{n-1} onto S^{n-1}, so its degree is -1, and this contradicts what we just deduced from the existence of the homotopy (4.8.8). □

From this argument we can deduce another interesting fact about the sphere S^{n-1} when $n-1$ is even. For $v \in S^{n-1}$ the tangent space to S^{n-1} at v is just the space,

$$T_v S^{n-1} = \{(v, w) \mid w \in \mathbf{R}^n \text{ and } v \cdot w = 0\},$$

so a vector field on S^{n-1} can be viewed as a function $g \colon S^{n-1} \to \mathbf{R}^n$ with the property

$$g(v) \cdot v = 0$$

for all $v \in S^{n-1}$. If this function is non-zero at all points, then, letting h be the function (4.8.6), and arguing as above, we're led to a contradiction. Hence we conclude the following theorem.

Theorem 4.8.10. *If $n-1$ is even and $\boldsymbol{v}$ is a vector field on the sphere S^{n-1}, then there exists a point $p \in S^{n-1}$ at which $\boldsymbol{v}(p) = 0$.*

Note that if $n-1$ is odd this statement is *not* true. The vector field

$$x_1 \frac{\partial}{\partial x_2} - x_2 \frac{\partial}{\partial x_1} + \cdots + x_{2n-1} \frac{\partial}{\partial x_{2n}} - x_{2n} \frac{\partial}{\partial x_{2n-1}}$$

is a counterexample. It is nowhere vanishing and at $p \in S^{n-1}$ is tangent to S^{n-1}.

Application 4.8.11 (The Jordan–Brouwer separation theorem). Let X be a compact oriented $(n-1)$-dimensional submanifold of $\mathbf{R}^n$. In this subsection of § 4.8 we'll outline a proof of the following theorem (leaving the details as a string of exercises).

Theorem 4.8.12. *If X is connected, the complement of $\mathbf{R}^n \smallsetminus X$ of X has exactly two connected components.*

This theorem is known as the Jordan–Brouwer separation theorem (and in two dimensions as the Jordan curve theorem). For simple, easy to visualize, submanifolds of $\mathbf{R}^n$ like the $(n-1)$-sphere this result is obvious, and for this reason it's easy to be misled into thinking of it as being a trivial (and not very interesting) result. However, for submanifolds of $\mathbf{R}^n$ like the curve in $\mathbf{R}^2$ depicted in Figure 4.8.1 it's much less obvious. (In ten seconds or less: is the point p in this figure inside this curve or outside?)

To determine whether a point $p \in \mathbf{R}^n \smallsetminus X$ is inside X or outside X, one needs a topological invariant to detect the difference; such an invariant is provided by the *winding number.*

Definition 4.8.13. For $p \in \mathbf{R}^n \smallsetminus X$ let

$$\gamma_p \colon X \to S^{n-1}$$

be the map

$$\gamma_p(x) = \frac{x-p}{|x-p|}.$$

The ***winding number*** $W(X, p)$ of X about p is the degree

$$W(X, p) := \deg(\gamma_p).$$

We show below that $W(X, p) = 0$ if p is outside, and if p is inside X, then $W(X, p) = \pm 1$ (depending on the orientation of X). Hence the winding number tells us which of the two components of $\mathbf{R}^n \setminus X$ the point p is contained in.

Exercises for §4.8

Exercise 4.8.i. Let U be a connected component of $\mathbf{R}^n \setminus X$. Show that if p_0 and p_1 are in U, $W(X, p_0) = W(X, p_1)$.

Hints:

- First suppose that the line segment,

$$p_t = (1-t)p_0 + tp_1\,, \quad 0 \le t \le 1$$

lies in U. Conclude from the homotopy invariance of degree that

$$W(X, p_0) = W(X, p_t) = W(X, p_1).$$

- Show that there exists a sequence of points $q_1, \ldots, q_N \in U$, with $q_1 = p_0$ and $q_N = p_1$, such that the line segment joining q_i to q_{i+1} is in U.

Exercise 4.8.ii. Let X be a connected $(n-1)$-dimensional submanifold of $\mathbf{R}^n$. Show that $\mathbf{R}^n \setminus X$ has at most two connected components.

Hints:

- Show that if q is in X there exists a small ε-ball $B_\varepsilon(q)$ centered at q such that $B_\varepsilon(q) - X$ has two components. (See Theorem 4.2.15).
- Show that if p is in $\mathbf{R}^n \setminus X$, there exists a sequence $q_1, \ldots, q_N \in \mathbf{R}^n \setminus X$ such that $q_1 = p$, $q_N \in B_\varepsilon(q)$ and the line segments joining q_i to q_{i+1} are in $\mathbf{R}^n \setminus X$.

Exercise 4.8.iii. For $v \in S^{n-1}$, show that $x \in X$ is in $\gamma_p^{-1}(v)$ if and only if x lies on the ray

$$p + tv, \quad 0 < t < \infty. \tag{4.8.14}$$

Exercise 4.8.iv. Let $x \in X$ be a point on this ray. Show that

$$(d\gamma_p)_x \colon T_pX \to T_vS^{n-1} \tag{4.8.15}$$

is bijective if and only if $v \notin T_pX$, i.e., if and only if the ray (4.8.14) is *not* tangent to X at x.

Hint: $\gamma_p \colon X \to S^{n-1}$ is the composition of the maps

$$\tau_p \colon X \to \mathbf{R}^n \setminus \{0\}, \quad x \mapsto x - p,$$

and

$$\pi \colon \mathbf{R}^n \setminus \{0\} \to S^{n-1}, \quad y \mapsto \frac{y}{|y|}.$$

Show that if $\pi(y) = v$, then the kernel of $(d\pi)_g$ is the one-dimensional subspace of $\mathbf{R}^n$ spanned by v. Conclude that if $y = x - p$ and $v = y/|y|$ the composite map

$$(d\gamma_p)_x = (d\pi)_y \circ (d\tau_p)_x$$

is bijective if and only if $v \notin T_x X$.

Exercise 4.8.v. From Exercises 4.8.iii and 4.8.iv deduce that v is a regular value of γ_p if and only if the ray (4.8.14) intersects X in a finite number of points and at each point of intersection is *not* tangent to X at that point.

Exercise 4.8.vi. In Exercise 4.8.v show that the map (4.8.15) is orientation preserving if the orientations of $T_x X$ and v are compatible with the standard orientation of $T_p\mathbf{R}^n$. (See Exercise 1.6.v.)

Exercise 4.8.vii. Conclude that $\deg(\gamma_p)$ counts (with orientations) the number of points where the ray (4.8.14) intersects X.

Exercise 4.8.viii. Let $p_1 \in \mathbf{R}^n \setminus X$ be a point on the ray (4.8.14). Show that if $v \in S^{n-1}$ is a regular value of γ_p, it is a regular value of γ_{p_1} and show that the number

$$\deg(\gamma_p) - \deg(\gamma_{p_1}) = W(X, p) - W(X, p_1)$$

counts (with orientations) the number of points on the ray lying between p and p_1.

Hint: Exercises 4.8.v and 4.8.vii.

Exercise 4.8.ix. Let $x \in X$ be a point on the ray (4.8.14). Suppose $x = p + tv$. Show that if ε is a small positive number and

$$p_\pm = p + (t \pm \varepsilon)v$$

then

$$W(X, p_+) = W(X, p_-) \pm 1,$$

and from Exercise 4.8.i conclude that p_+ and p_- lie in different components of $\mathbf{R}^n \setminus X$. In particular conclude that $\mathbf{R}^n \setminus X$ has exactly two components.

Exercise 4.8.x. Finally show that if p is very large the difference

$$\gamma_p(x) - \frac{p}{|p|}, \quad x \in X,$$

is very small, i.e., γ_p is *not* surjective and hence the degree of γ_p is zero. Conclude that for $p \in \mathbf{R}^n \setminus X$, p is in the unbounded component of $\mathbf{R}^n \setminus X$ if $W(X, p) = 0$ and in the bounded component if $W(X, p) = \pm 1$ (the "$\pm$" depending on the orientation of X).

Remark 4.8.16. The proof of Jordan–Brouwer sketched above gives us an effective way of deciding whether the point p in Figure 4.8.1 is inside X or outside X. Draw a non-tangential ray from p. If it intersects X in an even number of points, p is outside X and if it intersects X is an odd number of points, p is inside.

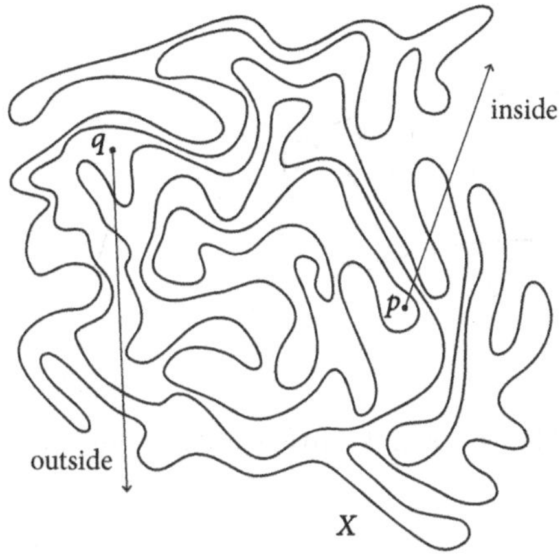

Figure 4.8.1. A Jordan curve with p inside of the curve and q outside.

Application 4.8.17 (The Gauss–Bonnet theorem). Let $X \subset \mathbf{R}^n$ be a compact, connected, oriented $(n-1)$-dimensional submanifold. By the Jordan–Brouwer theorem X is the boundary of a bounded smooth domain, so for each $x \in X$ there exists a unique outward pointing unit normal vector n_x. The Gauss map

$$\gamma \colon X \to S^{n-1}$$

is the map $x \mapsto n_x$. Let σ be the Riemannian volume form of S^{n-1}, or, in other words, the restriction to S^{n-1} of the form,

$$\sum_{i=1}^{n}(-1)^{i-1} x_i dx_1 \wedge \cdots \wedge \widehat{dx_i} \wedge \cdots \wedge dx_n,$$

and let σ_X be the Riemannian volume form of X. Then for each $p \in X$

$$(\gamma^* \sigma)_p = K(p)(\sigma_X)_q, \tag{4.8.18}$$

where $K(p)$ is the ***scalar curvature*** of X at p. This number measures the extent to which "X is curved" at p. For instance, if X_a is the circle, $|x| = a$ in $\mathbf{R}^2$, the Gauss map is the map $p \to p/a$, so for all p we have $K_a(p) = 1/a$, reflecting the fact that for $a < b$, X_a is more curved than X_b.

The scalar curvature can also be negative. For instance for surfaces X in $\mathbf{R}^3$, $K(p)$ is positive at p if X is ***convex*** at p and negative if X is ***convex–concave*** at p. (See Figures 4.8.2 and 4.8.3. The surface in Figure 4.8.2 is convex at p, and the surface in Figure 4.8.3 is convex–concave.)

Let $\mathrm{vol}(S^{n-1})$ be the Riemannian volume of the $(n-1)$-sphere, i.e., let

$$\mathrm{vol}(S^{n-1}) = \frac{2\pi^{n/2}}{\Gamma(n/2)}$$

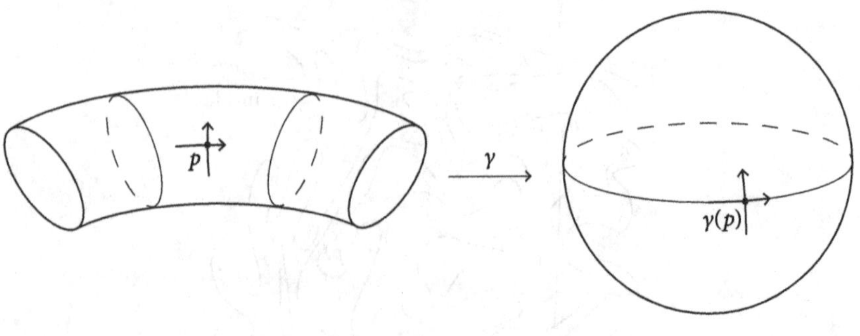

Figure 4.8.2. Positive scalar curvature at p.

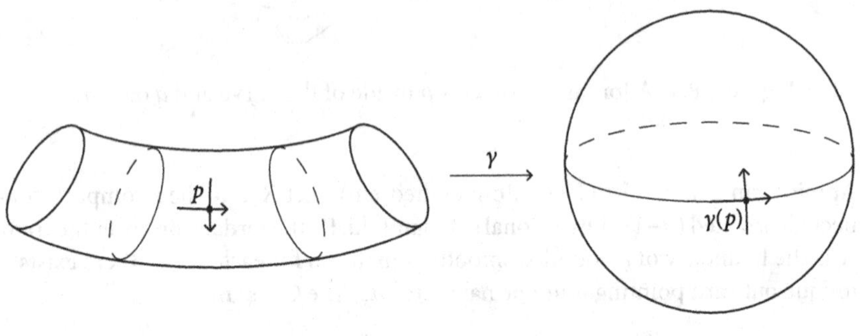

Figure 4.8.3. Negative scalar curvature at p.

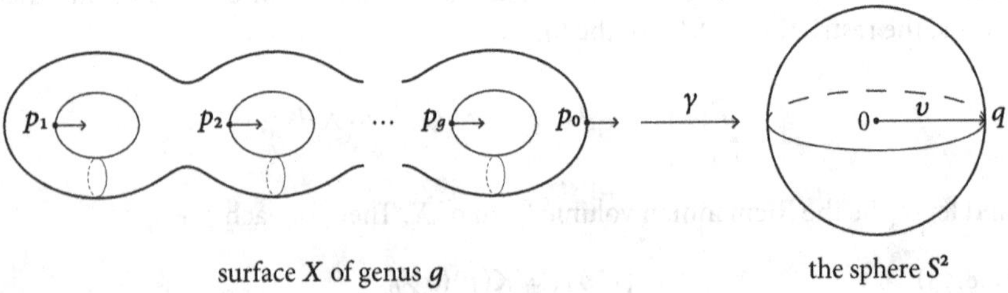

Figure 4.8.4. The Gauss map from a surface of genus g to the sphere.

(where Γ is the Gamma function). Then by (4.8.18) the quotient

$$\frac{\int K\sigma_X}{\mathrm{vol}(S^{n-1})} \tag{4.8.19}$$

is the degree of the Gauss map, and hence is a topological invariant of the surface of X. For $n = 3$ the ***Gauss–Bonnet theorem*** asserts that this topological invariant is just $1 - g$ where g is the ***genus*** of X or, in other words, the "number of holes". Figure 4.8.4 gives a pictorial proof of this result. (Notice that at the points $p_1, \dots, p_g$ the surface X is convex–concave so the scalar curvature at these points is negative, i.e., the Gauss map is orientation reversing. On the other hand, at the point p_0 the surface is convex, so the Gauss map at this point is orientation preserving.)

4.9. The index of a vector field

Let D be a bounded smooth domain in $\mathbf{R}^n$ and $\boldsymbol{v} = \sum_{i=1}^n v_i \partial/\partial x_i$ a C^∞ vector field defined on the closure $\overline{D}$ of D. We will define below a topological invariant which, for "generic" $\boldsymbol{v}$'s, counts (with appropriate $\pm$-signs) the number of zeros of $\boldsymbol{v}$. To avoid complications we'll assume $\boldsymbol{v}$ has *no* zeros on the boundary of D.

Definition 4.9.1. Let ∂D be the boundary of D and let

$$f_{\boldsymbol{v}} : \partial D \to S^{n-1}$$

be the mapping

$$p \mapsto \frac{\boldsymbol{v}(p)}{|\boldsymbol{v}(p)|},$$

where $\boldsymbol{v}(p) = (v_1(p), \dots, v_n(p))$. The ***index*** of $\boldsymbol{v}$ with respect to D is by definition the degree of $f_{\boldsymbol{v}}$.

We'll denote this index by $\operatorname{ind}(\boldsymbol{v}, D)$ and as a first step in investigating its properties we'll prove the following theorem.

Theorem 4.9.2. *Let D_1 be a smooth domain in $\mathbf{R}^n$ whose closure is contained in D. Then*

$$\operatorname{ind}(\boldsymbol{v}, D) = \operatorname{ind}(\boldsymbol{v}, D_1)$$

provided that $\boldsymbol{v}$ has no zeros in the set $D \setminus D_1$.

Proof. Let $W = D \setminus \overline{D}_1$. Then the map $f_{\boldsymbol{v}} : \partial D \to S^{n-1}$ extends to a C^∞ map

$$F : W \to S^{n-1}, \quad p \mapsto \frac{\boldsymbol{v}(p)}{|\boldsymbol{v}(p)|}.$$

Moreover,

$$\partial W = X \cup X_1^-,$$

where X is the boundary of D with its natural boundary orientation and X_1^- is the boundary of D_1 with its boundary orientation reversed. Let ω be an element of $\Omega^{n-1}(S^{n-1})$ whose integral over S^{n-1} is 1. Then if $f = f_{\boldsymbol{v}}$ and $f_1 = (f_1)_{\boldsymbol{v}}$ are the restrictions of F to X and X_1 we get from Stokes' theorem (and the fact that $d\omega = 0$) the identity

$$\begin{aligned} 0 = \int_W F^\star d\omega &= \int_W dF^\star \omega \\ &= \int_X f^\star \omega - \int_{X_1} f_1^\star \omega = \deg(f) - \deg(f_1). \end{aligned}$$

Hence

$$\operatorname{ind}(v, D) = \deg(f) = \deg(f_1) = \operatorname{ind}(v, D). \qquad \square$$

Suppose now that v has a finite number of isolated zeros $p_1, \ldots, p_k$ in D. Let $B_\varepsilon(p_i)$ be the open ball of radius ε with center at p_i. By making ε small enough we can assume that each of these balls is contained in D and that they're mutually disjoint. We will define the ***local index*** $\operatorname{ind}(v, p_i)$ of v at p_i to be the index of v with respect to $B_\varepsilon(p_i)$. By Theorem 4.9.2 these local indices are unchanged if we replace ε by a smaller ε, and by applying this theorem to the domain

$$D_1 = \bigcup_{i=1}^{k} B_{p_i}(\varepsilon)$$

we get, as a corollary of the theorem, the formula

$$\operatorname{ind}(v, D) = \sum_{i=1}^{k} \operatorname{ind}(v, p_i), \tag{4.9.3}$$

which computes the global index of v with respect to D in terms of these local indices.

Let's now see how to compute these local indices. Let's first suppose that the point $p = p_i$ is at the origin of $\mathbf{R}^n$. Then near $p = 0$

$$v = v_{\text{lin}} + v',$$

where

$$v = v_{\text{lin}} = \sum_{1 \le i,j \le n} a_{i,j} x_i \frac{\partial}{\partial x_j}$$

the $a_{i,j}$'s being constants and

$$v_i = \sum_{j=1}^{n} f_{ij} \frac{\partial}{\partial x_j},$$

where the f_{ij}'s vanish to second order near zero, i.e., satisfy

$$|f_{i,j}(x)| \le C|x|^2 \tag{4.9.4}$$

for some constant $C > 0$.

Definition 4.9.5. We'll say that the point $p = 0$ is a ***non-degenerate zero*** of v if the matrix $A = (a_{i,j})$ is non-singular.

This means in particular that

$$\sum_{j=1}^{n} \left| \sum_{i=1}^{n} a_{i,j} x_i \right| \ge C_1 |x| \tag{4.9.6}$$

form some constant $C_1 > 0$. Thus by (4.9.4) and (4.9.6) the vector field

$$v_t = v_{\text{lin}} + t v', \quad 0 \le t \le 1$$

has no zeros in the ball $B_\varepsilon(0)$, other than the point $p = 0$ itself, provided that ε is small enough. Therefore if we let X_ε be the boundary of $B_\varepsilon(0)$ we get a homotopy

$$F\colon X_\varepsilon \times [0,1] \to S^{n-1}, \quad (x,t) \mapsto f_{v_t}(x)$$

between the maps $f_{v_{\text{lin}}}\colon X_\varepsilon \to S^{n-1}$ and $f_v\colon X_\varepsilon \to S^{n-1}$, thus by Theorem 4.7.16 we see that $\deg(f_v) = \deg(f_{v_{\text{lin}}})$. Therefore,

$$\operatorname{ind}(v,p) = \operatorname{ind}(v_{\text{lin}}, p). \tag{4.9.7}$$

We have assumed in the above discussion that $p = 0$, but by introducing at p a translated coordinate system for which p is the origin, these comments are applicable to any zero p of v. More explicitly if $c_1, \dots, c_n$ are the coordinates of p then as above

$$v = v_{\text{lin}} + v',$$

where

$$v_{\text{lin}} := \sum_{1\le i,j\le n} a_{i,j}(x_i - c_i)\frac{\partial}{\partial x_j}$$

and v' vanishes to second order at p, and if p is a non-degenerate zero of q, i.e., if the matrix $A = (a_{i,j})$ is non-singular, then exactly the same argument as before shows that

$$\operatorname{ind}(v,p) = \operatorname{ind}(v_{\text{lin}}, p).$$

We will next prove the following theorem.

Theorem 4.9.8. *If p is a non-degenerate zero of v, the local index* $\operatorname{ind}(v,p)$ *is $+1$ or -1 depending on whether the determinant of the matrix, $A = (a_{i,j})$ is positive or negative.*

Proof. As above we can, without loss of generality, assume that $p = 0$. By (4.9.7) we can assume $v = v_{\text{lin}}$. Let D be the domain defined by

$$\sum_{j=1}^{n}\left(\sum_{i=1}^{n} a_{i,j}x_i\right)^2 < 1, \tag{4.9.9}$$

and let $X := \partial D$ be its boundary. Since the only zero of v_{lin} inside this domain is $p = 0$ we get from (4.9.3) and (4.9.4) that

$$\operatorname{ind}(v,p) = \operatorname{ind}(v_{\text{lin}}, p) = \operatorname{ind}(v_{\text{lin}}, D)\,.$$

Moreover, the map

$$f_{v_{\text{lin}}}\colon X \to S^{n-1}$$

is, in view of (4.9.9), just the linear map $v \mapsto Av$, restricted to X. In particular, since A is a diffeomorphism, this mapping is a diffeomorphism as well, so the degree of this map is $+1$ or -1 depending on whether this map is orientation preserving or not. To decide which of these alternatives is true let $\omega = \sum_{i=1}^{n}(-1)^i x_i dx_1 \wedge \cdots \wedge \widehat{dx_i} \wedge \cdots dx_n$ be the Riemannian volume form on S^{n-1} then

$$\int_X f_{\boldsymbol{v}_{\text{lin}}}^{\star}\omega = \deg(f_{\boldsymbol{v}_{\text{lin}}})\,\text{vol}(S^{n-1}).$$

Since $\boldsymbol{v}$ is the restriction of A to X. By Stokes' theorem this is equal to

$$\begin{aligned}\int_D A^{\star}d\omega &= n\int_D A^{\star}dx_1 \wedge \cdots \wedge dx_n \\ &= n\det(A)\int_D dx_1 \wedge \cdots \wedge dx_n \\ &= n\det(A)\,\text{vol}(D)\end{aligned}$$

which gives us the formula

$$\deg(f_{\boldsymbol{v}_{\text{lin}}}, D) = \frac{n \det A\,\text{vol}(D)}{\text{vol}(S^{n-1})}. \qquad \square$$

We'll briefly describe the various types of non-degenerate zeros that can occur for vector fields in two dimensions. To simplify this description a bit we'll assume that the matrix

$$A = \begin{pmatrix} a_{1,1} & a_{1,2} \\ a_{2,1} & a_{2,2} \end{pmatrix}$$

is diagonalizable and that its eigenvalues are not purely imaginary. Then the following five scenarios can occur:

(1) *The eigenvalues of A are real and positive.* In this case the integral curves of $\boldsymbol{v}_{\text{lin}}$ in a neighborhood of p look like the curves in Figure 4.9.1 and hence, in a small neighborhood of p, the integral sums of $\boldsymbol{v}$ itself look approximately like the curve in Figure 4.9.1.

(2) *The eigenvalues of A are real and negative.* In this case the integral curves of $\boldsymbol{v}_{\text{lin}}$ look like the curves in Figure 4.9.2, but the arrows are pointing into p, rather than out of p, i.e.,

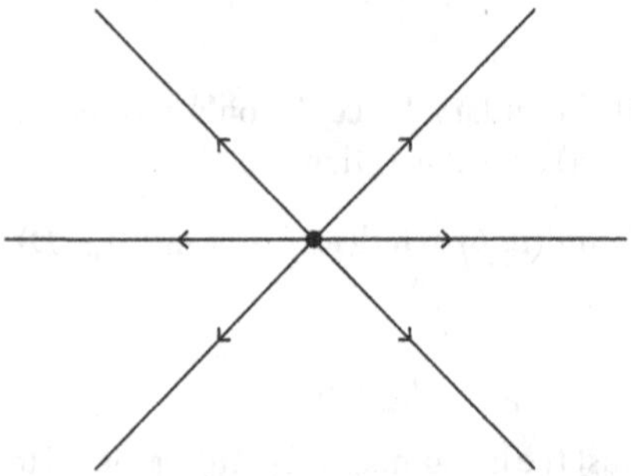

Figure 4.9.1. Real and positive eigenvalues.

(3) *The eigenvalues of A are real, but one is positive and one is negative.* In this case the integral curves of $\boldsymbol{v}_{\text{lin}}$ look like the curves in Figure 4.9.3.

(4) *The eigenvalues of A are complex and of the form $a \pm \sqrt{-b}$, with a positive.* In this case the integral curves of $\boldsymbol{v}_{\text{lin}}$ are spirals going out of p as in Figure 4.9.4.

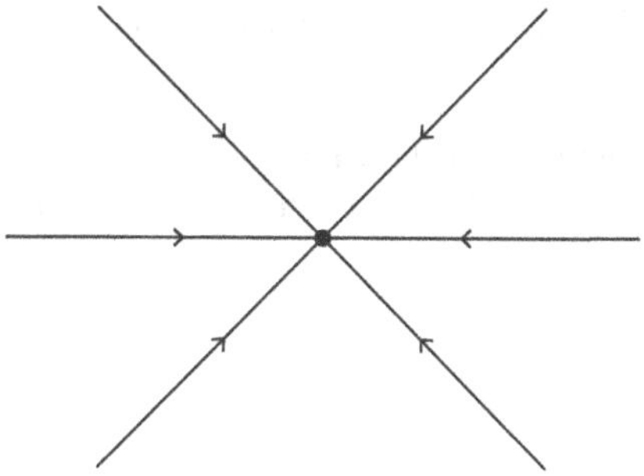

Figure 4.9.2. Real and negative eigenvalues.

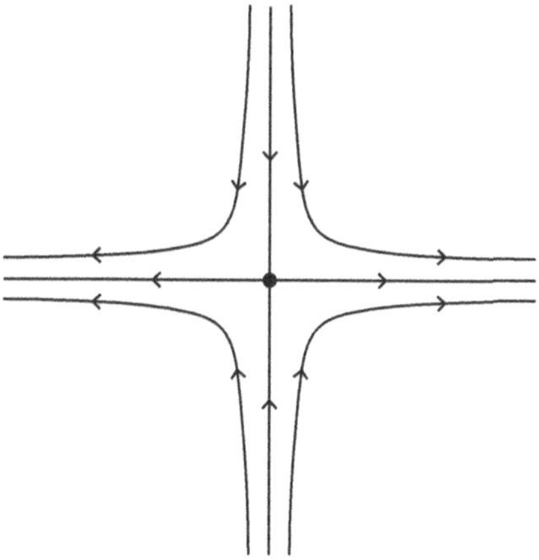

Figure 4.9.3. Real eigenvalues, one positive and one negative.

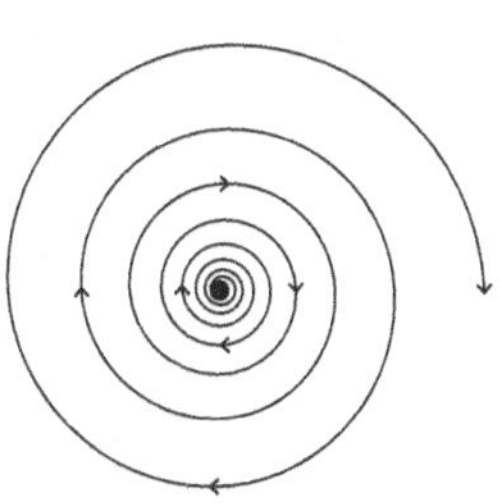

Figure 4.9.4. Complex eigenvalues of the form $a \pm \sqrt{-b}$, with a positive.

(5) *The eigenvalues of A are complex and of the form* $a \pm \sqrt{-b}$, *with a negative.* In this case the integral curves are as in Figure 4.9.4 but are spiraling into p rather than out of p.

Definition 4.9.10. Zeros of $\boldsymbol{v}$ of types (1) and (4) are called ***sources***; those of types (2) and (5) are called ***sinks*** and those of type (3) are called ***saddles***.

Thus in particular the two-dimensional version of equation (4.9.3) is as follows.

Theorem 4.9.11. *If the vector field* $\boldsymbol{v}$ *on* D *has only non-degenerate zeros, then* $\operatorname{ind}(\boldsymbol{v}, D)$ *is equal to the number of sources plus the number of sinks minus the number of saddles.*

Chapter 5

Cohomology via Forms

5.1. The de Rham cohomology groups of a manifold

In the last four chapters we've frequently encountered the question: When is a closed k-form on an open subset of $\mathbf{R}^N$ (or, more generally on a submanifold of $\mathbf{R}^N$) exact? To investigate this question more systematically than we've done heretofore, let X be an n-manifold and let

$$Z^k(X) := \{\omega \in \Omega^k(X) \mid d\omega = 0\}$$

and

$$B^k(X) := d(\Omega^{k-1}(X)) \subset \Omega^k(X)$$

be the vector spaces of closed and exact k-forms. Since $B^k(X)$ is a vector subspace of $Z^k(X)$, we can form the quotient space

$$H^k(X) := Z^k(X)/B^k(X),$$

and the dimension of this space is a measure of the extent to which closed forms fail to be exact. We will call this space the ***kth de Rham cohomology group of the manifold*** X. Since the vector spaces $Z^k(X)$ and $B^k(X)$ are both infinite dimensional in general, there is no guarantee that this quotient space is finite dimensional, however, we'll show later in this chapter that it is in a lot of interesting cases.

The spaces $H^k(X)$ also have compactly supported counterparts. Namely let

$$Z_c^k(X) := \{\omega \in \Omega_c^k(X) \mid d\omega = 0\}$$

and

$$B_c^k(X) := d(\Omega_c^{k-1}(X)) \subset \Omega_c^k(X).$$

Then as above $B_c^k(X)$ is a vector subspace of $Z_c^k(X)$ and the vector space quotient

$$H_c^k(X) := Z_c^k(X)/B_c^k(X)$$

is the ***kth compactly supported de Rham cohomology group of*** X.

Given a closed k-form $\omega \in Z^k(X)$ we will denote by $[\omega]$ the image of ω in the quotient space $H^k(X) = Z^k(X)/B^k(X)$ and call $[\omega]$ the ***cohomology class*** of ω. We will also use the same notation for compactly supported cohomology. If ω is in

$Z_c^k(X)$ we'll denote by $[\omega]$ the cohomology class of ω in the quotient space $H_c^k(X) = Z_c^k(X)/B_c^k(X)$.

We've already computed some cohomology groups of manifolds in the previous chapters (although we didn't explicitly describe these computations as "computing cohomology"). We'll make a list below of some of the properties that we have already discovered about the de Rham cohomology of a manifold X:

(1) If X is connected, $H^0(X) = \mathbf{R}$. To see this, note that a closed 0-form is a function $f \in C^\infty(X)$ having the property $df = 0$, and if X is connected the only such functions are constants.

(2) If X is connected and non-compact $H_c^0(X) = 0$. To see this, note that if f is in $C_0^\infty(X)$ and X is non-compact, f has to be zero at some point, and hence if $df = 0$, then f has to be identically zero.

(3) If X is n-dimensional,

$$\Omega^k(X) = \Omega_c^k(X) = 0$$

for $k < 0$ or $k > n$, hence

$$H^k(X) = H_c^k(X) = 0$$

for $k < 0$ or $k > n$.

(4) If X is an oriented, connected n-manifold, the integration operation is a linear map

$$\int_X : \Omega_c^n(X) \to \mathbf{R} \tag{5.1.1}$$

and, by Theorem 4.8.2, the kernel of this map is $B_c^n(X)$. Moreover, in degree n, $Z_c^n(X) = \Omega_c^n(X)$ and hence by the definition of $H_c^k(X)$ as the quotient $Z_c^k(X)/B_c^k(X)$, we get from (5.1.1) a bijective map

$$I_X : H_c^n(X) \xrightarrow{\simeq} \mathbf{R}. \tag{5.1.2}$$

In other words $H_c^n(X) = \mathbf{R}$.

(5) Let U be a star-shaped open subset of $\mathbf{R}^n$. In Exercises 2.6 iv–2.6 vii we sketched a proof of the assertion: For $k > 0$ every closed form $\omega \in Z^k(U)$ is exact. Translating this assertion into cohomology language, we showed that $H^k(U) = 0$ for $k > 0$.

(6) Let $U \subset \mathbf{R}^n$ be an open rectangle. In Exercises 3.2 iv–3.2 vii we sketched a proof of the assertion: If $\omega \in \Omega_c^k(U)$ is closed and $0 < k < n$, then $\omega = d\mu$ for some $(k-1)$-form $\mu \in \Omega_c^{k-1}(U)$. Hence we showed $H_c^k(U) = 0$ for $k < n$.

(7) *Poincaré lemma for manifolds:* Let X be an n-manifold and $\omega \in Z^k(X)$, $k > 0$, a closed k-form. Then for every point $p \in X$ there exist a neighborhood U of p and a $(k-1)$-form $\mu \in \Omega^{k-1}(U)$ such that $\omega = d\mu$ on U. To see this note that for open subsets of $\mathbf{R}^n$ we proved this result in §2.4 and since X is locally diffeomorphic at p to an open subset of $\mathbf{R}^n$ this result is true for manifolds as well.

(8) Let X be the unit sphere S^n in $\mathbf{R}^{n+1}$. Since S^n is compact, connected and oriented $H^0(S^n) = H^n(S^n) = \mathbf{R}$.

We will show that for $k \neq, 0, n$ we have

$$H^k(S^n) = 0.$$

To see this let $\omega \in \Omega^k(S^n)$ be a closed k-form and let $p = (0, \ldots, 0, 1) \in S^n$ be the "north pole" of S^n. By the Poincaré lemma there exist a neighborhood U of p in S^n and a $(k-1)$-form $\mu \in \Omega^{k-1}(U)$ with $\omega = d\mu$ on U. Let $\rho \in C_0^\infty(U)$ be a "bump function" which is equal to 1 on a neighborhood U_0 of U in p. Then

$$\omega_1 = \omega - d\rho\mu \tag{5.1.3}$$

is a closed k-form with compact support in $S^n \setminus \{p\}$. However, stereographic projection gives one a diffeomorphism

$$\phi\colon \mathbf{R}^n \to S^n \setminus \{p\}$$

(see Exercise 5.1.i), and hence $\phi^*\omega_1$ is a closed compactly supported k-form on $\mathbf{R}^n$ with support in a large rectangle. Thus by (5.1.3) $\phi^*\omega = d\nu$, for some $\nu \in \Omega_c^{k-1}(\mathbf{R}^n)$, and by (5.1.3)

$$\omega = d(\rho\mu + (\phi^{-1})^*\nu) \tag{5.1.4}$$

with $(\phi^{-1})^*\nu \in \Omega_c^{k-1}(S^n \setminus \{p\}) \subset \Omega^k(S^n)$, so we've proved that for $0 < k < n$ every closed k-form on S^n is exact.

We will next discuss some "pullback" operations in de Rham theory. Let X and Y be manifolds and $f: X \to Y$ a C^∞ map. For $\omega \in \Omega^k(Y)$, $df^*\omega = f^*d\omega$, so if ω is closed, $f^*\omega$ is as well. Moreover, if $\omega = d\mu$, $f^*\omega = df^*\mu$, so if ω is exact, $f^*\omega$ is as well. Thus we have linear maps

$$f^*: Z^k(Y) \to Z^k(X)$$

and

$$f^*: B^k(Y) \to B^k(X) \tag{5.1.5}$$

and composing $f^*: Z^k(Y) \to Z^k(X)$ with the projection

$$\pi\colon Z^k(X) \to Z^k(X)/B^k(X)$$

we get a linear map

$$Z^k(Y) \to H^k(X). \tag{5.1.6}$$

In view of (5.1.5), $B^k(Y)$ is in the kernel of this map, so by Proposition 1.2.9 one gets an induced linear map

$$f^\sharp: H^k(Y) \to H^k(Y),$$

such that $f^\sharp \circ \pi$ is the map (5.1.6). In other words, if ω is a closed k-form on Y, then $f^\sharp$ has the defining property

$$f^\sharp[\omega] = [f^*\omega]. \tag{5.1.7}$$

This "pullback" operation on cohomology satisfies the following chain rule: Let Z be a manifold and $g: Y \to Z$ a C^∞ map. Then if ω is a closed k-form on Z

$$(g \circ f)^* \omega = f^* g^* \omega$$

by the chain rule for pullbacks of forms, and hence by (5.1.7)

$$(g \circ f)^\sharp[\omega] = f^\sharp(g^\sharp[\omega]). \tag{5.1.8}$$

The discussion above carries over verbatim to the setting of compactly supported de Rham cohomology: If $f: X \to Y$ is a proper C^∞ map f induces a pullback map on cohomology

$$f^\sharp: H_c^k(Y) \to H_c^k(X)$$

and if $f: X \to Y$ and $g: Y \to Z$ are proper C^∞ maps then the chain rule (5.1.8) holds for compactly supported de Rham cohomology as well as for ordinary de Rham cohomology. Notice also that if $f: X \to Y$ is a diffeomorphism, we can take Z to be X itself and g to be f^{-1}, and in this case the chain rule tells us that the induced maps $f^\sharp: H^k(Y) \to H^k(Y)$ and $f^\sharp: H_c^k(Y) \to H_c^k(X)$ are bijections, i.e., $H^k(X)$ and $H^k(Y)$ and $H_c^k(X)$ and $H_c^k(Y)$ are isomorphic as vector spaces.

We will next establish an important fact about the pullback operation $f^\sharp$; we'll show that it's a ***homotopy invariant*** of f. Recall that two C^∞ maps

$$f_0, f_1: X \to Y \tag{5.1.9}$$

are *homotopic* if there exists a C^∞ map

$$F: X \times [0,1] \to Y$$

with the properties $F(p,0) = f_0(p)$ and $F(p,1) = f_1(p)$ for all $p \in X$. We will prove the following theorem.

Theorem 5.1.10. *If the maps (5.1.9) are homotopic, then, for the maps they induce on cohomology*

$$f_0^\sharp = f_1^\sharp. \tag{5.1.11}$$

Our proof of this will consist of proving this for an important special class of homotopies, and then by "pullback" tricks deducing this result for homotopies in general. Let $\boldsymbol{v}$ be a complete vector field on X and let

$$f_t: X \to X, \quad -\infty < t < \infty$$

be the one-parameter group of diffeomorphisms that $\boldsymbol{v}$ generates. Then

$$F: X \times [0,1] \to X, \qquad F(p,t) := f_t(p),$$

is a homotopy between f_0 and f_1, and we'll show that for this homotopic pair (5.1.11) is true.

Proof of a special case of Theorem 5.1.10. Recall that for $\omega \in \Omega^k(X)$

$$\frac{d}{dt} f_t^* \omega \bigg|_{t=0} = L_{\boldsymbol{v}} = \iota_{\boldsymbol{v}} d\omega + d\iota_{\boldsymbol{v}}\omega$$

and more generally for all t we have

$$\begin{aligned}\frac{d}{dt}f_t^\star\omega &= \frac{d}{ds}f_{s+t}^\star\omega\Big|_{s=0} = \frac{d}{ds}(f_s\circ f_t)^\star\omega\Big|_{s=0}\\ &= \frac{d}{ds}f_t^\star f_s^\star\omega\Big|_{s=0} = f_t^\star\frac{d}{ds}f_s^\star\omega\Big|_{s=0}\\ &= f_t^\star L_v\omega\\ &= f_t^\star\iota_v d\omega + df_t^\star\iota_v\omega.\end{aligned}$$

Thus if we set

$$Q_t\omega := f_t^\star\iota_v\omega$$

we get from this computation:

$$\frac{d}{dt}f_t^\star\omega = dQ_t\omega + Q_t d\omega$$

and integrating over $0\le t\le 1$:

$$f_1^\star\omega - f_0^\star\omega = dQ\omega + Qd\omega, \tag{5.1.12}$$

where $Q\colon \Omega^k(Y)\to\Omega^{k-1}(X)$ is the operator

$$Q\omega = \int_0^1 Q_t\omega dt.$$

The identity (5.1.11) is an easy consequence of this "chain homotopy" identity. If ω is in $Z^k(X)$, then $d\omega=0$, so

$$f_1^\star\omega - f_0^\star\omega = dQ\omega$$

and

$$f_1^\sharp[\omega] - f_0^\sharp[\omega] = [f_1^\star\omega - f_0^\star\omega] = 0. \qquad \square$$

We'll now describe how to extract from this result a proof of Theorem 5.1.10 for *any* pair of homotopic maps. We'll begin with the following useful observation.

Proposition 5.1.13. *If $f_0, f_1\colon X\to Y$ are homotopic C^∞ mappings, then there exists a C^∞ map*

$$F\colon X\times\mathbf{R}\to Y$$

such that the restriction of F to $X\times[0,1]$ is a homotopy between f_0 and f_1.

Proof. Let $\rho\in C_0^\infty(\mathbf{R})$, $\rho\ge 0$, be a bump function which is supported on the interval $[\frac14,\frac34]$ and is positive at $t=\frac12$. Then

$$\chi(t) = \frac{\int_{-\infty}^t \rho(s)ds}{\int_{-\infty}^\infty \rho(s)ds}$$

is a function which is zero on the interval $(-\infty, \frac{1}{4}]$, is 1 on the interval $[\frac{3}{4}, \infty)$, and, for all t, lies between 0 and 1. Now let

$$G\colon X \times [0,1] \to Y$$

be a homotopy between f_0 and f_1 and let $F\colon X \times \mathbf{R} \to Y$ be the map

$$F(x,t) = G(x, \chi(t)). \tag{5.1.14}$$

This is a C^∞ map and since

$$F(x,1) = G(x,\chi(1)) = G(x,1) = f_1(x),$$

and

$$F(x,0) = G(x,\chi(0)) = G(x,0) = f_0(x),$$

it gives one a homotopy between f_0 and f_1. □

We're now in position to deduce Theorem 5.1.10 from the version of this result that we proved above.

Proof of Theorem 5.1.10. Let

$$\gamma_t\colon X \times \mathbf{R} \to X \times \mathbf{R}, \quad -\infty < t < \infty$$

be the one-parameter group of diffeomorphisms

$$\gamma_t(x,a) = (x, a+t)$$

and let $v = \partial/\partial t$ be the vector field generating this group. For a k-form $\mu \in \Omega^k(X \times \mathbf{R})$, we have by (5.1.12) the identity

$$\gamma_1^\star \mu - \gamma_0^\star \mu = d\Gamma\mu + \Gamma d\mu,$$

where

$$\Gamma\mu = \int_0^1 \gamma_t^\star(\iota_{\partial/\partial t}\mu)dt. \tag{5.1.15}$$

Now let F, as in Proposition 5.1.13, be a C^∞ map

$$F\colon X \times \mathbf{R} \to Y$$

whose restriction to $X \times [0,1]$ is a homotopy between f_0 and f_1. Then for $\omega \in \Omega^k(Y)$

$$\gamma_1^\star F^\star\omega - \gamma_0^\star F^\star\omega = d\Gamma F^\star\mu + \Gamma F^\star d\mu$$

by the identity (5.1.14). Now let $\iota\colon X \to X \times \mathbf{R}$ be the inclusion $p \mapsto (p,0)$, and note that

$$(F \circ \gamma_1 \circ \iota)(p) = F(p,1) = f_1(p)$$

and

$$(F \circ \gamma_0 \circ \iota)(p) = F(p,0) = f_0(p),$$

i.e.,

$$F \circ \gamma_1 \circ \iota = f_1$$

and

$$F \circ \gamma_0 \circ \iota = f_0.$$

Thus

$$\iota^\star(\gamma_1^\star F^\star \omega - \gamma_0^\star F^\star \omega) = f_1^\star \omega - f_0^\star \omega$$

and on the other hand by (5.1.15)

$$\iota^\star(\gamma_1^\star F^\star \omega - \gamma_0^\star F^\star \omega) = d\iota^\star \Gamma F^\star \omega + \iota^\star \Gamma F^\star d\omega.$$

Letting

$$Q\colon \Omega^k(Y) \to \Omega^{k-1}(X)$$

be the "chain homotopy" operator

(5.1.16) $$Q\omega := \iota^\star \Gamma F^\star \omega$$

we can write the identity above more succinctly in the form

(5.1.17) $$f_1^\star \omega - f_0^\star \omega = dQ\omega + Qd\omega$$

and from this deduce, exactly as we did earlier, the identity (5.1.11). □

This proof can easily be adapted to the compactly supported setting. Namely the operator (5.1.16) is defined by the integral

(5.1.18) $$Q\omega = \int_0^1 \iota^\star \gamma_t^\star (\iota_{\partial/\partial t} F^\star \omega) dt.$$

Hence if ω is supported on a set A in Y, the integrand of (5.1.17) at t is supported on the set

(5.1.19) $$\{\, p \in X \mid F(p,t) \in A \,\},$$

and hence $Q\omega$ is supported on the set

$$\pi(F^{-1}(A) \cap X \times [0,1]),$$

where $\pi\colon X \times [0,1] \to X$ is the projection map $\pi(p,t) = p$.

Suppose now that f_0 and f_1 are proper mappings and

$$G\colon X \times [0,1] \to Y$$

a *proper* homotopy between f_0 and f_1, i.e., a homotopy between f_0 and f_1 which is proper as a C^∞ map. Then if F is the map (5.1.14) its restriction to $X \times [0,1]$ is also a proper map, so this restriction is also a proper homotopy between f_0 and f_1. Hence if ω is in $\Omega_c^k(Y)$ and A is its support, the set (5.1.19) is compact, so $Q\omega$ is in $\Omega_c^{k-1}(X)$. Therefore all summands in the "chain homotopy" formula (5.1.17) are compactly supported. Thus we've proved the following.

Theorem 5.1.20. *If $f_0, f_1 : X \to Y$ are proper C^∞ maps which are homotopic via a proper homotopy, the induced maps on cohomology*

$$f_0^\sharp, f_1^\sharp : H_c^k(Y) \to H_c^k(X)$$

are the same.

We'll conclude this section by noting that the cohomology groups $H^k(X)$ are equipped with a natural product operation. Namely, suppose $\omega_i \in \Omega^{k_i}(X)$, $i = 1, 2$, is a closed form and that $c_i = [\omega_i]$ is the cohomology class represented by ω_i. We can then define a product cohomology class $c_1 \cdot c_2$ in $H^{k_1+k_2}(X)$ by the recipe

$$c_1 \cdot c_2 := [\omega_1 \wedge \omega_2]. \tag{5.1.21}$$

To show that this is a legitimate definition we first note that since ω_2 is closed

$$d(\omega_1 \wedge \omega_2) = d\omega_1 \wedge \omega_2 + (-1)^{k_1}\omega_1 \wedge d\omega_2 = 0,$$

so $\omega_1 \wedge \omega_2$ is closed and hence does represent a cohomology class. Moreover if we replace ω_1 by another representative $\omega_1 + d\mu_1 = \omega'$ of the cohomology class c_1, then

$$\omega_1' \wedge \omega_2 = \omega_1 \wedge \omega_2 + d\mu_1 \wedge \omega_2.$$

But since ω_2 is closed,

$$\begin{aligned} d\mu_1 \wedge \omega_2 &= d(\mu_1 \wedge \omega_2) + (-1)^{k_1}\mu_1 \wedge d\omega_2 \\ &= d(\mu_1 \wedge \omega_2) \end{aligned}$$

so

$$\omega_1' \wedge \omega_2 = \omega_1 \wedge \omega_2 + d(\mu_1 \wedge \omega_2)$$

and $[\omega_1' \wedge \omega_2] = [\omega_1 \wedge \omega_2]$. Similarly (5.1.21) is unchanged if we replace ω_2 by $\omega_2 + d\mu_2$, so the definition of (5.1.21) depends neither on the choice of ω_1 nor ω_2 and hence is an intrinsic definition as claimed.

There is a variant of this product operation for compactly supported cohomology classes, and we'll leave for you to check that it's also well defined. Suppose c_1 is in $H_c^{k_1}(X)$ and c_2 is in $H^{k_2}(X)$ (i.e., c_1 is a compactly supported class and c_2 is an ordinary cohomology class). Let ω_1 be a representative of c_1 in $\Omega_c^{k_1}(X)$ and ω_2 a representative of c_2 in $\Omega^{k_2}(X)$. Then $\omega_1 \wedge \omega_2$ is a closed form in $\Omega_c^{k_1+k_2}(X)$ and hence defines a cohomology class

$$c_1 \cdot c_2 := [\omega_1 \wedge \omega_2] \tag{5.1.22}$$

in $H_c^{k_1+k_2}(X)$. We'll leave for you to check that this is intrinsically defined. We'll also leave for you to check that (5.1.22) is intrinsically defined if the roles of c_1 and c_2 are reversed, i.e., if c_1 is in $H^{k_1}(X)$ and c_2 in $H_c^{k_2}(X)$ and that the products (5.1.21) and (5.1.22) both satisfy

$$c_1 \cdot c_2 = (-1)^{k_1 k_2} c_2 \cdot c_1.$$

Finally we note that if Y is another manifold and $f : X \to Y$ a C^∞ map then for $\omega_1 \in \Omega^{k_1}(Y)$ and $\omega_2 \in \Omega^{k_2}(Y)$

$$f^*(\omega_1 \wedge \omega_2) = f^*\omega_1 \wedge f^*\omega_2$$

by (2.6.8) and hence if ω_1 and ω_2 are closed and $c_i = [\omega_i]$, then

(5.1.23) $$f^\sharp(c_1 \cdot c_2) = f^\sharp c_1 \cdot f^\sharp c_2.$$

Exercises for §5.1

Exercise 5.1.i (stereographic projection). Let $p \in S^n$ be the point $p = (0, \dots, 0, 1)$. Show that for every point $x = (x_1, \dots, x_{n+1})$ of $S^n \setminus \{p\}$ the ray

$$tx + (1-t)p\,, \ t > 0$$

intersects the plane $x_{n+1} = 0$ in the point

$$\gamma(x) = \frac{1}{1 - x_{n+1}}(x_1, \dots, x_n)$$

and that the map $\gamma\colon S^n \setminus \{p\} \to \mathbf{R}^n$ is a diffeomorphism.

Exercise 5.1.ii. Show that the operator

$$Q_t : \Omega^k(Y) \to \Omega^{k-1}(X)$$

in the integrand of equation (5.1.18), i.e., the operator

$$Q_t\omega = \iota^\star \gamma_t^\star (\iota_{\partial/\partial t}) F^\star \omega$$

has the following description. Let p be a point of X and let $q = f_t(p)$. The curve $s \mapsto f_s(p)$ passes through q at time $s = t$. Let $v(q) \in T_qY$ be the tangent vector to this curve at t. Show that

(5.1.24) $$(Q_t\omega)(p) = (df_t^\star)_p \iota_{v(q)} \omega_q.$$

Exercise 5.1.iii. Let U be a star-shaped open subset of $\mathbf{R}^n$, i.e., a subset of $\mathbf{R}^n$ with the property that for every $p \in U$ the ray $\{\, tp \,|\, 0 \le t \le 1 \,\}$ is in U.

(1) Let $\boldsymbol{v}$ be the vector field

$$\boldsymbol{v} = \sum_{i=1}^n x_i \frac{\partial}{\partial x_i}$$

and $\gamma_t : U \to U$ the map $p \mapsto tp$. Show that for every k-form $\omega \in \Omega^k(U)$

$$\omega = dQ\omega + Qd\omega,$$

where

$$Q\omega = \int_0^1 \gamma_t^\star \iota_{\boldsymbol{v}} \omega \frac{dt}{t}.$$

(2) Show that if $\omega = \sum_I a_I(x) dx_I$ then

(5.1.25) $$Q\omega = \sum_{I,r} \left(\int_0^1 t^{k-1} (-1)^{r-1} x_{i_r} a_I(tx) dt \right) dx_{I_r},$$

where $dx_{I_r} \coloneqq dx_{i_1} \wedge \cdots \wedge \widehat{dx_{i_r}} \wedge \cdots \wedge dx_{i_k}$

Exercise 5.1.iv. Let X and Y be oriented connected n-manifolds, and $f\colon X \to Y$ a proper map. Show that the linear map L defined by the commutative square

$$\begin{array}{ccc} H^n_c(Y) & \xrightarrow{f^\sharp} & H^n_c(X) \\ {\scriptstyle I_Y}\downarrow\wr & & \wr\downarrow{\scriptstyle I_X} \\ \mathbf{R} & \xrightarrow[L]{} & \mathbf{R} \end{array}$$

is multiplication by $\deg(f)$.

Exercise 5.1.v. A ***homotopy equivalence*** between X and Y is a pair of maps $f\colon X \to Y$ and $g\colon Y \to X$ such that $g \circ f \simeq \mathrm{id}_X$ and $f \circ g \simeq \mathrm{id}_Y$. Show that if X and Y are homotopy equivalent their cohomology groups are the same "up to isomorphism", i.e., f and g induce inverse isomorphisms $f^\sharp\colon H^k(Y) \to H^k(X)$ and $g^\sharp\colon H^k(X) \to H^k(Y)$.

Exercise 5.1.vi. Show that $\mathbf{R}^n \smallsetminus \{0\}$ and S^{n-1} are homotopy equivalent.

Exercise 5.1.vii. What are the cohomology groups of the n-sphere with two points deleted?

Hint: The n-sphere with one point deleted is homeomorphic to $\mathbf{R}^n$.

Exercise 5.1.viii. Let X and Y be manifolds and $f_0, f_1, f_2\colon X \to Y$ three C^∞ maps. Show that if f_0 and f_1 are homotopic and f_1 and f_2 are homotopic then f_0 and f_2 are homotopic.

Hint: The homotopy (5.1.7) has the property that

$$F(p,t) = f_t(p) = f_0(p)$$

for $0 \le t \le \frac{1}{4}$ and

$$F(p,t) = f_t(p) = f_1(p)$$

for $\frac{3}{4} \le t < 1$. Show that two homotopies with these properties: a homotopy between f_0 and f_1 and a homotopy between f_1 and f_2, are easy to "glue together" to get a homotopy between f_0 and f_2.

Exercise 5.1.ix.

(1) Let X be an n-manifold. Given points $p_0, p_1, p_2 \in X$, show that if p_0 can be joined to p_1 by a C^∞ curve $\gamma_0\colon [0,1] \to X$, and p_1 can be joined to p_2 by a C^∞ curve $\gamma_1\colon [0,1] \to X$, then p_0 can be joined to p_2 by a C^∞ curve, $\gamma\colon [0,1] \to X$.

Hint: A C^∞ curve, $\gamma\colon [0,1] \to X$, joining p_0 to p_2 can be thought of as a homotopy between the maps

$$\gamma_{p_0}\colon * \to X\,, \quad * \mapsto p_0$$

and

$$\gamma_{p_1} : * \to X\,, \quad * \mapsto p_1$$

where $*$ is the zero-dimensional manifold consisting of a single point.

(2) Show that if a manifold X is connected it is arc-wise connected: any two points can by joined by a C^∞ curve.

Exercise 5.1.x. Let X be a connected n-manifold and $\omega \in \Omega^1(X)$ a closed 1-form.

(1) Show that if $\gamma\colon [0,1] \to X$ is a C^∞ curve there exist a partition:

$$0 = a_0 < a_1 < \cdots < a_n = 1$$

of the interval $[0,1]$ and open sets $U_1, \dots, U_n$ in X such that $\gamma([a_{i-1}, a_i]) \subset U_i$ and $\omega|_{U_i}$ is exact.

(2) In part (1) show that there exist functions $f_i \in C^\infty(U_i)$ such that $\omega|_{U_i} = df_i$ and $f_i(\gamma(a_i)) = f_{i+1}(\gamma(a_i))$.

(3) Show that if p_0 and p_1 are the endpoints of γ

$$f_n(p_1) - f_1(p_0) = \int_0^1 \gamma^\star \omega.$$

(4) Let

$$\gamma_s\colon [0,1] \to X\,, \quad 0 \le s \le 1 \tag{5.1.26}$$

be a homotopic family of curves with $\gamma_s(0) = p_0$ and $\gamma_s(1) = p_1$. Prove that the integral

$$\int_0^1 \gamma_s^\star \omega$$

is independent of s_0.

Hint: Let s_0 be a point on the interval, $[0,1]$. For we have $\gamma = \gamma_{s_0}$ choose a_i's and f_i's as in parts (1)–(2) and show that for s close to s_0, we have $\gamma_s[a_{i-1}, a_i] \subset U_i$.

(5) A connected manifold X is ***simply connected*** if for any two curves $\gamma_0, \gamma_1 : [0,1] \to X$ with the same endpoints p_0 and p_1, there exists a homotopy (5.1.22) with $\gamma_s(0) = p_0$ and $\gamma_s(1) = p_1$, i.e., γ_0 can be smoothly deformed into γ_1 by a family of curves all having the same endpoints. Prove the following theorem.

Theorem 5.1.27. *If X is simply-connected, then $H^1(X) = 0$.*

Exercise 5.1.xi. Show that the product operation (5.1.21) is associative and satisfies left and right distributive laws.

Exercise 5.1.xii. Let X be a compact oriented $2n$-dimensional manifold. Show that the map

$$B\colon H^n(X) \times H^n(X) \to \mathbf{R}$$

defined by

$$B(c_1, c_2) = I_X(c_1 \cdot c_2)$$

is a bilinear form on $H^n(X)$ and that B is symmetric if n is even and alternating if n is odd.

5.2. The Mayer–Vietoris sequence

In this section we'll develop some techniques for computing cohomology groups of manifolds. (These techniques are known collectively as "diagram chasing" and the mastering of these techniques is more akin to becoming proficient in checkers or chess or the Sunday acrostics in the *New York Times* than in the areas of mathematics to which they're applied.) Let $C^0, C^1, C^2, \dots$ be vector spaces and $d : C^i \to C^{i+1}$ a linear map. The sequence of vector spaces and maps

$$C^0 \xrightarrow{d} C^1 \xrightarrow{d} C^2 \xrightarrow{d} \cdots \tag{5.2.1}$$

is called a ***cochain complex***, or simply a ***complex***, if $d^2 = 0$, i.e., if for $a \in C^k$, $d(da) = 0$. For instance if X is a manifold the de Rham complex

$$\Omega^0(X) \xrightarrow{d} \Omega^1(X) \xrightarrow{d} \Omega^2(X) \xrightarrow{d} \cdots \tag{5.2.2}$$

is an example of a complex, and the complex of compactly supported de Rham forms

$$\Omega^0_c(X) \xrightarrow{d} \Omega^1_c(X) \xrightarrow{d} \Omega^2_c(X) \xrightarrow{d} \cdots \tag{5.2.3}$$

is another example. One defines the ***cohomology groups*** of the complex (5.2.1) in exactly the same way that we defined the cohomology groups of the complexes (5.2.2) and (5.2.3) in §5.1. Let

$$Z^k := \ker(d : C^k \to C^{k+1}) = \{a \in C^k \mid da = 0\}$$

and

$$B^k := dC^{k-1}$$

i.e., let a be in B^k if and only if $a = db$ for some $b \in C^{k-1}$. Then $da = d^2 b = 0$, so B^k is a vector subspace of Z^k, and we define $H^k(C)$ — the ***kth cohomology group*** of the complex (5.2.1) — to be the quotient space

$$H^k(C) := Z^k / B^k. \tag{5.2.4}$$

Given $c \in Z^k$ we will, as in §5.1, denote its image in $H^k(C)$ by $[c]$ and we'll call c a ***representative*** of the cohomology class $[c]$.

We will next assemble a small dictionary of "diagram chasing" terms.

Definition 5.2.5. Let $V_0, V_1, V_2, \dots$ be vector spaces and $\alpha_i : V_i \to V_{i+1}$ linear maps. The sequence

$$V_0 \xrightarrow{\alpha_0} V_1 \xrightarrow{\alpha_1} V_2 \xrightarrow{\alpha_2} \cdots.$$

is an ***exact sequence*** if, for each i, the kernel of α_{i+1} is equal to the image of α_i.

For example the sequence (5.2.1) is exact if $Z_i = B_i$ for all i, or, in other words, if $H^i(C) = 0$ for all i. A simple example of an exact sequence that we'll encounter below is a sequence of the form

$$0 \longrightarrow V_1 \xrightarrow{\alpha_1} V_2 \xrightarrow{\alpha_2} V_3 \longrightarrow 0,$$

i.e., a five-term exact sequence whose first and last terms are the vector space, $V_0 = V_4 = 0$, and hence $\alpha_0 = \alpha_3 = 0$. This sequence is exact if and only if the following conditions hold:

(1) α_1 is injective,
(2) the kernel of α_2 equals the image of α_1, and
(3) α_2 is surjective.

We will call an exact sequence of this form a ***short exact sequence***. (We'll also encounter below an even shorter example of an exact sequence, namely a sequence of the form

$$0 \longrightarrow V_1 \xrightarrow{\alpha_1} V_2 \longrightarrow 0.$$

This is an exact sequence if and only if α_1 is bijective.)

Another basic notion in the theory of diagram chasing is the notion of a commutative diagram. The square diagram of vector spaces and linear maps

$$\begin{array}{ccc} A & \xrightarrow{f} & B \\ {\scriptstyle i}\downarrow & & \downarrow{\scriptstyle j} \\ C & \xrightarrow[g]{} & D \end{array}$$

commutes if $j \circ f = g \circ i$, and a more complicated diagram of vector spaces and linear maps like the diagram

$$\begin{array}{ccccc} A_1 & \longrightarrow & A_2 & \longrightarrow & A_3 \\ \downarrow & & \downarrow & & \downarrow \\ B_1 & \longrightarrow & B_2 & \longrightarrow & B_3 \\ \downarrow & & \downarrow & & \downarrow \\ C_1 & \longrightarrow & C_2 & \longrightarrow & C_3 \end{array}$$

commutes if every sub-square in the diagram commutes.

We now have enough "diagram chasing" vocabulary to formulate the Mayer–Vietoris theorem. For $r = 1, 2, 3$ let

(5.2.6) $$C_r^0 \xrightarrow{d} C_r^1 \xrightarrow{d} C_r^2 \xrightarrow{d} \cdots$$

be a complex and, for fixed k, let

(5.2.7) $$0 \longrightarrow C_1^k \xrightarrow{i} C_2^k \xrightarrow{j} C_3^k \longrightarrow 0$$

be a short exact sequence. Assume that the diagram below commutes:

$$\begin{array}{ccccccccc}
 & & \vdots & & \vdots & & \vdots & & \\
 & & \downarrow & & \downarrow & & \downarrow & & \\
0 & \longrightarrow & C_1^{k-1} & \xrightarrow{i} & C_2^{k-1} & \xrightarrow{j} & C_3^{k-1} & \longrightarrow & 0 \\
 & & \downarrow d & & \downarrow d & & \downarrow d & & \\
0 & \longrightarrow & C_1^{k} & \xrightarrow{i} & C_2^{k} & \xrightarrow{j} & C_3^{k} & \longrightarrow & 0 \\
 & & \downarrow d & & \downarrow d & & \downarrow d & & \\
0 & \longrightarrow & C_1^{k+1} & \xrightarrow[i]{} & C_2^{k+1} & \xrightarrow[j]{} & C_3^{k+1} & \longrightarrow & 0 \\
 & & \downarrow & & \downarrow & & \downarrow & & \\
 & & \vdots & & \vdots & & \vdots & &
\end{array}$$

That is, assume that in the left-hand squares, $di = id$, and in the right-hand squares, $dj = jd$.

The Mayer–Vietoris theorem addresses the following question: *If one has information about the cohomology groups of two of the three complexes* (5.2.6), *what information about the cohomology groups of the third can be extracted from this diagram*? Let us first observe that the maps i and j give rise to mappings between these cohomology groups. Namely, for $r = 1, 2, 3$ let Z_r^k be the kernel of the map $d\colon C_r^k \to C_r^{k+1}$, and B_r^k the image of the map $d\colon C_r^{k-1} \to C_r^k$. Since $id = di$, i maps B_1^k into B_2^k and Z_1^k into Z_2^k, therefore by (5.2.4) it gives rise to a linear mapping

$$i_\sharp\colon H^k(C_1) \to H^k(C_2).$$

Similarly since $jd = dj$, j maps B_2^k into B_3^k and Z_2^k into Z_3^k, and so by (5.2.4) gives rise to a linear mapping

$$j_\sharp\colon H^k(C_2) \to H^k(C_3)\,.$$

Moreover, since $j \circ i = 0$ the image of $i_\sharp$ is contained in the kernel of $j_\sharp$. We'll leave as an exercise the following sharpened version of this observation.

Proposition 5.2.8. *The kernel of $j_\sharp$ equals the image of $i_\sharp$, i.e., the three term sequence*

$$H^k(C_1) \xrightarrow{i_\sharp} H^k(C_2) \xrightarrow{j_\sharp} H^k(C_3) \tag{5.2.9}$$

is exact.

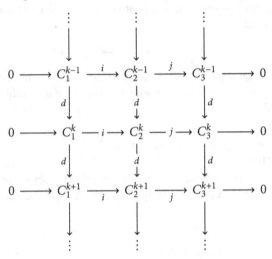

Since (5.2.7) is a short exact sequence one is tempted to conjecture that (5.2.9) is also a short exact sequence (which, if it were true, would tell us that the cohomology groups of any two of the complexes (5.2.6) completely determine the cohomology groups of the third). Unfortunately, this is not the case. To see how this conjecture can be violated let us try to show that the mapping $j_\sharp$ is surjective. Let c_3^k be an element of Z_3^k representing the cohomology class $[c_3^k]$ in $H^3(C_3)$. Since (5.2.7) is exact there exists a c_2^k in C_2^k which gets mapped by j onto c_3^k, and if c_3^k were in Z_2^k this would imply

$$j_\sharp[c_2^k] = [jc_2^k] = [c_3^k],$$

i.e., the cohomology class $[c_3^k]$ would be in the image of $j_\sharp$. However, since there's no reason for c_2^k to be in Z_2^k, there's also no reason for $[c_3^k]$ to be in the image of $j_\sharp$. What we can say, however, is that $jdc_2^k = d\,jc_2^k = dc_3^k = 0$ since c_3^k is in Z_3^k. Therefore by the exactness of (5.2.7) in degree $k+1$ there exists a unique element c_1^{k+1} in C_1^{k+1} with property

(5.2.10) $$dc_2^k = ic_1^{k+1}.$$

Moreover, since

$$0 = d(dc_2^k) = di(c_1^{k+1}) = i\,dc_1^{k+1}$$

and i is injective, $dc_1^{k+1} = 0$, i.e.,

(5.2.11) $$c_1^{k+1} \in Z_1^{k+1}.$$

Thus via (5.2.10) and (5.2.11) we've converted an element c_3^k of Z_3^k into an element c_1^{k+1} of Z_1^{k+1} and hence set up a correspondence

(5.2.12) $$Z_3^k \ni c_3^k \mapsto c_1^{k+1} \in Z_1^{k+1}.$$

Unfortunately this correspondence isn't, strictly speaking, a map of Z_3^k into Z_1^{k+1}; the c_1^k in (5.2.12) isn't determined by c_3^k alone but also by the choice we made of c_2^k. Suppose, however, that we make another choice of a c_2^k with the property $j(c_2^k) = c_3^k$. Then the difference between our two choices is in the kernel of j and hence, by the exactness of (5.2.6) at level k, is in the image of i. In other words, our two choices are related by

$$(c_2^k)_{\text{new}} = (c_2^k)_{\text{old}} + i(c_1^k)$$

for some c_1^k in C_1^k, and hence by (5.2.10)

$$(c_1^{k+1})_{\text{new}} = (c_1^{k+1})_{\text{old}} + dc_1^k.$$

Therefore, even though the correspondence (5.2.12) isn't strictly speaking a map it does give rise to a well-defined map

$$Z_3^k \to H^{k+1}(C_1), \quad c_3^k \mapsto [c_3^{k+1}].$$

Moreover, if c_3^k is in B_3^k, i.e., $c_3^k = dc_3^{k-1}$ for some $c_3^{k-1} \in C_3^{k-1}$, then by the exactness of (5.2.6) at level $k-1$, $c_3^{k-1} = j(c_2^{k-1})$ for some $c_2^{k-1} \in C_2^{k-1}$ and hence $c_3^k = j(dc_2^{k-2})$. In other words we can take the c_2^k above to be dc_2^{k-1} in which case the c_1^{k+1} in equation (5.2.10) is just zero. Thus the map (5.2.12) maps B_3^k to zero and hence by Proposition 1.2.9 gives rise to a well-defined map

$$\delta : H^k(C_3) \to H^{k+1}(C_1)$$

mapping $[c_3^k] \mapsto [c_1^{k+1}]$. We will leave it as an exercise to show that this mapping measures the failure of the arrow $j_\sharp$ in the exact sequence (5.2.9) to be surjective (and hence the failure of this sequence to be a short exact sequence at its right end).

Proposition 5.2.13. *The image of the map $j_\sharp : H^k(C_2) \to H^k(C_3)$ is equal to the kernel of the map $\delta : H^k(C_3) \to H^{k+1}(C_1)$.*

Hint: Suppose that in the correspondence (5.2.12), c_1^{k+1} is in B_1^{k+1}. Then $c_1^{k+1} = dc_1^k$ for some c_1^k in C_1^k. Show that $j(c_2^k - i(c_1^k)) = c_3^k$ and $d(c_2^k - i(c_1^k)) = 0$, i.e., $c_2^k - i(c_1^k)$ is in Z_2^k and hence $j_\sharp[c_2^k - i(c_1^k)] = [c_3^k]$.

Let us next explore the failure of the map $i_\sharp : H^{k+1}(C_1) \to H^{k+1}(C_2)$, to be injective. Let c_1^{k+1} be in Z_1^{k+1} and suppose that its cohomology class, $[c_1^{k+1}]$, gets mapped by $i_\sharp$ into zero. This translates into the statement

$$i(c_1^{k+1}) = dc_2^k \tag{5.2.14}$$

for some $c_2^k \in C_2^k$. Moreover since $dc_2^k = i(c_1^{k+1})$, we have $j(dc_2^k) = 0$. But if

$$c_3^k := j(c_2^k) \tag{5.2.15}$$

then $dc_3^k = dj(c_2^k) = j(dc_2^k) = j(i(c_1^{k+1})) = 0$, so c_3^k is in Z_3^k, and by (5.2.14), (5.2.15) and the definition of δ

$$[c_1^{k+1}] = \delta[c_3^k].$$

In other words the kernel of the map $i_\sharp : H^{k+1}(C_1) \to H^{k+1}(C_2)$ is contained in the image of the map $\delta : H^k(C_3) \to H^{k+1}(C_1)$. We will leave it as an exercise to show that this argument can be reversed to prove the converse assertion and hence to prove the following proposition.

Proposition 5.2.16. *The image of the map $\delta : H^k(C_1) \to H^{k+1}(C_1)$ is equal to the kernel of the map $i_\sharp : H^{k+1}(C_1) \to H^{k+1}(C_2)$.*

Putting together Propositions 5.2.8, 5.2.13 and 5.2.16 we obtain the main result of this section: the ***Mayer–Vietoris theorem***. The sequence of cohomology groups and linear maps

$$\cdots \xrightarrow{\delta} H^k(C_1) \xrightarrow{i_\sharp} H^k(C_2) \xrightarrow{j_\sharp} H^k(C_3) \xrightarrow{\delta} H^{k+1}(C_1) \xrightarrow{i_\sharp} \cdots \tag{5.2.17}$$

is exact.

Remark 5.2.18. To see an illuminating real-world example of an application of these ideas, we strongly recommend that our readers stare at Exercise 5.2.v with gives a method for computing the de Rham cohomology of the n-sphere S^n in terms of the de Rham cohomologies of $S^n \setminus \{(0, \ldots, 0, 1)\}$ and $S^n \setminus \{(0, \ldots, 0, -1)\}$ via an exact sequence of the form (5.2.17).

Remark 5.2.19. In view of the "$\cdots$"s this sequence can be a *very* long sequence and is commonly referred to as the ***long exact sequence in cohomology*** associated to the short exact sequence of complexes (5.2.7).

Before we discuss the applications of this result, we will introduce some vector space notation. Given vector spaces V_1 and V_2, we'll denote by $V_1 \oplus V_2$ the vector space sum of V_1 and V_2, i.e., the set of all pairs

$$(u_1, u_2), \quad u_i \in V_i$$

with the addition operation

$$(u_1, u_2) + (v_1 + v_2) := (u_1 + v_1, u_2 + v_2)$$

and the scalar multiplication operation

$$\lambda(u_1, u_2) := (\lambda u_1, \lambda u_2).$$

Now let X be a manifold and let U_1 and U_2 be open subsets of X. Then one has a linear map

$$i\colon \Omega^k(U_1 \cup U_2) \to \Omega^k(U_1) \oplus \Omega^k(U_2)$$

defined by

$$\omega \mapsto (\omega|_{U_1}, \omega|_{U_2}), \tag{5.2.20}$$

where $\omega|_{U_i}$ is the restriction of ω to U_i. Similarly one has a linear map

$$j\colon \Omega^k(U_1) \oplus \Omega^k(U_2) \to \Omega^k(U_1 \cap U_2)$$

defined by

$$(\omega_1, \omega_2) \mapsto \omega_1|_{U_1 \cap U_2} - \omega_2|_{U_1 \cap U_2}.$$

We claim the following theorem.

Theorem 5.2.21. *The sequence*

$$0 \longrightarrow \Omega^k(U_1 \cup U_2) \xrightarrow{\ i\ } \Omega^k(U_1) \oplus \Omega^k(U_2) \xrightarrow{\ j\ } \Omega^k(U_1 \cap U_2) \longrightarrow 0$$

is a short exact sequence.

Proof. If the right-hand side of (5.2.20) is zero, ω itself has to be zero so the map (5.2.20) is injective. Moreover, if the right-hand side of (5.2) is zero, ω_1 and ω_2 are equal on the overlap, $U_1 \cap U_2$, so we can glue them together to get a C^∞ k-form on $U_1 \cup U_2$ by setting $\omega = \omega_1$ on U_1 and $\omega = \omega_2$ on U_2. Thus by (5.2.20) $i(\omega) = (\omega_1, \omega_2)$, and this shows that the kernel of j is equal to the image of i. Hence to complete the proof we only have to show that j is surjective, i.e., that every form ω on $\Omega^k(U_1 \cap U_2)$ can be written as a difference, $\omega_1|_{U_1 \cap U_2} - \omega_2|_{U_1 \cap U_2}$, where ω_1 is in $\Omega^k(U_1)$ and ω_2 is in $\Omega^k(U_2)$. To prove this we'll need the following variant of the partition of unity theorem.

Theorem 5.2.22. *There exist functions, $\varphi_1, \varphi_2 \in C^\infty(U_1 \cup U_2)$, such that support ϕ_α is contained in U_α and $\phi_1 + \phi_2 = 1$.*

Before proving this let us use it to complete our proof of Theorem 5.2.21. Given $\omega \in \Omega^k(U_1 \cap U_2)$ let

$$\omega_1 := \begin{cases} \phi_2\omega & \text{on } U_1 \cap U_2, \\ 0 & \text{on } U_1 \setminus U_1 \cap U_2, \end{cases} \tag{5.2.23}$$

and let

$$\omega_2 := \begin{cases} -\phi_1\omega & \text{on } U_1 \cap U_2, \\ 0 & \text{on } U_2 \setminus U_1 \cap U_2. \end{cases} \tag{5.2.24}$$

Since ϕ_2 is supported on U_2 the form defined by (5.2.23) is C^∞ on U_1 and since ϕ_1 is supported on U_1 the form defined by (5.2.24) is C^∞ on U_2 and since $\phi_1 + \phi_2 = 1$, $\omega_1 - \omega_2 = (\phi_1 + \phi_2)\omega = \omega$ on $U_1 \cap U_2$. □

Proof of Theorem 5.2.22. Let $\rho_i \in C_0^\infty(U_1 \cup U_2)$, $i = 1, 2, 3, \ldots$, be a partition of unity subordinate to the cover $\{U_1, U_2\}$ of $U_1 \cup U_2$ and let ϕ_1 be the sum of the ρ_i's with support on U_1 and ϕ_2 the sum of the remaining ρ_i's. It's easy to check (using part (b) of Theorem 4.6.1) that ϕ_α is supported in U_α and (using part (a) of Theorem 4.6.1) that $\phi_1 + \phi_2 = 1$. □

Now let

$$0 \longrightarrow C_1^0 \xrightarrow{d} C_1^1 \xrightarrow{d} C_1^2 \xrightarrow{d} \cdots \tag{5.2.25}$$

be the de Rham complex of $U_1 \cup U_2$, let

$$0 \longrightarrow C_3^0 \xrightarrow{d} C_3^1 \xrightarrow{d} C_3^2 \xrightarrow{d} \cdots \tag{5.2.26}$$

be the de Rham complex of $U_1 \cap U_2$, and let

$$0 \longrightarrow C_2^0 \xrightarrow{d} C_2^1 \xrightarrow{d} C_2^2 \xrightarrow{d} \cdots \tag{5.2.27}$$

be the vector space direct sum of the de Rham complexes of U_1 and U_2, i.e., the complex whose kth term is

$$C_2^k = \Omega^k(U_1) \oplus \Omega^k(U_2)$$

with $d\colon C_2^k \to C_2^{k+1}$ defined to be the map $d(\mu_1, \mu_2) = (d\mu_1, d\mu_2)$. Since $C_1^k = \Omega^k(U_1 \cup U_2)$ and $C_3^k = \Omega^k(U_1 \cap U_2)$ we have, by Theorem 5.2.21, a short exact sequence

$$0 \longrightarrow C_1^k \xrightarrow{i} C_2^k \xrightarrow{j} C_3^k \longrightarrow 0,$$

and it's easy to see that i and j commute with the d's:

$$di = id \quad \text{and} \quad dj = jd.$$

Hence we're exactly in the situation to which Mayer–Vietoris applies. Since the cohomology groups of the complexes (5.2.25) and (5.2.26) are the de Rham cohomology groups $H^k(U_1 \cup U_2)$ and $H^k(U_1 \cap U_2)$, and the cohomology groups of the complex (5.2.27) are the vector space direct sums $H^k(U_1) \oplus H^k(U_2)$, we obtain from the abstract Mayer–Vietoris theorem, the following de Rham theoretic version of Mayer–Vietoris.

Theorem 5.2.28. *Letting $U = U_1 \cup U_2$ and $V = U_1 \cap U_2$ one has a long exact sequence in de Rham cohomology:*

$$\cdots \xrightarrow{\delta} H^k(U) \xrightarrow{i_\sharp} H^k(U_1) \oplus H^k(U_2) \xrightarrow{j_\sharp} H^k(V) \xrightarrow{\delta} H^{k+1}(U) \xrightarrow{i_\sharp} \cdots .$$

This result also has an analogue for compactly supported de Rham cohomology. Let

$$i\colon \Omega^k_c(U_1 \cap U_2) \to \Omega^k_c(U_1) \oplus \Omega^k_c(U_2) \tag{5.2.29}$$

be the map

$$i(\omega) := (\omega_1, \omega_2),$$

where

$$\omega_\alpha := \begin{cases} \omega & \text{on } U_1 \cap U_2, \\ 0 & \text{on } U_\alpha \smallsetminus U_1 \cap U_2. \end{cases} \tag{5.2.30}$$

(Since ω is compactly supported on $U_1 \cap U_2$ the form defined by equation (5.2.30) is a C^∞ form and is compactly supported on U_α.) Similarly, let

$$j\colon \Omega^k_c(U_1) \oplus \Omega^k_c(U_2) \to \Omega^k_c(U_1 \cup U_2)$$

be the map

$$j(\omega_1, \omega_2) := \tilde{\omega}_1 - \tilde{\omega}_2,$$

where

$$\tilde{\omega}_\alpha := \begin{cases} \omega_\alpha & \text{on } U_\alpha, \\ 0 & \text{on } (U_1 \cup U_2) \smallsetminus U_\alpha. \end{cases}$$

As above it's easy to see that i is injective and that the kernel of j is equal to the image of i. Thus if we can prove that j is surjective we'll have proved the following.

Theorem 5.2.31. *The sequence*

$$0 \longrightarrow \Omega^k_c(U_1 \cap U_2) \xrightarrow{i} \Omega^k_c(U_1) \oplus \Omega^k_c(U_2) \xrightarrow{j} \Omega^k_c(U_1 \cup U_2) \longrightarrow 0$$

is a short exact sequence.

Proof. To prove the surjectivity of j we mimic the proof above. Given $\omega \in \Omega^k_c(U_1 \cup U_2)$, let

$$\omega_1 := \phi_1 \omega|_{U_1}$$

and

$$\omega_2 := -\phi_2 \omega|_{U_2}.$$

Then by (5.2.30) we have that $\omega = j(\omega_1, \omega_2)$. □

Thus, applying Mayer–Vietoris to the compactly supported versions of the complexes (5.2.6), we obtain the following theorem.

Theorem 5.2.32. *Letting $U = U_1 \cap U_2$ and $V = U_1 \cup U_2$ there exists a long exact sequence in compactly supported de Rham cohomology*

$$\cdots \xrightarrow{\delta} H^k_c(U) \xrightarrow{i_\sharp} H^k_c(U_1) \oplus H^k_c(U_2) \xrightarrow{j_\sharp} H^k_c(V) \xrightarrow{\delta} H^{k+1}_c(U) \xrightarrow{i_\sharp} \cdots.$$

Exercises for §5.2

Exercise 5.2.i. Prove Proposition 5.2.8.

Exercise 5.2.ii. Prove Proposition 5.2.13.

Exercise 5.2.iii. Prove Proposition 5.2.16.

Exercise 5.2.iv. Show that if U_1, U_2 and $U_1 \cap U_2$ are non-empty and connected, the first segment of the Mayer–Vietoris sequence is a short exact sequence

$$0 \longrightarrow H^0(U_1 \cup U_2) \xrightarrow{i_\sharp} H^0(U_1) \oplus H^0(U_2) \xrightarrow{j_\sharp} H^0(U_1 \cap U_2) \longrightarrow 0.$$

Exercise 5.2.v. Let $X = S^n$ and let U_1 and U_2 be the open subsets of S^n obtained by removing from S^n the points, $p_1 = (0, \ldots, 0, 1)$ and $p_2 = (0, \ldots, 0, -1)$.

(1) Using stereographic projection show that U_1 and U_2 are diffeomorphic to $\mathbf{R}^n$.
(2) Show that $U_1 \cup U_2 = S^n$ and $U_1 \cap U_2$ is homotopy equivalent to S^{n-1}. (See Exercise 5.1.v.)
Hint: $U_1 \cap U_2$ is diffeomorphic to $\mathbf{R}^n \smallsetminus \{0\}$.
(3) Deduce from the Mayer–Vietoris sequence that $H^{i+1}(S^n) = H^i(S^{n-1})$ for $i \geq 1$.
(4) Using part (3) give an inductive proof of a result that we proved by other means in §5.1: $H^k(S^n) = 0$ for $1 \leq k < n$.

Exercise 5.2.vi. Using the Mayer–Vietoris sequence of Exercise 5.2.v with cohomology replaced by compactly supported cohomology show that

$$H^k_c(\mathbf{R}^n \smallsetminus \{0\}) \cong \begin{cases} \mathbf{R}, & k = 1, n, \\ 0 & \text{otherwise.} \end{cases}$$

Exercise 5.2.vii. Let n a positive integer and let

$$0 \longrightarrow V_1 \xrightarrow{f_1} V_2 \longrightarrow \cdots \longrightarrow V_{n-1} \xrightarrow{f_{n-1}} V_n \longrightarrow 0$$

be an exact sequence of finite-dimensional vector spaces. Prove that $\sum_{i=1}^n \dim(V_i) = 0$.

5.3. Cohomology of good covers

In this section we will show that for compact manifolds (and for lots of other manifolds besides) the de Rham cohomology groups which we defined in § 5.1 are finite-dimensional vector spaces and thus, in principle, "computable" objects. A key ingredient in our proof of this fact is the notion of a *good cover* of a manifold.

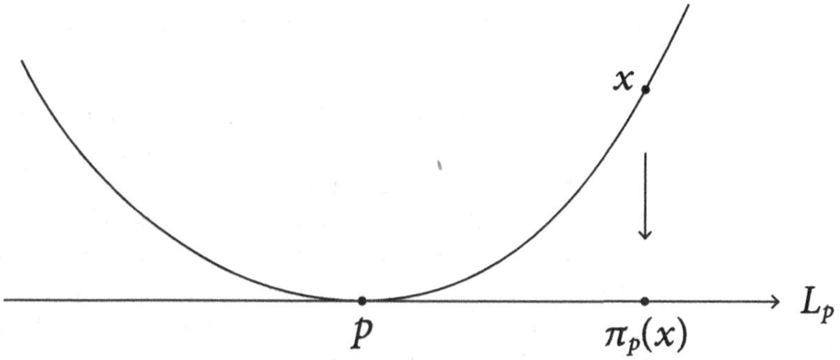

Figure 5.3.1. The orthogonal projection of X onto L_p.

Definition 5.3.1. Let X be an n-manifold, and let $\mathcal{U} = \{U_\alpha\}_{\alpha \in I}$ be an open cover of X. Then $\mathcal{U}$ is a ***good cover*** if for every finite set of indices $\alpha_1, \dots, \alpha_k \in I$ the intersection $U_{\alpha_1} \cap \cdots \cap U_{\alpha_k}$ is either empty or is diffeomorphic to $\mathbf{R}^n$.

One of our first goals in this section will be to show that good covers exist. We will sketch below a proof of the following.

Theorem 5.3.2. *Every manifold admits a good cover.*

The proof involves an elementary result about open convex subsets of $\mathbf{R}^n$.

Proposition 5.3.3. *A bounded open convex subset $U \subset \mathbf{R}^n$ is diffeomorphic to $\mathbf{R}^n$.*

A proof of this will be sketched in Exercises 5.3.i–5.3.iv. One immediate consequence of Proposition 5.3.3 is an important special case of Theorem 5.3.2.

Theorem 5.3.4. *Every open subset U of $\mathbf{R}^n$ admits a good cover.*

Proof. For each $p \in U$ let U_p be an open convex neighborhood of p in U (for instance an ε-ball centered at p). Since the intersection of any two convex sets is again convex the cover $\{U_p\}_{p \in U}$ is a good cover by Proposition 5.3.3. □

For manifolds the proof of Theorem 5.3.2 is somewhat trickier. The proof requires a manifold analogue of the notion of convexity and there are several serviceable candidates. The one we will use is the following. Let $X \subset \mathbf{R}^N$ be an n-manifold and for $p \in X$ let T_pX be the tangent space to X at p. Recalling that T_pX sits inside $T_p\mathbf{R}^N$ and that

$$T_p\mathbf{R}^N = \{(p, v) \mid v \in \mathbf{R}^N\}.$$

We get a map

$$T_pX \hookrightarrow T_p\mathbf{R}^N \to \mathbf{R}^N, \quad (p, x) \mapsto p + x,$$

and this map maps T_pX bijectively onto an n-dimensional "affine" subspace L_p of $\mathbf{R}^N$ which is tangent to X at p. Let $\pi_p \colon X \to L_p$ be, as Figure 5.3.1, the orthogonal projection of X onto L_p.

Definition 5.3.5. An open subset V of X is ***convex*** if for every $p \in V$ the map $\pi_p : X \to L_p$ maps V diffeomorphically onto a convex open subset of L_p.

It's clear from this definition of convexity that the intersection of two open convex subsets of X is an open convex subset of X and that every open convex subset of X is diffeomorphic to $\mathbf{R}^n$. Hence to prove Theorem 5.3.2 it suffices to prove that every point $p \in X$ is contained in an open convex subset U_p of X. Here is a sketch of how to prove this. In Figure 5.3.1 let $B^\varepsilon(p)$ be the ball of radius ε about p in L_p centered at p. Since L_p and T_p are tangent at p the derivative of π_p at p is just the identity map, so for ε small π_p maps a neighborhood $U^\varepsilon(p)$ of p in X diffeomorphically onto $B^\varepsilon(p)$. We claim the following:

Proposition 5.3.6. *For ε small, $U^\varepsilon(p)$ is a convex subset of X.*

Intuitively this assertion is pretty obvious: if q is in $U^\varepsilon(p)$ and ε is small the map

$$B^\varepsilon(p) \xrightarrow{\pi_p^{-1}} U^\varepsilon(p) \xrightarrow{\pi_q} L_q$$

is to order ε^2 equal to the identity map, so it's intuitively clear that its image is a slightly warped, but still convex, copy of $B^\varepsilon(p)$. We won't, however, bother to write out the details that are required to make this proof rigorous.

A good cover is a particularly good "good cover" if it is a finite cover. We'll codify this property in the definition below.

Definition 5.3.7. An n-manifold X is said to have ***finite topology*** if X admits a finite covering by open sets $U_1, \dots, U_N$ with the property that for every multi-index, $I = (i_1, \dots, i_k)$, $1 \le i_1 \le i_2 \cdots < i_K \le N$, the set

$$U_I := U_{i_1} \cap \cdots \cap U_{i_k} \tag{5.3.8}$$

is either empty or is diffeomorphic to $\mathbf{R}^n$.

If X is a compact manifold and $\mathcal{U} = \{U_\alpha\}_{\alpha \in I}$ is a good cover of X then by the Heine–Borel theorem we can extract from $\mathcal{U}$ a finite subcover $U_1, \dots, U_N$, where

$$U_i := U_{\alpha_i}\,, \quad \text{for } \alpha_1, \dots, \alpha_N \in I,$$

hence we conclude the following theorem.

Theorem 5.3.9. *Every compact manifold has finite topology.*

More generally, for any manifold X, let C be a compact subset of X. Then by Heine–Borel we can extract from the cover $\mathcal{U}$ a finite subcollection $U_1, \dots, U_N$, where

$$U_i := U_{\alpha_i}, \quad \text{for } \alpha_1, \dots, \alpha_N \in I,$$

that covers C, hence letting $U := U_1 \cup \cdots \cup U_N$, we've proved the following.

Theorem 5.3.10. *If X is an n-manifold and C a compact subset of X, then there exists an open neighborhood U of C in X with finite topology.*

We can in fact even strengthen this further. Let U_0 be any open neighborhood of C in X. Then in the theorem above we can replace X by U_0 to conclude the following.

Theorem 5.3.11. *Let X be a manifold, C a compact subset of X, and U_0 an open neighborhood of C in X. Then there exists an open neighborhood U of C in X, U contained in U_0, so that U has finite topology.*

We will justify the term "finite topology" by devoting the rest of this section to proving.

Theorem 5.3.12. *Let X be an n-manifold. If X has finite topology, the de Rham cohomology groups $H^k(X)$, for $k = 0, \dots, n$, and the compactly supported de Rham cohomology groups $H_c^k(X)$, for $k = 0, \dots, n$, are finite-dimensional vector spaces.*

The basic ingredients in the proof of this will be the Mayer–Vietoris techniques that we developed in §5.2 and the following elementary result about vector spaces.

Lemma 5.3.13. *Let V_1, V_2, and V_3 be vector spaces and*

$$V_1 \xrightarrow{\alpha} V_2 \xrightarrow{\beta} V_3 \tag{5.3.14}$$

an exact sequence of linear maps. Then if V_1 and V_3 are finite dimensional, so is V_2.

Proof. Since V_3 is finite dimensional, $\operatorname{im}(\beta)$ is of finite dimension k. Hence there exist vectors $v_1, \dots, v_k \in V_2$ such that

$$\operatorname{im}(\beta) = \operatorname{span}(\beta(v_1), \dots, \beta(v_k)).$$

Now let $v \in V_2$. Then $\beta(v)$ is a linear combination $\beta(v) = \sum_{i=1}^k c_i\beta(v_i)$, where $c_1, \dots, c_k \in \mathbf{R}$. So

$$v' := v - \sum_{i=1}^{k} c_i v_i \tag{5.3.15}$$

is in the kernel of β and hence, by the exactness of (5.3.14), in the image of α. But V_1 is finite dimensional, so $\operatorname{im}(\alpha)$ is finite dimensional. Letting $v_{k+1}, \dots, v_m$ be a basis of $\operatorname{im}(\alpha)$ we can by (5.3.15) write v as a sum $v = \sum_{i=1}^m c_i v_i$. In other words $v_1, \dots, v_m$ is a basis of V_2. □

Proof of Theorem 5.3.12. Our proof will be by induction on the number of open sets in a good cover of X. More specifically, let

$$\mathcal{U} = \{U_1, \dots, U_N\}$$

be a good cover of X. If $N = 1$, $X = U_1$ and hence X is diffeomorphic to $\mathbf{R}^n$, so $H^k(X) = 0$ for $k > 0$, and $H^0(X) \cong \mathbf{R}$, so the theorem is certainly true in this case. Let us now prove it's true for arbitrary N by induction. Let $U := U_2 \cup \cdots \cup U_N$. Then U is a submanifold of X, and, thinking of U as a manifold in its own right, $\{U_2, \dots, U_N\}$ is a good cover of U involving only $N - 1$ sets. Hence the cohomology groups of U are finite dimensional by the induction hypothesis. The manifold

$U \cap U_1$ has a good cover given by $\{U \cap U_2, \ldots, U \cap U_N\}$, so by the induction hypothesis the cohomology groups of $U \cap U_1$ are finite dimensional. To prove that the theorem is true for X we note that $X = U_1 \cup U$, and the Mayer–Vietoris sequence gives an exact sequence

$$H^{k-1}(U_1 \cap U) \xrightarrow{\delta} H^k(X) \xrightarrow{i_\sharp} H^k(U_1) \oplus H^k(U).$$

Since the right- and left-hand terms are finite dimensional it follows from Lemma 5.3.13 that the middle term is also finite dimensional. □

The proof works practically verbatim for compactly supported cohomology. For $N = 1$

$$H_c^k(X) = H_c^k(U_1) = H_c^k(\mathbf{R}^n)$$

so all the cohomology groups of $H^k(X)$ are finite dimensional in this case, and the induction "$N - 1$" ⇒ "N" follows from the exact sequence

$$H_c^k(U_1) \oplus H_c^k(U) \xrightarrow{j_\sharp} H_c^k(X) \xrightarrow{\delta} H_c^{k+1}(U_1 \cap U).$$

Remark 5.3.16. A careful analysis of the proof above shows that the dimensions of the vector spaces $H^k(X)$ are determined by the intersection properties of the open sets U_i, i.e., by the list of multi-indices I for which the intersections (5.3.8) are non-empty.

This collection of multi-indices is called the ***nerve*** of the cover $\mathcal{U}$, and this remark suggests that there should be a cohomology theory which has as input the nerve of $\mathcal{U}$ and as output cohomology groups which are isomorphic to the de Rham cohomology groups. Such a theory does exist, and we further address it in §5.8. (A nice account of it can also be found in [13, Chapter 5].)

Exercises for §5.3

Exercise 5.3.i. Let U be a bounded open subset of $\mathbf{R}^n$. A continuous function

$$\psi : U \to [0, \infty)$$

is called an ***exhaustion function*** if it is proper as a map of U into $[0, \infty)$; i.e., if, for every $a > 0$, $\psi^{-1}([0, a])$ is compact. For $x \in U$ let

$$d(x) = \inf\{\, |x - y| \mid y \in \mathbf{R}^n - U \,\},$$

i.e., let $d(x)$ be the "distance" from x to the boundary of U. Show that $d(x) > 0$ and that $d(x)$ is continuous as a function of x. Conclude that $\psi_0 = 1/d$ is an exhaustion function.

Exercise 5.3.ii. Show that there exists a C^∞ exhaustion function $\phi_0 : U \to [0, \infty)$ with the property $\phi_0 \geq \psi_0^2$ where ψ_0 is the exhaustion function in Exercise 5.3.i.

Hints: For $i = 2, 3, \ldots$ let

$$C_i = \left\{ x \in U \;\middle|\; \frac{1}{i} \leq d(x) \leq \frac{1}{i-1} \right\}$$

and

$$U_i = \left\{ x \in U \;\middle|\; \frac{1}{i+1} < d(x) < \frac{1}{i-2} \right\}.$$

Let $\rho_i \in C_0^\infty(U_i)$, $\rho_i \geq 0$, be a "bump" function which is identically one on C_i and let $\phi_0 = \sum_{i=2}^{\infty} i^2 \rho_i + 1$.

Exercise 5.3.iii. Let U be a bounded open convex subset of $\mathbf{R}^n$ containing the origin. Show that there exists an exhaustion function

$$\psi : U \to \mathbf{R}, \quad \psi(0) = 1,$$

having the property that ψ is a ***monotonically increasing*** function of t along the ray tx, $0 \leq t \leq 1$, for all points $x \in U$.

Hints:

- Let $\rho(x)$, $0 \leq \rho(x) \leq 1$, be a C^∞ function which is one outside a small neighborhood of the origin in U and is zero in a still smaller neighborhood of the origin. Modify the function ϕ_0, in the previous exercise by setting $\phi(x) = \rho(x)\phi_0(x)$ and let

$$\psi(x) = \int_0^1 \phi(sx)\frac{ds}{s} + 1.$$

Show that for $0 \leq t \leq 1$

$$\frac{d\psi}{dt}(tx) = \phi(tx)/t$$

and conclude from equation (5.3.15) that ψ is monotonically increasing along the ray, tx, $0 \leq t \leq 1$.

- Show that for $0 < \varepsilon < 1$,

$$\psi(x) \geq \varepsilon\phi(y),$$

where y is a point on the ray tx, $0 \leq t \leq 1$ a distance less than $\varepsilon|x|$ from X.

- Show that there exist constants, C_0 and C_1, $C_1 > 0$ such that

$$\psi(x) = \frac{C_1}{d(x)} + C_0.$$

(Take ε to be equal to $\frac{1}{2}d(x)/|x|$.)

Exercise 5.3.iv. Show that every bounded open convex subset U of $\mathbf{R}^n$ is diffeomorphic to $\mathbf{R}^n$.

Hints:

- Let $\psi(x)$ be the exhaustion function constructed in Exercise 5.3.iii and let

$$f : U \to \mathbf{R}^n$$

be the map: $f(x) = \psi(x)x$. Show that this map is a bijective map of U onto $\mathbf{R}^n$.

› Show that for $x \in U$ and $v \in \mathbf{R}^n$

$$(df)_x v = \psi(x)v + d\psi_x(v)x$$

and conclude that df_x is bijective at x, i.e., that f is locally a diffeomorphism of a neighborhood of x in U onto a neighborhood of $f(x)$ in $\mathbf{R}^n$.

› Putting these together show that f is a diffeomorphism of U onto $\mathbf{R}^n$.

Exercise 5.3.v. Let $U \subset \mathbf{R}$ be the union of the open intervals, $k < x < k+1$ for k an integer. Show that U *does not* have finite topology.

Exercise 5.3.vi. Let $V \subset \mathbf{R}^2$ be the open set obtained by deleting from $\mathbf{R}^2$ the points, $p_n = (0, n)$, for every integer n. Show that V *does not* have finite topology.

Hint: Let γ_n be a circle of radius $\frac{1}{2}$ centered about the point p_n. Using Exercises 2.1.viii and 2.1.ix show that there exists a closed smooth 1-form ω_n on V with the property that $\int_{\gamma_n} \omega_n = 1$ and $\int_{\gamma_m} \omega_n = 0$ for $m \neq n$.

Exercise 5.3.vii. Let X be an n-manifold and $\mathcal{U} = \{U_1, U_2\}$ a good cover of X. What are the cohomology groups of X if the nerve of this cover is:

(1) $\{1\}, \{2\}$;
(2) $\{1\}, \{2\}, \{1,2\}$.

Exercise 5.3.viii. Let X be an n-manifold and $\mathcal{U} = \{U_1, U_2, U_3\}$ a good cover of X. What are the cohomology groups of X if the nerve of this cover is:

(1) $\{1\}, \{2\}, \{3\}$;
(2) $\{1\}, \{2\}, \{3\}, \{1,2\}$;
(3) $\{1\}, \{2\}, \{3\}, \{1,2\}, \{1,3\}$;
(4) $\{1\}, \{2\}, \{3\}, \{1,2\}, \{1,3\}, \{2,3\}$;
(5) $\{1\}, \{2\}, \{3\}, \{1,2\}, \{1,3\}, \{2,3\}, \{1,2,3\}$.

Exercise 5.3.ix. Let S^1 be the unit circle in $\mathbf{R}^3$ parameterized by arc length: $(x, y) = (\cos\theta, \sin\theta)$. Let U_1 be the set of points $(\cos\theta, \sin\theta) \in S^1$ with $0 < \theta < \frac{2\pi}{3}$, U_2 the set of points $(\cos\theta, \sin\theta) \in S^1$ with $\frac{\pi}{2} < \theta < \frac{3\pi}{2}$, and U_3 the set of points $(\cos\theta, \sin\theta) \in S^1$ with $-\frac{2\pi}{3} < \theta < \frac{\pi}{3}$.

(1) Show that the U_i's are a good cover of S^1.
(2) Using the previous exercise compute the cohomology groups of S^1.

Exercise 5.3.x. Let S^2 be the unit 2-sphere in $\mathbf{R}^3$. Show that the sets

$$U_i = \{(x_1, x_2, x_3) \in S^2 \mid x_i > 0\}$$

$i = 1, 2, 3$ and

$$U_i = \{(x_1, x_2, x_3) \in S^2 \mid x_{i-3} < 0\},$$

$i = 4, 5, 6$, are a good cover of S^2. What is the nerve of this cover?

Exercise 5.3.xi. Let X and Y be manifolds. Show that if they both have finite topology, their product $X \times Y$ does as well.

Exercise 5.3.xii.

(1) Let X be a manifold and let $U_1, \ldots, U_N$ be a good cover of X. Show that $U_1 \times \mathbf{R}, \ldots, U_N \times \mathbf{R}$ is a good cover of $X \times \mathbf{R}$ and that the nerves of these two covers are the same.

(2) By Remark 5.3.16,

$$H^k(X \times \mathbf{R}) \cong H^k(X).$$

Verify this directly using homotopy techniques.

(3) More generally, show that for all $\ell \geq 0$

$$H^k(X \times \mathbf{R}^\ell) \cong H^k(X)$$

(a) by concluding that this has to be the case in view of Remark 5.3.16,
(b) and by proving this directly using homotopy techniques.

5.4. Poincaré duality

In this chapter we've been studying two kinds of cohomology groups: the ordinary de Rham cohomology groups H^k and the compactly supported de Rham cohomology groups H^k_c. It turns out that these groups are closely related. In fact if X is a connected oriented n-manifold and has finite topology, then $H^{n-k}_c(X)$ is the vector space dual of $H^k(X)$. We give a proof of this later in this section, however, before we do we need to review some basic linear algebra.

Given finite-dimensional vector spaces V and W, a ***bilinear pairing*** between V and W is a map

$$B \colon V \times W \to \mathbf{R}, \tag{5.4.1}$$

which is linear in each of its factors. In other words, for fixed $w \in W$, the map

$$\ell_w \colon V \to \mathbf{R}\,,\ v \mapsto B(v,w)$$

is linear, and for $v \in V$, the map

$$\ell_v \colon W \to \mathbf{R}\,,\ w \mapsto B(v,w)$$

is linear. Therefore, from the pairing (5.4.1) one gets a map

$$L_B \colon W \to V^\star\,,\ w \mapsto \ell_w \tag{5.4.2}$$

and since $\ell_{w_1} + \ell_{w_2}(v) = B(v, w_1 + w_2) = \ell_{w_1+w_2}(v)$, this map is linear. We'll say that (5.4.1) is a ***non-singular pairing*** if (5.4.2) is bijective. Notice, by the way, that the roles of V and W can be reversed in this definition. Letting $B^\sharp(w,v) := B(v,w)$ we get an analogous linear map

$$L_{B^\sharp} \colon V \to W^\star$$

and in fact

$$(L_{B^\sharp}(v))(w) = (L_B(w))(v) = B(v,w). \tag{5.4.3}$$

Thus if

$$\mu : V \to (V^\star)^\star$$

is the canonical identification of V with $(V^\star)^\star$ given by the recipe

$$\mu(v)(\ell) = \ell(v)$$

for $v \in V$ and $\ell \in V^\star$, we can rewrite equation (5.4.3) more suggestively in the form

$$L_{B^\natural} = L_B^\star \mu \tag{5.4.4}$$

i.e., L_B and $L_{B^\natural}$ are just the transposes of each other. In particular L_B is bijective if and only if $L_{B^\natural}$ is bijective.

Let us now apply these remarks to de Rham theory. Let X be a connected, oriented n-manifold. If X has finite topology the vector spaces $H_c^{n-k}(X)$ and $H^k(X)$ are both finite dimensional. We will show that there is a natural bilinear pairing between these spaces, and hence by the discussion above, a natural linear mapping of $H^k(X)$ into the vector space dual of $H_c^{n-k}(X)$. To see this let c_1 be a cohomology class in $H_c^{n-k}(X)$ and c_2 a cohomology class in $H^k(X)$. Then by (5.1.22) their product $c_1 \cdot c_2$ is an element of $H_c^n(X)$, and so by (5.1.2) we can define a pairing between c_1 and c_2 by setting

$$B(c_1, c_2) := I_X(c_1 \cdot c_2). \tag{5.4.5}$$

Notice that if $\omega_1 \in \Omega_c^{n-k}(X)$ and $\omega_2 \in \Omega^k(X)$ are closed forms representing the cohomology classes c_1 and c_2, respectively, then by (5.1.22) this pairing is given by the integral

$$B(c_1, c_2) := \int_X \omega_1 \wedge \omega_2 .$$

We'll next show that this bilinear pairing is non-singular in one important special case.

Proposition 5.4.6. *If X is diffeomorphic to $\mathbf{R}^n$ the pairing defined by (5.4.5) is non-singular.*

Proof. To verify this there is very little to check. The vector spaces $H^k(\mathbf{R}^n)$ and $H_c^{n-k}(\mathbf{R}^n)$ are zero except for $k = 0$, so all we have to check is that the pairing

$$H_c^n(X) \times H^0(X) \to \mathbf{R}$$

is non-singular. To see this recall that every compactly supported n-form is closed and that the only closed 0-forms are the constant functions, so at the level of forms, the pairing (5.4.5) is just the pairing

$$\Omega^n(X) \times \mathbf{R} \to \mathbf{R}, \quad (\omega, c) \mapsto c \int_X \omega,$$

and this is zero if and only if c is zero or ω is in $d\Omega_c^{n-1}(X)$. Thus at the level of cohomology this pairing is non-singular. □

We will now show how to prove this result in general.

Theorem 5.4.7 (Poincaré duality). *Let X be an oriented, connected n-manifold having finite topology. Then the pairing (5.4.5) is non-singular.*

The proof of this will be very similar in spirit to the proof that we gave in the last section to show that if X has finite topology its de Rham cohomology groups are finite dimensional. Like that proof, it involves Mayer–Vietoris plus some elementary diagram-chasing. The "diagram-chasing" part of the proof consists of the following two lemmas.

Lemma 5.4.8. *Let V_1, V_2 and V_3 be finite-dimensional vector spaces, and let*

$$V_1 \xrightarrow{\alpha} V_2 \xrightarrow{\beta} V_3$$

be an exact sequence of linear mappings. Then the sequence of transpose maps

$$V_3^\star \xrightarrow{\beta^\star} V_2^\star \xrightarrow{\alpha^\star} V_1^\star$$

is exact.

Proof. Given a vector subspace W_2 of V_2, let

$$W_2^\perp = \{\ell \in V_2^\star \mid \ell(w) = 0 \text{ for all } w \in W\}.$$

We'll leave for you to check that if W_2 is the kernel of β, then $W_2^\perp$ is the image of $\beta^\star$ and that if W_2 is the image of α, $W_2^\perp$ is the kernel of $\alpha^\star$. Hence if $\ker(\beta) = \operatorname{im}(\alpha)$, $\operatorname{im}(\beta^\star) = \ker(\alpha^\star)$. □

Lemma 5.4.9 (5-lemma). *Consider a commutative diagram of vector spaces*

$$\begin{array}{ccccccccc} B_1 & \xrightarrow{\beta_1} & B_2 & \xrightarrow{\beta_2} & B_3 & \xrightarrow{\beta_3} & B_4 & \xrightarrow{\beta_4} & B_5 \\ \downarrow{\scriptstyle\gamma_1} & & \downarrow{\scriptstyle\gamma_2} & & \downarrow{\scriptstyle\gamma_3} & & \downarrow{\scriptstyle\gamma_4} & & \downarrow{\scriptstyle\gamma_5} \\ A_1 & \xrightarrow[\alpha_1]{} & A_2 & \xrightarrow[\alpha_2]{} & A_3 & \xrightarrow[\alpha_3]{} & A_4 & \xrightarrow[\alpha_4]{} & A_5\,. \end{array} \tag{5.4.10}$$

satisfying the following conditions:

(1) *All the vector spaces are finite dimensional.*
(2) *The two rows are exact.*
(3) *The linear maps, γ_1, γ_2, γ_4, and γ_5 are bijections.*

Then the map γ_3 is a bijection.

Proof. We'll show that γ_3 is surjective. Given $a_3 \in A_3$ there exists a $b_4 \in B_4$ such that $\gamma_4(b_4) = \alpha_3(a_3)$ since γ_4 is bijective. Moreover, $\gamma_5(\beta_4(b_4)) = \alpha_4(\alpha_3(a_3)) = 0$, by the exactness of the top row. Therefore, since γ_5 is bijective, $\beta_4(b_4) = 0$, so by the exactness of the bottom row $b_4 = \beta_3(b_3)$ for some $b_3 \in B_3$, and hence

$$\alpha_3(\gamma_3(b_3)) = \gamma_4(\beta_3(b_3)) = \gamma_4(b_4) = \alpha_3(a_3).$$

Thus $\alpha_3(a_3 - \gamma_3(b_3)) = 0$, so by the exactness of the top row

$$a_3 - \gamma_3(b_3) = \alpha_2(a_2)$$

for some $a_2 \in A_2$. Hence by the bijectivity of γ_2 there exists a $b_2 \in B_2$ with $a_2 = \gamma_2(b_2)$, and hence

$$a_3 - \gamma_3(b_3) = \alpha_2(a_2) = \alpha_2(\gamma_2(b_2)) = \gamma_3(\beta_2(b_2)).$$

Thus finally

$$a_3 = \gamma_3(b_3 + \beta_2(b_2)).$$

Since a_3 was any element of A_3 this proves the surjectivity of γ_3.

One can prove the injectivity of γ_3 by a similar diagram-chasing argument, but one can also prove this with less duplication of effort by taking the transposes of all the arrows in (5.4.10) and noting that the same argument as above proves the surjectivity of $\gamma_3^\star : A_3^\star \to B_3^\star$. □

To prove Theorem 5.4.7 we apply these lemmas to the diagram below. In this diagram U_1 and U_2 are open subsets of X, $M = U_1 \cup U_2$, we write $U_{1,2} := U_1 \cap U_2$, and the vertical arrows are the mappings defined by the pairing (5.4.5). We will leave for you to check that this is a commutative diagram "up to sign". (To make it commutative one has to replace some of the vertical arrows by their negatives.) This is easy to check except for the commutative square on the extreme left. To check that this square commutes, some serious diagram-chasing is required.

$$\begin{array}{ccccccccc}
\cdots \longrightarrow & H_c^{k-1}(M) & \longrightarrow & H_c^k(U_{1,2}) & \longrightarrow & H_c^k(U_1) \oplus H_c^k(U_2) & \longrightarrow & H_c^k(M) & \longrightarrow \cdots \\
 & \downarrow & & \downarrow & & \downarrow & & \downarrow & \\
\cdots \to & H^{n-(k-1)}(M)^\star & \to & H^{n-k}(U_{1,2})^\star & \to & H^{n-k}(U_1)^\star \oplus H^{n-k}(U_2)^\star & \to & H^{n-k}(M)^\star & \to \cdots
\end{array}$$

By Mayer–Vietoris the top row of the diagram is exact and by Mayer–Vietoris and Lemma 5.4.8 the bottom row of the diagram is exact. Hence we can apply the "five lemma" to the diagram to conclude the following.

Lemma 5.4.11. *If the maps*

$$H^k(U) \to H_c^{n-k}(U)^\star$$

defined by the pairing (5.4.5) are bijective for U_1, U_2 and $U_{1,2} = U_1 \cap U_2$, they are also bijective for $M = U_1 \cup U_2$.

Thus to prove Theorem 5.4.7 we can argue by induction as in § 5.3. Let $U_1, U_2, \ldots, U_N$ be a good cover of X. If $N = 1$, then $X = U_1$ and, hence, since U_1 is diffeomorphic to $\mathbf{R}^n$, the map (5.4.12) is bijective by Proposition 5.4.6. Now let us assume the theorem is true for manifolds involving good covers by k open sets where k is less than N. Let $U' = U_1 \cup \cdots \cup U_{N-1}$ and $U'' = U_N$. Since

$$U' \cap U'' = U_1 \cap U_N \cup \cdots \cup U_{N-1} \cap U_N$$

it can be covered by a good cover by k open sets, $k < N$, and hence the hypotheses of the lemma are true for U', U'' and $U' \cap U''$. Thus the lemma says that (5.4.12) is bijective for the union X of U' and U''. □

Exercises for §5.4

Exercise 5.4.i (proper pushforward for compactly supported de Rham cohomology). Let X be an m-manifold, Y an n-manifold and $f : X \to Y$ a C^∞ map. Suppose that both of these manifolds are oriented and connected and have finite topology. Show that there exists a unique linear map

$$f_! : H_c^{m-k}(X) \to H_c^{n-k}(Y) \tag{5.4.12}$$

with the property

$$B_Y(f_! c_1, c_2) = B_X(c_1, f^\sharp c_2) \tag{5.4.13}$$

for all $c_1 \in H_c^{m-k}(X)$ and $c_2 \in H^k(Y)$. (In this formula B_X is the bilinear pairing (5.4.5) on X and B_Y is the bilinear pairing (5.4.5) on Y.)

Exercise 5.4.ii. Suppose that the map f in Exercise 5.4.i is proper. Show that there exists a unique linear map

$$f_! : H^{m-k}(X) \to H^{n-k}(Y)$$

with the property

$$B_Y(c_1, f_! c_2) = (-1)^{k(m-n)} B_X(f^\sharp c_1, c_2)$$

for all $c_1 \in H_c^k(Y)$ and $c_2 \in H^{m-k}(X)$, and show that, if X and Y are compact, this mapping is the same as the mapping $f_!$ in Exercise 5.4.i.

Exercise 5.4.iii. Let U be an open subset of $\mathbf{R}^n$ and let $f : U \times \mathbf{R} \to U$ be the projection, $f(x,t) = x$. Show that there is a unique linear mapping

$$f_\star : \Omega_c^{k+1}(U \times \mathbf{R}) \to \Omega_c^k(U)$$

with the property

$$\int_U f_\star \mu \wedge \nu = \int_{U \times \mathbf{R}} \mu \wedge f^\star \nu \tag{5.4.14}$$

for all $\mu \in \Omega_c^{k+1}(U \times \mathbf{R})$ and $\nu \in \Omega^{n-k}(U)$.

Hint: Let $x_1, \ldots, x_n$ and t be the standard coordinate functions on $\mathbf{R}^n \times \mathbf{R}$. By Exercise 2.3.v every $(k+1)$-form $\omega \in \Omega_c^{k+1}(U \times \mathbf{R})$ can be written uniquely in "reduced form" as a sum

$$\omega = \sum_I f_I dt \wedge dx_I + \sum_J g_J dx_J$$

over multi-indices, I and J, which are strictly increasing. Let

$$f_\star \omega = \sum_I \left(\int_{\mathbf{R}} f_I(x,t) dt \right) dx_I.$$

Exercise 5.4.iv. Show that the mapping $f_\star$ in Exercise 5.4.iii satisfies $f_\star d\omega = d f_\star \omega$.

Exercise 5.4.v. Show that if ω is a closed compactly supported $(k+1)$-form on $U \times \mathbf{R}$ then

$$[f_\star \omega] = f_![\omega],$$

where $f_!$ is the mapping (5.4.13) and $f_\star$ the mapping (5.4.14).

Exercise 5.4.vi.

(1) Let U be an open subset of $\mathbf{R}^n$ and let $f : U \times \mathbf{R}^\ell \to U$ be the projection, $f(x,t) = x$. Show that there is a unique linear mapping

$$f_\star : \Omega_c^{k+\ell}(U \times \mathbf{R}^\ell) \to \Omega_c^k(U)$$

with the property

$$\int_U f_\star \mu \wedge \nu = \int_{U \times \mathbf{R}^\ell} \mu \wedge f^\star \nu$$

for all $\mu \in \Omega_c^{k+\ell}(U \times \mathbf{R}^\ell)$ and $\nu \in \Omega^{n-k}(U)$.
Hint: Exercise 5.4.iii plus induction on ℓ.

(2) Show that for $\omega \in \Omega_c^{k+\ell}(U \times \mathbf{R}^\ell)$

$$d f_\star \omega = f_\star d\omega.$$

(3) Show that if ω is a closed, compactly supported $(k+\ell)$-form on $X \times \mathbf{R}^\ell$

$$[f_\star \omega] = f_![\omega],$$

where $f_! : H_c^{k+\ell}(U \times \mathbf{R}^\ell) \to H_c^k(U)$ is the map (5.4.13).

Exercise 5.4.vii. Let X be an n-manifold and Y an m-manifold. Assume X and Y are compact, oriented and connected, and orient $X \times Y$ by giving it its natural product orientation. Let

$$f : X \times Y \to Y$$

be the projection map $f(x, y) = y$. Given

$$\omega \in \Omega^m(X \times Y)$$

and $p \in Y$, let

$$f_\star\omega(p) := \int_X \iota_p^\star \omega, \tag{5.4.15}$$

where $\iota_p \colon X \to X \times Y$ is the inclusion map $\iota_p(x) = (x, p)$.

(1) Show that the function $f_\star\omega$ defined by equation (5.4.15) is C^∞, i.e., is in $\Omega^0(Y)$.
(2) Show that if ω is closed this function is constant.
(3) Show that if ω is closed

$$[f_\star\omega] = f_![\omega],$$

where $f_! \colon H^n(X \times Y) \to H^0(Y)$ is the map (5.4.13).

Exercise 5.4.viii.

(1) Let X be an n-manifold which is compact, connected and oriented. Combining Poincaré duality with Exercise 5.3.xii show that

$$H_c^{k+\ell}(X \times \mathbf{R}^\ell) \cong H_c^k(X).$$

(2) Show, moreover, that if $f \colon X \times \mathbf{R}^\ell \to X$ is the projection $f(x, a) = x$, then

$$f_! \colon H_c^{k+\ell}(X \times \mathbf{R}^\ell) \to H_c^k(X)$$

is a bijection.

Exercise 5.4.ix. Let X and Y be as in Exercise 5.4.i. Show that the pushforward operation (5.4.13) satisfies the ***projection formula***

$$f_!(c_1 \cdot f^\sharp c_2) = f_!(c_1) \cdot c_2,$$

for $c_1 \in H_c^k(X)$ and $c_2 \in H^\ell(Y)$.

5.5. Thom classes and intersection theory

Let X be a connected, oriented n-manifold. If X has finite topology its cohomology groups are finite dimensional, and since the bilinear pairing B defined by (5.4.5) is non-singular we get from this pairing bijective linear maps

$$L_B \colon H_c^{n-k}(X) \to H^k(X)^\star \tag{5.5.1}$$

and

$$L_B^\star \colon H^{n-k}(X) \to H_c^k(X)^\star. \tag{5.5.2}$$

In particular, if $\ell \colon H^k(X) \to \mathbf{R}$ is a linear function (i.e., an element of $H^k(X)^\star$), then by (5.5.1) we can convert ℓ into a cohomology class

$$L_B^{-1}(\ell) \in H_c^{n-k}(X), \tag{5.5.3}$$

and similarly if $\ell_c \colon H_c^k(X) \to \mathbf{R}$ is a linear function, we can convert it by (5.5.2) into a cohomology class

$$(L_B^\star)^{-1}(\ell_c) \in H^{n-k}(X).$$

One way that linear functions like this arise in practice is by integrating forms over submanifolds of X. Namely let Y be a closed oriented k-dimensional submanifold of X. Since Y is oriented, we have by (5.1.2) an integration operation in cohomology

$$I_Y \colon H_c^k(Y) \to \mathbf{R},$$

and since Y is closed the inclusion map ι_Y of Y into X is proper, so we get from it a pullback operation on cohomology

$$\iota_Y^\sharp \colon H_c^k(X) \to H_c^k(Y)$$

and by composing these two maps, we get a linear map $\ell_Y = I_Y \circ \iota_Y^\sharp \colon H_c^k(X) \to \mathbf{R}$. The cohomology class

$$T_Y \coloneqq L_B^{-1}(\ell_Y) \in H_c^k(X)$$

associated with ℓ_Y is called the ***Thom class*** of the manifold Y and has the defining property

$$B(T_Y, c) = I_Y(\iota_Y^\sharp c) \tag{5.5.4}$$

for $c \in H_c^k(X)$. Let us see what this defining property looks like at the level of forms. Let $\tau_Y \in \Omega^{n-k}(X)$ be a closed k-form representing T_Y. Then by (5.4.5), the formula (5.5.4) for $c = [\omega]$ becomes the integral formula

$$\int_X \tau_Y \wedge \omega = \int_Y \iota_Y^\star \omega. \tag{5.5.5}$$

In other words, for every closed form $\omega \in \Omega_c^{n-k}(X)$, the integral of ω over Y is equal to the integral over X of $\tau_Y \wedge \omega$. A closed form τ_Y with this "reproducing" property is called a ***Thom form*** for Y. Note that if we add to τ_Y an exact $(n-k)$-form $\mu \in d\Omega^{n-k-1}(X)$, we get another representative $\tau_Y + \mu$ of the cohomology class T_Y, and hence another form with this reproducing property. Also, since equation (5.5.5) is a direct translation into form language of equation (5.5.4), any closed $(n-k)$-form τ_Y with the reproducing property (5.5.5) is a representative of the cohomology class T_Y.

These remarks make sense as well for compactly supported cohomology. Suppose Y is compact. Then from the inclusion map we get a pullback map

$$\iota_Y^\sharp \colon H^k(X) \to H^k(Y)$$

and since Y is compact, the integration operation I_Y is a map $H^k(Y) \to \mathbf{R}$, so the composition of these two operations is a map

$$\ell_Y \colon H^k(X) \to \mathbf{R},$$

which by (5.5.3) gets converted into a cohomology class

$$T_Y = L_B^{-1}(\ell_Y) \in H_c^{n-k}(X).$$

Moreover, if $\tau_Y \in \Omega_c^{n-k}(X)$ is a closed form, it represents this cohomology class if and only if it has the reproducing property

$$\int_X \tau_Y \wedge \omega = \int_Y \iota_Y^\star \omega \tag{5.5.6}$$

for closed forms $\omega \in \Omega^{n-k}(X)$. (There's a subtle difference, however, between formulas (5.5.5) and (5.5.6). In (5.5.5) ω has to be closed *and* compactly supported and in (5.5.6) it just has to be closed.)

As above we have a lot of latitude in our choice of τ_Y: we can add to it any element of $d\Omega_c^{n-k-1}(X)$. One consequence of this is the following.

Theorem 5.5.7. *Given a neighborhood U of Y in X, there exists a closed form $\tau_Y \in \Omega_c^{n-k}(U)$ with the reproducing property*

$$\int_U \tau_Y \wedge \omega = \int_Y \iota_Y^\star \omega \tag{5.5.8}$$

for all closed forms $\omega \in \Omega^k(U)$.

Hence, in particular, τ_Y has the reproducing property (5.5.6) for all closed forms $\omega \in \Omega^{n-k}(X)$. This result shows that the Thom form τ_Y can be chosen to have support in an *arbitrarily small neighborhood* of Y.

Proof of Theorem 5.5.7. By Theorem 5.3.9 we can assume that U has finite topology and hence, in our definition of τ_Y, we can replace the manifold X by the open submanifold U. This gives us a Thom form τ_Y with support in U and with the reproducing property (5.5.8) for closed forms $\omega \in \Omega^{n-k}(U)$. □

Let us see what Thom forms actually look like in concrete examples. Suppose Y is defined globally by a system of ℓ independent equations, i.e., suppose there exist an open neighborhood O of Y in X, a C^∞ map $f : O \to \mathbf{R}^\ell$, and a bounded open convex neighborhood V of the origin in $\mathbf{R}^n$ satisfying the following properties.

Properties 5.5.9.

(1) The origin is a regular value of f.
(2) $f^{-1}(\overline{V})$ is closed in X.
(3) $Y = f^{-1}(0)$.

Then by (1) and (3) Y is a closed submanifold of O and by (2) it's a closed submanifold of X. Moreover, it has a natural orientation: For every $p \in Y$ the map

$$df_p : T_pX \to T_0\mathbf{R}^\ell$$

is surjective, and its kernel is T_pY, so from the standard orientation of $T_0\mathbf{R}^\ell$ one gets an orientation of the quotient space

$$T_pX/T_pY,$$

and hence since T_pX is oriented, one gets by Theorem 1.9.9 an orientation on T_pY. (See Example 4.4.5.) Now let μ be an element of $\Omega_c^\ell(X)$. Then $f^\star\mu$ is supported in $f^{-1}(\overline{V})$ and hence by property (2) we can extend it to X by setting it equal to zero outside O. We will prove the following.

Theorem 5.5.10. *If $\int_V \mu = 1$, then $f^\star\mu$ is a Thom form for Y.*

To prove this we'll first prove that if $f^\star\mu$ has property (5.5.5) for some choice of μ it has this property for every choice of μ.

Lemma 5.5.11. *Let μ_1 and μ_2 be forms in $\Omega_c^\ell(V)$ such that $\int_V \mu_1 = \int_V \mu_2 = 1$. Then for every closed k-form $\nu \in \Omega_c^k(X)$ we have*

$$\int_X f^\star\mu_1 \wedge \nu = \int_X f^\star\mu_2 \wedge \nu.$$

Proof. By Theorem 3.2.2, $\mu_1 - \mu_2 = d\beta$ for some $\beta \in \Omega_c^{\ell-1}(V)$, hence, since $d\nu = 0$

$$(f^\star\mu_1 - f^\star\mu_2) \wedge \nu = df^\star\beta \wedge \nu = d(f^\star\beta \wedge \nu).$$

Therefore, by Stokes theorem, the integral over X of the expression on the left is zero. □

Now suppose $\mu = \rho(x_1, \dots, x_\ell)dx_1 \wedge \cdots \wedge dx_\ell$, for ρ in $C_0^\infty(V)$. For $t \le 1$ let

$$\mu_t = t^\ell \rho\left(\frac{x_1}{t}, \dots, \frac{x_\ell}{t}\right) dx_1 \wedge \cdots \wedge dx_\ell .$$

This form is supported in the convex set tV, so by Lemma 5.5.11

$$\int_X f^\star\mu_t \wedge \nu = \int_X f^\star\mu \wedge \nu$$

for all closed forms $\nu \in \Omega_c^k(X)$. Hence to prove that $f^\star\mu$ has the property (5.5.5), it suffices to prove that

$$\lim_{t\to 0} \int f^\star\mu_t \wedge \nu = \int_Y \iota_Y^\star \nu. \tag{5.5.12}$$

We prove this by proving a stronger result.

Lemma 5.5.13. *The assertion (5.5.12) is true for every k-form $\nu \in \Omega_c^k(X)$.*

Proof. The canonical submersion theorem (see Theorem B.17) says that for every $p \in Y$ there exist a neighborhood U_p of p in Y, a neighborhood, W of 0 in $\mathbf{R}^n$, and an orientation-preserving diffeomorphism $\psi: (W, 0) \to (U_p, p)$ such that

$$f \circ \psi = \pi, \tag{5.5.14}$$

where $\pi: \mathbf{R}^n \to \mathbf{R}^\ell$ is the canonical submersion, $\pi(x_1,\dots,x_n) = (x_1,\dots,x_\ell)$. Let $\mathcal{U}$ be the cover of O by the open sets $O \setminus Y$ and the U_p's. Choosing a partition of unity subordinate to this cover it suffices to verify (5.5.12) for ν in $\Omega_c^k(O \setminus Y)$ and ν in $\Omega_c^k(U_p)$. Let us first suppose ν is in $\Omega_c^k(O \setminus Y)$. Then $f(\operatorname{supp}\nu)$ is a compact subset of $\mathbf{R}^\ell \setminus \{0\}$ and hence for t small $f(\operatorname{supp}\nu)$ is disjoint from tV, and both sides of (5.5.12) are zero. Next suppose that ν is in $\Omega_c^k(U_p)$. Then $\psi^\star\nu$ is a compactly supported k-form on W so we can write it as a sum

$$\psi^\star\nu = \sum_I h_I(x)dx_I, \quad h_I \in C_0^\infty(W)$$

the I's being strictly increasing multi-indices of length k. Let $I_0 = (\ell+1, \ell_2+2, \dots, n)$. Then

$$\pi^\star\mu_t \wedge \psi^\star\nu = t^\ell\rho\left(\frac{x_1}{t},\dots,\frac{x_\ell}{t}\right)h_{I_0}(x_1,\dots,x_n)dx_r \wedge \cdots \wedge dx_n \tag{5.5.15}$$

and by (5.5.14)

$$\psi^\star(f^\star\mu_t \wedge \nu) = \pi^\star\mu_t \wedge \psi^\star\nu$$

and hence since ψ is orientation preserving

$$\begin{aligned}\int_X f^\star\mu_t \wedge \nu = \int_{U_p} f^\star\mu_t \wedge \nu &= t^\ell \int_{\mathbf{R}^n} \rho\left(\frac{x_1}{t},\dots,\frac{x_\ell}{t}\right)h_{I_0}(x_1,\dots,x_n)dx \\ &= \int_{\mathbf{R}^n} \rho(x_1,\dots,x_\ell)h_{I_0}(tx_1,\dots,tx_\ell,x_{\ell+1},\dots,x_n)dx\end{aligned}$$

and the limit of this expression as t tends to zero is

$$\int \rho(x_1,\dots,x_\ell)h_{I_0}(0,\dots,0,x_{\ell+1},\dots,x_n)dx_1\cdots dx_n$$

or

$$\int h_I(0,\dots,0,x_{\ell+1},\dots,x_n)\,dx_{\ell+1}\cdots dx_n. \tag{5.5.16}$$

This, however, is just the integral of $\psi^\star\nu$ over the set $\pi^{-1}(0) \cap W$. By (5.5.12), ψ maps $\pi^{-1}(0)\cap W$ diffeomorphically onto $Y \cap U_p$ and by our recipe for orienting Y this diffeomorphism is an orientation-preserving diffeomorphism, so the integral (5.5.16) is equal to $\int_Y \nu$. □

We'll now describe some applications of Thom forms to *topological intersection theory*. Let Y and Z be closed, oriented submanifolds of X of dimensions k and ℓ where $k + \ell = n$, and let us assume one of them (say Z) is compact. We will show below how to define an *intersection number* $I(Y, Z)$, which on the one hand will be a topological invariant of Y and Z and on the other hand will actually count, with appropriate $\pm$-signs, the number of points of intersection of Y and Z when they intersect non-tangentially. (Thus this notion is similar to the notion of the degree $\deg(f)$ for a C^∞ mapping f. On the one hand $\deg(f)$ is a topological invariant of f. It's unchanged if we deform f by a homotopy. On the other hand if q is a regular value of f, then $\deg(f)$ counts with appropriate $\pm$-signs the number of points in the set $f^{-1}(q)$.)

We'll first give the topological definition of this intersection number. This is by the formula

$$I(Y,Z) = B(T_Y, T_Z), \tag{5.5.17}$$

where $T_Y \in H^\ell(X)$ and $T_Z \in H^k_c(X)$ and B is the bilinear pairing (5.4.5). If $\tau_Y \in \Omega^\ell(X)$ and $\tau_Z \in \Omega^k_c(X)$ are Thom forms representing T_Y and T_Z, (5.5.17) can also be defined as the integral

$$I(Y,Z) = \int_X \tau_Y \wedge \tau_Z$$

or by (5.5.8), as the integral over Y,

$$I(Y,Z) = \int_Y \iota_Y^\star \tau_Z \tag{5.5.18}$$

or, since $\tau_Y \wedge \tau_Z = (-1)^{k\ell} \tau_Z \wedge \tau_Y$, as the integral over Z

$$I(X,Y) = (-1)^{k\ell} \int_Z \iota_Z^\star \tau_Y.$$

In particular

$$I(Y,Z) = (-1)^{k\ell} I(Z,Y).$$

As a test case for our declaring $I(Y,Z)$ to be the intersection number of Y and Z we will first prove the following proposition.

Proposition 5.5.19. *If Y and Z **do not** intersect, then $I(Y,Z) = 0$.*

Proof. If Y and Z don't intersect then, since Y is closed, $U = X \smallsetminus Y$ is an open neighborhood of Z in X, therefore since Z is compact there exists by Theorem 5.5.7 a Thom form $\tau_Z \in \Omega^\ell_c(U)$. Thus $\iota_Y^\star \tau_Z = 0$, and so by (5.5.18) we have $I(Y,Z) = 0$. □

We'll next indicate how one computes $I(Y,Z)$ when Y and Z intersect "non-tangentially", or, to use terminology more in current usage, when their intersection is *transversal*. Recall that at a point of intersection $p \in Y \cap Z$, T_pY and T_pZ are vector subspaces of T_pX.

Definition 5.5.20. Y and Z intersect ***transversally*** if for every $p \in Y \cap Z$ we have

$$T_pY \cap T_pZ = 0.$$

Since $n = k + \ell = \dim T_pY + \dim T_pZ = \dim T_pX$, this condition is equivalent to

$$T_pX = T_pY \oplus T_pZ, \tag{5.5.21}$$

i.e., every vector $u \in T_pX$, can be written uniquely as a sum $u = v + w$, with $v \in T_pY$ and $w \in T_pZ$. Since X, Y and Z are oriented, their tangent spaces at p are oriented, and we'll say that these spaces are ***compatibly oriented*** if the orientations of the two sides of (5.5.21) agree. (In other words if $v_1, \dots, v_k$ is an oriented basis of T_pY and $w_1, \dots, w_\ell$ is an oriented basis of T_pZ, the n vectors, $v_1, \dots, v_k, w_1, \dots, w_\ell$, are an oriented basis of T_pX.) We will define the ***local intersection number*** $I_p(Y, Z)$ of Y and Z at p to be equal to $+1$ if X, Y, and Z are compatibly oriented at p and to be equal to -1 if they're not. With this notation we'll prove the following theorem.

Theorem 5.5.22. *If Y and Z intersect transversally, then $Y \cap Z$ is a finite set and*

$$I(Y, Z) = \sum_{p \in Y \cap Z} I_p(Y, Z).$$

To prove this we first need to show that transverse intersections look nice locally.

Theorem 5.5.23. *If Y and Z intersect transversally, then for every $p \in Y \cap Z$, there exist an open neighborhood V_p of p in X, an open neighborhood U_p of the origin in $\mathbf{R}^n$, and an orientation-preserving diffeomorphism $\psi_p \colon V_p \xrightarrow{\sim} U_p$ which maps $V_p \cap Y$ diffeomorphically onto the subset of U_p defined by the equations: $x_1 = \cdots = x_\ell = 0$, and maps $V \cap Z$ onto the subset of U_p defined by the equations: $x_{\ell+1} = \cdots = x_n = 0$.*

Proof. Since this result is a local result, we can assume that $X = \mathbf{R}^n$ and hence by Theorem 4.2.15 that there exist a neighborhood V_p of p in $\mathbf{R}^n$ and submersions $f \colon (V_p, p) \to (\mathbf{R}^\ell, 0)$ and $g \colon (V_p, p) \to (\mathbf{R}^k, 0)$ with the properties

$$V_p \cap Y = f^{-1}(0) \tag{5.5.24}$$

and

$$V_p \cap Z = g^{-1}(0). \tag{5.5.25}$$

Moreover, by (4.3.5) $T_pY = (df_p)^{-1}(0)$ and $T_pZ = (dg_p)^{-1}(0)$. Hence by (5.5.21), the equations

$$df_p(v) = dg_p(v) = 0 \tag{5.5.26}$$

for $v \in T_pX$ imply that $v = 0$. Now let $\psi_p \colon V_p \to \mathbf{R}^n$ be the map

$$(f, g) \colon V_p \to \mathbf{R}^\ell \times \mathbf{R}^k = \mathbf{R}^n.$$

Then by (5.5.26), $d\psi_p$ is bijective, therefore, shrinking V_p if necessary, we can assume that ψ_p maps V_p diffeomorphically onto a neighborhood U_p of the origin in $\mathbf{R}^n$, and hence by (5.5.24) and (5.5.25) ψ_p maps $V_p \cap Y$ onto the set: $x_1 = \cdots = x_\ell = 0$ and maps $V_p \cap Z$ onto the set: $x_{\ell+1} = \cdots = x_n = 0$. Finally, if ψ isn't orientation preserving, we can make it so by composing it with the involution $(x_1, \dots, x_n) \mapsto (x_1, x_2, \dots, x_{n-1}, -x_n)$. ☐

From this result we deduce the following theorem.

Theorem 5.5.27. *If Y and Z intersect transversally, their intersection is a finite set.*

Proof. By Theorem 5.5.23 the only point of intersection in V_p is p itself. Moreover, since Y is closed and Z is compact, $Y \cap Z$ is compact. Therefore, since the V_p's cover $Y \cap Z$ we can extract a finite subcover by compactness. However, since no two V_p's cover the same point of $Y \cap Z$, this cover must already be a finite subcover. □

We will now prove Theorem 5.5.22.

Proof of Theorem 5.2.22. Since Y is closed, the map $\iota_Y : Y \hookrightarrow X$ is proper, so by Theorem 3.4.7 there exists a neighborhood U of Z in X such that $U \cap Y$ is contained in the union of the open sets V_p above. Moreover by Theorem 5.5.7 we can choose τ_Z to be supported in U and by Theorem 5.3.2 we can assume that U has finite topology, so we're reduced to proving the theorem with X replaced by U and Y replaced by $Y \cap U$. Let

$$O \coloneqq U \cap \bigcup_{p \in X} V_p,$$

let $f : O \to \mathbf{R}^\ell$ be the map whose restriction to $V_p \cap U$ is $\pi \circ \psi_p$ where π is, as in equation (5.5.14), the canonical submersion of $\mathbf{R}^n$ onto $\mathbf{R}^\ell$, and finally let V be a bounded convex neighborhood of $\mathbf{R}^\ell$, whose closure is contained in the intersection of the open sets, $\pi \circ \psi_p(V_p \cap U)$. Then $f^{-1}(\overline{V})$ is a closed subset of U, so if we replace X by U and Y by $Y \cap U$, the data (f, O, V) satisfy Properties 5.5.9. Thus to prove Theorem 5.5.22 it suffices by Theorem 5.5.10 to prove this theorem with

$$\tau_Y = \sigma_p(Y) f^\star \mu$$

on $V_p \cap O$ where $\sigma_p(Y) = +1$ or -1 depending on whether the orientation of $Y \cap V_p$ in Theorem 5.5.10 coincides with the given orientation of Y or not. Thus

$$\begin{aligned} I(Y,Z) &= (-1)^{k\ell} I(Z,Y) \\ &= (-1)^{k\ell} \sum_{p \in X} \sigma_p(Y) \int_Z \iota_Z^\star f^\star \mu \\ &= (-1)^{k\ell} \sum_{p \in X} \sigma_p(Y) \int_Z \iota_Z^\star \psi_p^\star \pi^\star \mu \\ &= \sum_{p \in X} (-1)^{k\ell} \sigma_p(Y) \int_{Z \cap V_p} (\pi \circ \psi_p \circ \iota_Z)^\star \mu. \end{aligned}$$

But $\pi \circ \psi_p \circ \iota_Z$ maps an open neighborhood of p in $U_p \cap Z$ diffeomorphically onto V, and μ is compactly supported in V, so since $\int_V \mu = 1$,

$$\int_{Z \cap U_p} (\pi \circ \psi_p \circ \iota_Z)^\star \mu = \sigma_p(Z) \int_V \mu = \sigma_p(Z),$$

where $\sigma_p(Z) = +1$ or -1 depending on whether $\pi \circ \psi_p \circ \iota_Z$ is orientation preserving or not. Thus finally

$$I(Y,Z) = \sum_{p \in X} (-1)^{k\ell} \sigma_p(Y)\sigma_p(Z).$$

We will leave as an exercise the task of unraveling these orientations and showing that

$$(-1)^{k\ell}\sigma_p(Y)\sigma_p(Z) = I_p(Y,Z)$$

and hence that $I(Y,Z) = \sum_{p \in X} I_p(Y,Z)$. □

Exercises for §5.5

Exercise 5.5.i. Let X be a connected oriented n-manifold, W a connected oriented ℓ-dimensional manifold, $f: X \to W$ a C^∞ map, and Y a closed submanifold of X of dimension $k := n - \ell$. Suppose Y is a "level set" of the map f, i.e., suppose that q is a regular value of f and that $Y = f^{-1}(q)$. Show that if μ is in $\Omega_c^\ell(Z)$ and its integral over Z is 1, then one can orient Y so that $\tau_Y = f^\star \mu$ is a Thom form for Y.

Hint: Theorem 5.5.10.

Exercise 5.5.ii. In Exercise 5.5.i show that if $Z \subset X$ is a compact oriented ℓ-dimensional submanifold of X then

$$I(Y,Z) = (-1)^{k\ell} \deg(f \circ \iota_Z).$$

Exercise 5.5.iii. Let q_1 be another regular value of the map $f: X \to W$ and let $Y_1 = f^{-1}(q)$. Show that

$$I(Y,Z) = I(Y_1,Z).$$

Exercise 5.5.iv.

(1) Show that if q is a regular value of the map $f \circ \iota_Z : Z \to W$, then Z and Y intersect transversally.

(2) Show that this is an "if and only if" proposition: If Y and Z intersect transversally, then q is a regular value of the map $f \circ \iota_Z$.

Exercise 5.5.v. Suppose q is a regular value of the map $f \circ \iota_Z$. Show that p is in $Y \cap Z$ if and only if p is in the preimage $(f \circ \iota_Z)^{-1}(q)$ of q and that

$$I_p(X,Y) = (-1)^{k\ell}\sigma_p,$$

where σ_p is the orientation number of the map $f \circ \iota_Z$, at p, i.e., $\sigma_p = 1$ if $f \circ \iota_Z$ is orientation preserving at p and $\sigma_p = -1$ if $f \circ \iota_Z$ is orientation reversing at p.

Exercise 5.5.vi. Suppose the map $f: X \to W$ is proper. Show that there exists a neighborhood V of q in W having the property that *all* points of V are regular values of f.

Hint: Since q is a regular value of f there exists, for every $p \in f^{-1}(q)$ a neighborhood U_p of p on which f is a submersion. Conclude, by Theorem 3.4.7, that there exists a neighborhood V of q with $f^{-1}(V) \subset \bigcup_{p \in f^{-1}(q)} U_p$.

Exercise 5.5.vii. Show that in every neighborhood V_1 of q in V there exists a point q_1 whose preimage

$$Y_1 := f^{-1}(q_1)$$

intersects Z transversally. Conclude that one can "deform Y an arbitrarily small amount so that it intersects Z transversally".

Hint: Exercise 5.5.iv plus Sard's theorem.

Exercise 5.5.viii (Intersection theory for mappings). Let X be an oriented, connected n-manifold, Z a compact, oriented ℓ-dimensional submanifold, Y an oriented manifold of dimension $k := n - \ell$ and $f : Y \to X$ a proper C^∞ map. Define the ***intersection number*** of f with Z to be the integral

$$I(f, Z) := \int_Y f^\star \tau_Z.$$

(1) Show that $I(f, Z)$ is a homotopy invariant of f, i.e., show that if $f_0, f_1 : Y \to X$ are proper C^∞ maps and are properly homotopic, then

$$I(f_0, Z) = I(f_1, Z).$$

(2) Show that if Y is a closed submanifold of X of dimension $k = n - \ell$ and $\iota_Y : Y \to X$ is the inclusion map, then $I(\iota_Y, Z) = I(Y, Z)$.

Exercise 5.5.ix.

(1) Let X be an oriented connected n-manifold and let Z be a compact zero-dimensional submanifold consisting of a single point $z_0 \in X$. Show that if μ is in $\Omega_c^n(X)$ then μ is a Thom form for Z if and only if its integral is 1.

(2) Let Y be an oriented n-manifold and $f : Y \to X$ a C^∞ map. Show that for $Z = \{z_0\}$ as in part (1) we have $I(f, Z) = \deg(f)$.

5.6. The Lefschetz theorem

In this section we'll apply the intersection techniques that we developed in §5.5 to a concrete problem in dynamical systems: counting the number of fixed points of a differentiable mapping. The Brouwer fixed point theorem, which we discussed in §3.6, told us that a C^∞ map of the unit ball into itself has to have at least one fixed point. The Lefschetz theorem is a similar result for manifolds. It will tell us that a C^∞ map of a compact manifold into itself has to have a fixed point if a certain topological invariant of the map, its *global Lefschetz number*, is non-zero.

Before stating this result, we will first show how to translate the problem of counting fixed points of a mapping into an intersection number problem. Let X be an oriented compact n-manifold and $f : X \to X$ a C^∞ map. Define the ***graph of f*** in $X \times X$ to be the set

$$\Gamma_f := \{(x, f(x)) \mid x \in X\} \subset X \times X.$$

It's easy to see that this is an n-dimensional submanifold of $X \times X$ and that this manifold is diffeomorphic to X itself. In fact, in one direction, there is a C^∞ map

$$\gamma_f : X \to \Gamma_f, \quad x \mapsto (x, f(x)), \tag{5.6.1}$$

and, in the other direction, a C^∞ map

$$\pi \colon \Gamma_f \to X, \quad (x, f(x)) \mapsto x,$$

and it's obvious that these maps are inverses of each other and hence diffeomorphisms. We will orient Γ_f by requiring that γ_f and π be orientation-preserving diffeomorphisms.

An example of a graph is the graph of the identity map of X onto itself. This is the ***diagonal*** in $X \times X$

$$\Delta := \{(x, x) \mid x \in X\} \subset X \times X$$

and its intersection with Γ_f is the set

$$\Gamma_f \cap \Delta = \{(x, x) \mid f(x) = x\},$$

which is just the set of fixed points of f. Hence a natural way to count the fixed points of f is as the intersection number of Γ_f and Δ in $X \times X$. To do so we need these three manifolds to be oriented, but, as we noted above, Γ_f and Δ acquire orientations from the identifications (5.6.1) and, as for $X \times X$, we'll give it its natural orientation as a product of oriented manifolds. (See §4.5.)

Definition 5.6.2. The ***global Lefschetz number*** of f is the intersection number

$$L(f) := I(\Gamma_f, \Delta).$$

In this section we'll give two recipes for computing this number: one by topological methods and the other by making transversality assumptions and computing this number as a sum of local intersection numbers via Theorem 5.5.22. We first show what one gets from the transversality approach.

Definition 5.6.3. The map $f \colon X \to X$ is a ***Lefschetz map***, or simply ***Lefschetz***, if Γ_f and Δ intersect transversally.

Let us see what being Lefschetz entails. Suppose p is a fixed point of f. Then at the point $q = (p, p)$ of Γ_f

$$T_q(\Gamma_f) = (d\gamma_f)_p T_p X = \{(v, df_p(v)) \mid v \in T_p X\} \tag{5.6.4}$$

and, in particular, for the identity map,

$$T_q(\Delta) = \{(v, v) \mid v \in T_p X\}.$$

Therefore, if Δ and Γ_f are to intersect transversally, the intersection of $T_q(\Gamma_f) \cap T_q(\Delta)$ inside $T_q(X \times X)$ has to be the zero space. In other words if

$$(v, df_p(v)) = (v, v) \tag{5.6.5}$$

then $v = 0$. But the identity (5.6.5) says that v is a fixed point of df_p, so transversality at p amounts to the assertion

$$df_p(v) = v \iff v = 0,$$

or in other words the assertion that the map

$$(\mathrm{id}_{T_pX} - df_p)\colon T_pX \to T_pX \tag{5.6.6}$$

is bijective.

Proposition 5.6.7. *The local intersection number $I_p(\Gamma_f, \Delta)$ is 1 if (5.6.6) is orientation preserving and -1 otherwise.*

In other words $I_p(\Gamma_f, \Delta)$ is the sign of $\det(\mathrm{id}_{T_pX} - df_p)$.

Proof. To prove this let $e_1, \dots, e_n$ be an oriented basis of T_pX and let

$$df_p(e_i) = \sum_{j=1}^{n} a_{j,i} e_j. \tag{5.6.8}$$

Now set $v_i := (e_i, 0) \in T_q(X \times X)$ and $w_i := (0, e_i) \in T_q(X \times X)$. Then by the definition of the product orientation on $X \times X$, we have that $(v_1, \dots, v_n, w_1, \dots, w_n)$ is an oriented basis of $T_q(X \times X)$, and by (5.6.4)

$$\left(v_1 + \sum_{j=1}^{n} a_{j,i} w_j, \dots, v_n + \sum_{j=1}^{n} a_{j,n} w_j\right)$$

is an oriented basis of $T_q\Gamma_f$ and $v_1 + w_1, \dots, v_n + w_n$ is an oriented basis of $T_q\Delta$. Thus $I_p(\Gamma_f, \Delta) = \pm 1$, depending on whether or not the basis

$$\left(v_1 + \sum_{j=1}^{n} a_{j,i} w_j, \dots, v_n + \sum_{j=1}^{n} a_{j,n} w_j, v_1 + w_1, \dots, v_n + w_n\right)$$

of $T_q(X \times X)$ is compatibly oriented with the basis (5.6.8). Thus $I_p(\Gamma_f, \Delta) = \pm 1$, the sign depending on whether the determinant of the $2n \times 2n$ matrix relating these two bases:

$$P := \begin{pmatrix} \mathrm{id}_n & A \\ \mathrm{id}_n & \mathrm{id}_n \end{pmatrix}$$

is positive or negative, where $A = (a_{i,j})$. However, by the block matrix determinant formula, we see that to see that

$$\det(P) = \det(\mathrm{id}_n - A\,\mathrm{id}_n^{-1}\,\mathrm{id}_n)\det(\mathrm{id}_n) = \det(\mathrm{id}_n - A),$$

hence by (5.6.8) we see that $\det(P) = \det(\mathrm{id}_n - df_p)$. (If you are not familiar with this, it is easy to see that by elementary row operations P can be converted into the matrix

$$\begin{pmatrix} \mathrm{id}_n & A \\ 0 & \mathrm{id}_n - A \end{pmatrix},$$

Exercise 1.8.vii.) □

Let us summarize what we have shown so far.

Theorem 5.6.9. *A $f\colon X \to X$ is a Lefschetz map if and only if for every fixed point p of f the map*

$$\mathrm{id}_{T_pX} - df_p\colon T_pX \to T_pX$$

is bijective. Moreover for Lefschetz maps f we have

$$L(f) = \sum_{p=f(p)} L_p(f),$$

where $L_p(f) = +1$ if $\mathrm{id}_{T_pX} - df_p$ is orientation preserving and $L_p(f) = -1$ if $\mathrm{id}_{T_pX} - df_p$ is orientation reversing.

We'll next describe how to compute $L(f)$ as a topological invariant of f. Let ι_{Γ_f} be the inclusion map of Γ_f into $X \times X$ and let $T_\Delta \in H^n(X \times X)$ be the Thom class of Δ. Then by (5.5.18)

$$L(f) = I_{\Gamma_f}(\iota_{\Gamma_f}^\star T_\Delta)$$

and hence since the mapping $\gamma_f\colon X \to X \times X$ defined by (5.6.1) is an orientation preserving diffeomorphism of $X \xrightarrow{\sim} \Gamma_f$ we have

$$L(f) = I_X(\gamma_f^\star T_\Delta). \tag{5.6.10}$$

To evaluate the expression on the right we'll need to know some facts about the cohomology groups of product manifolds. The main result on this topic is the *Künneth theorem*, and we'll take up the formulation and proof of this theorem in § 5.7. First, however, we'll describe a result which follows from the Künneth theorem and which will enable us to complete our computation of $L(f)$.

Let π_1 and π_2 be the projection of $X \times X$ onto its first and second factors, i.e., for $i = 1, 2$ let

$$\pi_i\colon X \times X \to X$$

be the projection $\pi_i(x_1, x_2) \coloneqq x_i$. Then by equation (5.6.1) we have

$$\pi_1 \circ \gamma_f = \mathrm{id}_X \tag{5.6.11}$$

and

$$\pi_2 \circ \gamma_f = f. \tag{5.6.12}$$

Lemma 5.6.13. *If ω_1 and ω_2 are in $\Omega^n(X)$ then*

$$\int_{X\times X} \pi_1^\star\omega_1 \wedge \pi_2^\star\omega_2 = \left(\int_X \omega_1\right)\left(\int_X \omega_2\right). \tag{5.6.14}$$

Proof. By a partition of unity argument we can assume that ω_i has compact support in a parameterizable open set V_i. Let U_i be an open subset of $\mathbf{R}^n$ and $\phi_i\colon U_i \to V_i$ an orientation preserving diffeomorphism. Then

$$\phi_i^\star \omega = \rho_i dx_1 \wedge \cdots \wedge dx_n$$

with $\rho_i \in C_0^\infty(U_i)$, so the right-hand side of (5.6.14) is the product of integrals over $\mathbf{R}^n$:

$$\left(\int_{\mathbf{R}^n} \rho_1(x)dx\right)\left(\int_{\mathbf{R}^n} \rho_2(x)dx\right). \tag{5.6.15}$$

Moreover, since $X \times X$ is oriented by its product orientation, the map

$$\psi : U_1 \times U_2 \to V_1 \times V_2$$

is given by $(x, y) \mapsto (\phi_1(x)\phi_2(y))$ is an orientation-preserving diffeomorphism and since $\pi_i \circ \psi = \phi_i$

$$\begin{aligned}\psi^\star(\pi_1^\star\omega_1 \wedge \pi_2^\star\omega_2) &= \phi_1^\star\omega_1 \wedge \phi_2^\star\omega_2\\ &= \rho_1(x)\rho_2(y)dx_1 \wedge \cdots \wedge dx_n \wedge dy_1 \wedge \cdots \wedge dy_n\end{aligned}$$

and hence the left-hand side of (5.6.14) is the integral over $\mathbf{R}^{2n}$ of the function, $\rho_1(x)\rho_2(y)$, and therefore, by integration by parts, is equal to the product (5.6.15). □

As a corollary of this lemma we get a product formula for cohomology classes.

Lemma 5.6.16. *If c_1 and c_2 are in $H^n(X)$, then*

$$I_{X\times X}(\pi_1^\star c_1 \cdot \pi_2^\star c_2) = I_X(c_1)I_X(c_2).$$

Now let $d_k := \dim H^k(X)$ and note that since X is compact, Poincaré duality tells us that $d_k = d_\ell$ when $\ell = n - k$. In fact it tells us even more. Let $\mu_1^k, \ldots, \mu_{d_k}^k$ be a basis of $H^k(X)$. Then, since the pairing (5.4.5) is non-singular, there exists for $\ell = n - k$ a "dual" basis

$$\nu_j^\ell,\ \ j = 1, \ldots, d_\ell$$

of $H^\ell(X)$ satisfying

$$I_X(\mu_i^k \cdot \nu_j^\ell) = \delta_{i,j}. \tag{5.6.17}$$

Lemma 5.6.18. *The cohomology classes*

$$\pi_1^\sharp \nu_r^\ell \cdot \pi_2^\sharp \mu_s^k,\ \ k + \ell = n$$

for $k = 0, \ldots, n$ and $1 \le r, s \le d_k$, are a basis for $H^n(X \times X)$.

This is the corollary of the Künneth theorem that we alluded to above (and whose proof we'll give in §5.7). Using these results we prove the following.

Theorem 5.6.19. *The Thom class $T_\Delta \in H^n(X \times X)$ is given explicitly by the formula*

$$T_\Delta = \sum_{k+\ell=n} (-1)^\ell \sum_{i=1}^{d_k} \pi_1^\sharp \mu_i^k \cdot \pi_2^\sharp \nu_i^\ell. \tag{5.6.20}$$

Proof. We have to check that for every cohomology class $c \in H^n(X \times X)$, the class T_Δ defined by (5.6.20) has the reproducing property

$$I_{X\times X}(T_\Delta \cdot c) = I_\Delta(\iota_\Delta^\sharp c), \tag{5.6.21}$$

where ι_Δ is the inclusion map of Δ into $X \times X$. However the map

$$\gamma_\Delta \colon X \to X \times X, \quad x \mapsto (x, x)$$

is an orientation-preserving diffeomorphism of X onto Δ, so it suffices to show that

$$I_{X \times X}(T_\Delta \cdot c) = I_X(\gamma_\Delta^\sharp c) \tag{5.6.22}$$

and by Lemma 5.6.18 it suffices to verify (5.6.22) for c's of the form

$$c = \pi_1^\sharp \nu_r^\ell \cdot \pi_2^\sharp \mu_s^k.$$

The product of this class with a typical summand of (5.6.20), for instance, the summand

$$(-1)^{\ell'} \pi_1^\sharp \mu_i^{k'} \cdot \pi_2^\sharp \nu_i^{\ell'}, \quad k' + \ell' = n, \tag{5.6.23}$$

is equal,

$$\pi_1^\sharp \mu_i^{k'} \cdot \nu_r^\ell \cdot \pi_2^\sharp \mu_s^k \cdot \nu_i^{\ell'}.$$

Notice, however, that if $k \neq k'$ this product is zero: For $k < k'$, $k' + \ell$ is greater than $k + \ell$ and hence greater than n. Therefore

$$\mu_i^{k'} \cdot \nu_r^\ell \in H^{k'+\ell}(X)$$

is zero since X is of dimension n, and for $k > k'$, ℓ' is greater than ℓ and $\mu_s^k \cdot \nu_i^{\ell'}$ is zero for the same reason. Thus in taking the product of T_Δ with c we can ignore all terms in the sum except for the terms, $k' = k$ and $\ell' = \ell$. For these terms, the product of (5.6.23) with c is

$$(-1)^{k\ell} \pi_1^\sharp \mu_i^k \cdot \nu_r^\ell \cdot \pi_2^\sharp \mu_3^k \cdot \nu_i^\ell.$$

(Exercise: Check this. *Hint:* $(-1)^\ell (-1)^{\ell^2} = 1$.) Thus

$$T_\Delta \cdot c = (-1)^{k\ell} \sum_{i=1}^{d_k} \pi_1^\sharp \mu_i^k \cdot \nu_r^\ell \cdot \pi_2^\sharp \mu_s^k \cdot \nu_i^\ell$$

and hence by Lemma 5.6.13 and (5.6.17)

$$\begin{aligned} I_{X \times X}(T_\Delta \cdot c) &= (-1)^{k\ell} \sum_{i=1}^{d_k} I_X(\mu_i^k \cdot \nu_r^\ell) I_X(\mu_s^k \cdot \nu_i^\ell) \\ &= (-1)^{k\ell} \sum_{i=1}^{d_k} \delta_{i,r} \delta_{i,s} \\ &= (-1)^{k\ell} \delta_{r,s}. \end{aligned}$$

On the other hand for $c = \pi_1^\sharp \nu_r^\ell \cdot \pi_2^\sharp \mu_s^k$

$$\begin{aligned} \gamma_\Delta^\sharp c &= \gamma_\Delta^\sharp \pi_1^\sharp \nu_r^\ell \cdot \gamma_\Delta^\sharp \pi_2^\sharp \mu_s^k \\ &= (\pi_1 \cdot \gamma_\Delta)^\sharp \nu_r^\ell (\pi_2 \cdot \gamma_\Delta)^\sharp \mu_s^k \\ &= \nu_r^\ell \cdot \mu_s^k \end{aligned}$$

since

$$\pi_1 \cdot \nu_\Delta = \pi_2 \cdot \gamma_\Delta = \mathrm{id}_X .$$

So

$$I_X(\gamma_\Delta^\sharp c) = I_X(\nu_r^\ell \cdot \mu_s^k) = (-1)^{k\ell}\delta_{r,s}$$

by (5.6.17). Thus the two sides of (5.6.21) are equal. □

We're now in position to compute $L(f)$, i.e., to compute the expression $I_X(\gamma_f^\star T_\Delta)$ on the right-hand side of (5.6.10). Since $\nu_1^\ell, \dots, \nu_{d_\ell}^\ell$ is a basis of $H^\ell(X)$ the linear mapping

(5.6.24) $$f^\sharp : H^\ell(X) \to H^\ell(X)$$

can be described in terms of this basis by a matrix $(f_{i,j}^\ell)$ with the defining property

$$f^\sharp \nu_i^\ell = \sum_{j=1}^{d_\ell} f_{j,i}^\ell \nu_j^\ell .$$

Thus by equations (5.6.11), (5.6.12) and (5.6.20) we have

$$\begin{aligned}
\gamma_f^\sharp T_\Delta &= \gamma_f^\sharp \sum_{k+\ell=n} (-1)^\ell \sum_{i=1}^{d_k} \pi_1^\sharp \mu_i^k \cdot \pi_2^\sharp \nu_i^\ell \\
&= \sum_{k+\ell=n} (-1)^\ell \sum_{i=1}^{d_k} (\pi_1 \cdot \gamma_f)^\sharp \mu_i^k \cdot (\pi_2 \cdot \nu_f)^\sharp \nu_i^\ell \\
&= \sum_{k+\ell=n} (-1)^\ell \sum_{i=1}^{d_k} \mu_i^k \cdot f^\sharp \nu_i^\ell \\
&= \sum_{k+\ell=n} (-1)^\ell \sum_{i=1}^{d_k} f_{j,i}^\ell \mu_i^k \cdot \nu_j^\ell .
\end{aligned}$$

Thus by (5.6.17)

$$\begin{aligned}
I_X(\gamma_f^\sharp T_\Delta) &= \sum_{k+\ell=n} (-1)^\ell \sum_{1\le i,j\le d_\ell} f_{j,i}^\ell I_X(\mu_i^k \cdot \nu_j^\ell) \\
&= \sum_{k+\ell=n} (-1)^\ell \sum_{1\le i,j\le d_\ell} f_{j,i}^\ell \delta_{i,j} \\
&= \sum_{\ell=0}^{n} (-1)^\ell \sum_{i=1}^{d_\ell} f_{i,i}^\ell .
\end{aligned}$$

But $\sum_{i=1}^{d_\ell} f_{i,i}^\ell$ is just the trace of the linear mapping (5.6.24) (see Exercise 5.6.xii), so we end up with the following purely topological prescription of $L(f)$.

Theorem 5.6.25. *The Lefschetz number $L(f)$ is the alternating sum*

$$L(f) = \sum_{\ell=0}^{n} (-1)^\ell \operatorname{tr}(f^\sharp \,|\, H^\ell(X)),$$

where $\mathrm{tr}(f^\sharp \,|\, H^\ell(X))$ *is the trace of the map* $f^\sharp : H^\ell(X) \to H^\ell(X)$.

Exercises for §5.6

Exercise 5.6.i. Show that if $f_0 : X \to X$ and $f_1 : X \to X$ are homotopic C^∞ mappings $L(f_0) = L(f_1)$.

Exercise 5.6.ii.

(1) The ***Euler characteristic*** $\chi(X)$ of X is defined to be the intersection number of the diagonal with itself in $X \times X$, i.e., the "self-intersection" number

$$\chi(X) := I(\Delta, \Delta) = I_{X\times X}(T_\Delta, T_\Delta).$$

Show that if a C^∞ map $f : X \to X$ is homotopic to the identity, then $L(f) = \chi(X)$.

(2) Show that

$$\chi(X) = \sum_{\ell=0}^{n} (-1)^\ell \dim H^\ell(X).$$

(3) Show that if X is an odd-dimensional (compact) manifold, then $\chi(X) = 0$.

Exercise 5.6.iii.

(1) Let S^n be the unit n-sphere in $\mathbf{R}^{n+1}$. Show that if $g : S^n \to S^n$ is a C^∞ map, then

$$L(g) = 1 + (-1)^n \deg(g).$$

(2) Conclude that if $\deg(g) \neq (-1)^{n+1}$, then g has to have a fixed point.

Exercise 5.6.iv. Let f be a C^∞ mapping of the closed unit ball B^{n+1} into itself and let $g : S^n \to S^n$ be the restriction of f to the boundary of B^{n+1}. Show that if $\deg(g) \neq (-1)^{n+1}$ then the fixed point of f predicted by Brouwer's theorem can be taken to be a point on the boundary of B^{n+1}.

Exercise 5.6.v.

(1) Show that if $g : S^n \to S^n$ is the antipodal map $g(x) := -x$, then $\deg(g) = (-1)^{n+1}$.

(2) Conclude that the result in Exercise 5.6.iv is sharp. Show that the map

$$f : B^{n+1} \to B^{n+1}, \quad f(x) = -x,$$

has only one fixed point, namely the origin, and in particular has no fixed points on the boundary.

Exercise 5.6.vi. Let v be a vector field on a compact manifold X. Since X is compact, v generates a one-parameter group of diffeomorphisms

$$f_t : X \to X, \quad -\infty < t < \infty. \tag{5.6.26}$$

(1) Let Σ_t be the set of fixed points of f_t. Show that this set contains the set of zeros of v, i.e., the points $p \in X$ where $v(p) = 0$.

(2) Suppose that for some t_0, f_{t_0} is Lefschetz. Show that for all t, f_t maps Σ_{t_0} into itself.

(3) Show that for $|t| < \varepsilon$, ε small, the points of Σ_{t_0} are fixed points of f_t.
(4) Conclude that Σ_{t_0} is equal to the set of zeros of v.
(5) In particular, conclude that for all t the points of Σ_{t_0} are fixed points of f_t.

Exercise 5.6.vii.

(1) Let V be a finite-dimensional vector space and

$$F(t): V \to V, \quad -\infty < t < \infty$$

a one-parameter group of linear maps of V onto itself. Let $A = \frac{dF}{dt}(0)$ and show that $F(t) = \exp tA$. (See Exercise 2.2.viii.)

(2) Show that if $\mathrm{id}_V - F(t_0): V \to V$ is bijective for some t_0, then $A: V \to V$ is bijective.

Hint: Show that if $Av = 0$ for some $v \in V - \{0\}$, $F(t)v = v$.

Exercise 5.6.viii. Let $\boldsymbol{v}$ be a vector field on a compact manifold X and let (5.6.26) be the one-parameter group of diffeomorphisms generated by $\boldsymbol{v}$. If $\boldsymbol{v}(p) = 0$ then by part (1) of Exercise 5.6.vi, p is a fixed point of f_t for all t.

(1) Show that

$$(df_t): T_pX \to T_pX$$

is a one-parameter group of linear mappings of T_pX onto itself.

(2) Conclude from Exercise 5.6.vii that there exists a linear map

$$L_v(p): T_pX \to T_pX \tag{5.6.27}$$

with the property

$$\exp tL_v(p) = (df_t)_p.$$

Exercise 5.6.ix. Suppose f_{t_0} is a Lefschetz map for some t_0. Let $a = t_0/N$ where N is a positive integer. Show that f_a is a Lefschetz map.

Hints:

- Show that

$$f_{t_0} = f_a \circ \cdots \circ f_a = f_a^N$$

(i.e., f_a composed with itself N times).

- Show that if p is a fixed point of f_a, it is a fixed point of f_{t_0}.
- Conclude from Exercise 5.6.vi that the fixed points of f_a are the zeros of v.
- Show that if p is a fixed point of f_a,

$$(df_{t_0})_p = (df_a)_p^N.$$

- Conclude that if $(df_a)_p v = v$ for some $v \in T_pX \smallsetminus \{0\}$, then $(df_{t_0})_p v = v$.

Exercise 5.6.x. Show that for all t, $L(f_t) = \chi(X)$.

Hint: Exercise 5.6.ii.

Exercise 5.6.xi (Hopf theorem). A vector field $\boldsymbol{v}$ on a compact manifold X is a ***Lefschetz vector field*** if for some $t_0 \in \mathbf{R}$ the map f_{t_0} is a Lefschetz map.

(1) Show that if $\boldsymbol{v}$ is a Lefschetz vector field then it has a finite number of zeros and for each zero p the linear map (5.6.27) is bijective.
(2) For a zero p of $\boldsymbol{v}$ let $\sigma_p(\boldsymbol{v}) = +1$ if the map (5.6.27) is orientation preserving and let $\sigma_p(\boldsymbol{v}) = -1$ if it's orientation reversing. Show that

$$\chi(X) = \sum_{\boldsymbol{v}(p)=0} \sigma_p(\boldsymbol{v}).$$

Hint: Apply the Lefschetz theorem to f_a, $a = t_0/N$, N large.

Exercise 5.6.xii (review of the trace). For $A = (a_{i,j})$ an $n \times n$ matrix define

$$\operatorname{tr}(A) := \sum_{i=1}^{n} a_{i,i}.$$

(1) Show that if A and B are $n \times n$ matrices

$$\operatorname{tr}(AB) = \operatorname{tr}(BA).$$

(2) Show that if B is an invertible $n \times n$ matrix

$$\operatorname{tr}(BAB^{-1}) = \operatorname{tr}(A).$$

(3) Let V be an n-dimensional vector space and $L: V \to V$ a linear map. Fix a basis $v_1, \ldots, v_n$ of V and define the trace of L to be the trace of A where A is the defining matrix for L in this basis, i.e.,

$$Lv_i = \sum_{j=1}^{n} a_{j,i} v_j.$$

Show that this is an *intrinsic* definition not depending on the basis $v_1, \ldots, v_n$.

5.7. The Künneth theorem

Let X be an n-manifold and Y an r-dimensional manifold, both of these manifolds having finite topology. Let

$$\pi: X \times Y \to X$$

be the projection map $\pi(x, y) = x$ and

$$\rho: X \times Y \to Y$$

the projection map. Since X and Y have finite topology their cohomology groups are finite-dimensional vector spaces. For $0 \le k \le n$ let

$$\mu_i^k, \quad 1 \le i \le \dim H^k(X)$$

be a basis of $H^k(X)$ and for $0 \le \ell \le r$ let

$$v_j^\ell, \quad 1 \le j \le \dim H^\ell(Y)$$

be a basis of $H^\ell(Y)$. Then for $k + \ell = m$ the product $\pi^\sharp \mu_i^k \cdot \rho^\sharp v_j^\ell$ is in $H^m(X \times Y)$. The Künneth theorem asserts the following.

Theorem 5.7.1. *The product manifold $X \times Y$ has finite topology and hence the cohomology groups $H^m(X \times Y)$ are finite dimensional. Moreover, the products over $k + \ell = m$*

$$\pi^\sharp \mu_i^k \cdot \rho^\sharp v_j^\ell, \quad 0 \le i \le \dim H^k(X), \quad 0 \le j \le \dim H^\ell(Y), \tag{5.7.2}$$

are a basis for the vector space $H^m(X \times Y)$.

The fact that $X \times Y$ has finite topology is easy to verify. If $U_1, \ldots, U_M$ is a good cover of X and $V_1, \ldots, V_N$ is a good cover of Y, the product of these open sets $U_i \times V_j$, for $1 \le i \le M$ and $1 \le j \le N$, is a good cover of $X \times Y$: For every multi-index I, U_I is either empty or diffeomorphic to $\mathbf{R}^n$, and for every multi-index J, V_J is either empty or diffeomorphic to $\mathbf{R}^r$, hence for any product multi-index (I, J), the product $U_I \times V_J$ is either empty or diffeomorphic to $\mathbf{R}^n \times \mathbf{R}^r$. The tricky part of the proof is verifying that the products from (5.7.2) are a basis of $H^m(X \times Y)$, and to do this it will be helpful to state the theorem above in a form that avoids our choosing specified bases for $H^k(X)$ and $H^\ell(Y)$. To do so we'll need to generalize slightly the notion of a bilinear pairing between two vector space.

Definition 5.7.3. Let V_1, V_2 and W be finite-dimensional vector spaces. A map $B\colon V_1 \times V_2 \to W$ of sets is a ***bilinear map*** if it is linear in each of its factors, i.e., for $v_2 \in V_2$ the map

$$v_1 \in V_1 \mapsto B(v_1, v_2)$$

is a linear map of V_1 into W and for $v_1 \in V_1$ so is the map

$$v_2 \in V_2 \mapsto B(v_1, v_2).$$

It's clear that if B_1 and B_2 are bilinear maps of $V_1 \times V_2$ into W and λ_1 and λ_2 are real numbers the function

$$\lambda_1 B_1 + \lambda_2 B_2 : V_1 \times V_2 \to W$$

is also a bilinear map of $V_1 \times V_2$ into W, so the set of all bilinear maps of $V_1 \times V_2$ into W forms a vector space. In particular the set of all bilinear maps of $V_1 \times V_2$ into $\mathbf{R}$ is a vector space, and since this vector space will play an essential role in our intrinsic formulation of the Künneth theorem, we'll give it a name. We'll call it the ***tensor product*** of $V_1^\star$ and $V_2^\star$ and denote it by $V_1^\star \otimes V_2^\star$. To explain where this terminology comes from we note that if ℓ_1 and ℓ_2 are vectors in $V_1^\star$ and $V_2^\star$ then one can define a bilinear map

$$\ell_1 \otimes \ell_2 \colon V_1 \times V_2 \to \mathbf{R}$$

by setting $(\ell_1 \otimes \ell_2)(v_1, v_2) := \ell_1(v_1)\ell_2(v_2)$. In other words one has a tensor product map:

$$V_1^\star \times V_2^\star \to V_1^\star \otimes V_2^\star \tag{5.7.4}$$

mapping (ℓ_1, ℓ_2) to $\ell_1 \otimes \ell_2$. We leave for you to check that this is a bilinear map of $V_1^\star \times V_2^\star$ into $V_1^\star \otimes V_2^\star$ and to check as well the following proposition.

Proposition 5.7.5. *If $\ell_1^1, \dots, \ell_m^1$ is a basis of $V_1^\star$ and $\ell_1^2, \dots, \ell_n^2$ is a basis of $V_2^\star$ then $\ell_i^1 \otimes \ell_j^2$, for $1 \le i \le m$ and $1 \le j \le n$, is a basis of $V_1^\star \otimes V_2^\star$.*

Hint: If V_1 and V_2 are the same vector space you can find a proof of this in §1.3 and the proof is basically the same if they're different vector spaces.

Corollary 5.7.6. *Let V_1 and V_2 be finite-dimensional vector spaces. Then*

$$\dim(V_1^\star \otimes V_2^\star) = \dim(V_1^\star)\dim(V_2^\star) = \dim(V_1)\dim(V_2).$$

We'll now perform some slightly devious maneuvers with "duality" operations. First note that for any finite-dimensional vector space V, the *evaluation pairing*

$$V \times V^\star \to \mathbf{R}\,, \ (v, \ell) \mapsto \ell(v)$$

is a non-singular bilinear pairing, so, as we explained in § 5.4 it gives rise to a bijective linear mapping

$$V \to (V^\star)^\star. \tag{5.7.7}$$

Next note that if

$$L\colon V_1 \times V_2 \to W \tag{5.7.8}$$

is a bilinear mapping and $\ell\colon W \to \mathbf{R}$ a linear mapping (i.e., an element of $W^\star$), then the composition of ℓ and L is a bilinear mapping

$$\ell \circ L\colon V_1 \times V_2 \to \mathbf{R}$$

and hence by definition an element of $V_1^\star \otimes V_2^\star$. Thus from the bilinear mapping (5.7.8) we get a ***linear*** mapping

$$L^\sharp\colon W^\star \to V_1^\star \otimes V_2^\star. \tag{5.7.9}$$

We'll now define a notion of tensor product for the vector spaces V_1 and V_2 themselves.

Definition 5.7.10. The vector space $V_1 \otimes V_2$ is the vector space dual of $V_1^\star \otimes V_2^\star$, i.e., is the space

$$V_1 \otimes V_2 := (V_1^\star \otimes V_2^\star)^\star. \tag{5.7.11}$$

One implication of (5.7.11) is that there is a natural bilinear map

$$V_1 \times V_2 \to V_1 \otimes V_2. \tag{5.7.12}$$

(In (5.7.4) replace V_i by $V_i^\star$ and note that by (5.7.7) $(V_i^\star)^\star = V_i$.) Another is the following proposition.

Proposition 5.7.13. *Let L be a bilinear map of $V_1 \times V_2$ into W. Then there exists a unique linear map*

$$L^{\#} : V_1 \otimes V_2 \to W$$

with the property

$$L^{\#}(v_1 \otimes v_2) = L(v_1, v_2), \tag{5.7.14}$$

where $v_1 \otimes v_2$ is the image of (v_1, v_2) with respect to (5.7.12).

Proof. Let $L^{\#}$ be the transpose of the map $L^{\sharp}$ in (5.7.9) and note that by (5.7.7) $(W^{\star})^{\star} = W$. □

Notice that by Proposition 5.7.13 the property (5.7.14) is the *defining* property of $L^{\#}$; it uniquely determines this map. (This is in fact the whole point of the tensor product construction. Its purpose is to convert bilinear maps, which are not *linear*, into linear maps.)

After this brief digression (into an area of mathematics which some mathematicians unkindly refer to as "abstract nonsense"), let us come back to our motive for this digression: an intrinsic formulation of the Künneth theorem. As above let X and Y be manifolds of dimension n and r, respectively, both having finite topology. For $k + \ell = m$ one has a bilinear map

$$H^k(X) \times H^{\ell}(Y) \to H^m(X \times Y)$$

mapping (c_1, c_2) to $\pi^{\star} c_1 \cdot \rho^{\star} c_2$, and hence by Proposition 5.7.13 a ***linear*** map

$$H^k(X) \otimes H^{\ell}(Y) \to H^m(X \times Y). \tag{5.7.15}$$

Let

$$H_1^m(X \times Y) := \bigoplus_{k+\ell=m} H^k(X) \otimes H^{\ell}(Y).$$

The maps (5.7.15) can be combined into a single linear map

$$H_1^m(X \times Y) \to H^m(X \times Y) \tag{5.7.16}$$

and our intrinsic version of the Künneth theorem asserts the following.

Theorem 5.7.17. *The map (5.7.16) is bijective.*

Here is a sketch of how to prove this. (Filling in the details will be left as a series of exercises.) Let U be an open subset of X which has finite topology and let

$$\mathcal{H}_1^m(U) := \bigoplus_{k+\ell=m} H^k(U) \otimes H^{\ell}(Y)$$

and

$$\mathcal{H}_2^m(U) := H^m(U \times Y).$$

As we've just seen there's a Künneth map

$$\kappa : \mathcal{H}_1^m(U) \to \mathcal{H}_2^m(U).$$

Exercises for §5.7

Exercise 5.7.i. Let U_1 and U_2 be open subsets of X, both having finite topology, and let $U = U_1 \cup U_2$. Show that there is a long exact sequence:

$$\cdots \xrightarrow{\delta} \mathcal{H}_1^m(U) \longrightarrow \mathcal{H}_1^m(U_1) \oplus \mathcal{H}_1^m(U_2) \longrightarrow \mathcal{H}_1^m(U_1 \cap U_2) \xrightarrow{\delta} \mathcal{H}_1^{m+1}(U) \longrightarrow \cdots,$$

Hint: Take the usual Mayer–Vietoris sequence

$$\cdots \xrightarrow{\delta} H^k(U) \longrightarrow H^k(U_1) \oplus H^k(U_2) \longrightarrow H^k(U_1 \cap U_2) \xrightarrow{\delta} H^{k+1}(U) \longrightarrow \cdots,$$

tensor each term in this sequence with $H^\ell(Y)$, and sum over $k + \ell = m$.

Exercise 5.7.ii. Show that for $\mathcal{H}_2$ there is a similar sequence.
Hint: Apply Mayer–Vietoris to the open subsets $U_1 \times Y$ and $U_2 \times Y$ of M.

Exercise 5.7.iii. Show that the diagram below commutes:

$$\begin{array}{ccccccccc}
\cdots \xrightarrow{\delta} & \mathcal{H}_1^m(U) & \longrightarrow & \mathcal{H}_1^m(U_1) \oplus \mathcal{H}_1^m(U_2) & \longrightarrow & \mathcal{H}_1^m(U_1 \cap U_2) & \xrightarrow{\delta} & \mathcal{H}_1^{m+1}(U) & \longrightarrow \cdots \\
& \downarrow{\scriptstyle\kappa} & & \downarrow{\scriptstyle\kappa} & & \downarrow{\scriptstyle\kappa} & & \downarrow{\scriptstyle\kappa} & \\
\cdots \xrightarrow{\delta} & \mathcal{H}_2^m(U) & \longrightarrow & \mathcal{H}_2^m(U_1) \oplus \mathcal{H}_2^m(U_2) & \longrightarrow & \mathcal{H}_2^m(U_1 \cap U_2) & \xrightarrow{\delta} & \mathcal{H}_2^{m+1}(U) & \longrightarrow \cdots
\end{array}$$

(This looks hard but is actually very easy: just write down the definition of each arrow in the language of forms.)

Exercise 5.7.iv. Conclude from Exercise 5.7.iii that if the Künneth map is bijective for U_1, U_2 and $U_1 \cap U_2$ it is bijective for U.

Exercise 5.7.v. Prove the Künneth theorem by induction on the number of open sets in a good cover of X. To get the induction started, note that

$$H^k(\mathbf{R}^n \times Y) \cong H^k(Y).$$

(See Exercise 5.3.xi.)

5.8. Čech cohomology

We pointed out at the end of §5.3 that if a manifold X admits a finite good cover

$$\mathcal{U} = \{U_1, \ldots, U_d\},$$

then in principle its cohomology can be read off from the intersection properties of the U_i's. In this section we'll explain in more detail how to do this. More explicitly, we will show how to construct from the intersection properties of the U_i's a sequence of cohomology groups $H^k(\mathcal{U}; \mathbf{R})$, and we will prove that these *Čech cohomology groups* are isomorphic to the de Rham cohomology groups of X. (However we will leave for the reader a key ingredient in the proof: a diagram-chasing argument that is similar to the diagram-chasing arguments that were used to prove the Mayer–Vietoris theorem in §5.2 and the five lemma in §5.4.)

We will denote by $N^k(\mathcal{U})$ the set of all multi-indices

$$I = (i_0, \dots, i_k), \quad 1 \le i_0, \dots, i_k \le d$$

with the property that the intersection

$$U_I := U_{i_0} \cap \cdots \cap U_{i_k}$$

is non-empty. (For example, for $1 \le i \le d$, the multi-index $I = (i_0, \dots, i_k)$ with $i_0 = \cdots = i_k = i$ is in $N^k(\mathcal{U})$ since $U_I = U_i$.) The disjoint union $N(\mathcal{U}) := \coprod_{k \ge 0} N^k(\mathcal{U})$ is called the ***nerve*** of the cover $\mathcal{U}$ and $N^k(\mathcal{U})$ is the ***k-skeleton*** of the nerve $N(\mathcal{U})$. Note that if we delete i_r from a multi-index $I = (i_0, \dots, i_k)$ in $N^k(\mathcal{U})$, i.e., replace I by

$$I_r := (i_0, \dots, i_{r-1}, i_{r+1}, \dots, i_k),$$

then $I_r \in N^{k-1}(\mathcal{U})$.

We now associate a cochain complex to $N(\mathcal{U})$ by defining $C^k(\mathcal{U}; \mathbf{R})$ to be the finite dimensional vector space consisting of all (set-theoretic) maps $c \colon N^k(\mathcal{U}) \to \mathbf{R}$:

$$C^k(\mathcal{U}; \mathbf{R}) := \{c \colon N^k(\mathcal{U}) \to \mathbf{R}\},$$

with addition and scalar multiplication of functions. Define a coboundary operator

$$\delta \colon C^{k-1}(\mathcal{U}; \mathbf{R}) \to C^k(\mathcal{U}; \mathbf{R})$$

by setting

$$\delta(c)(I) := \sum_{r=0}^{k} (-1)^r c(I_r). \tag{5.8.1}$$

We claim that δ actually is a coboundary operator, i.e., $\delta \circ \delta = 0$. To see this, we note that for $c \in C^{k-1}(\mathcal{U}; \mathbf{R})$ and $I \in N^k(\mathcal{U})$, by (5.8.1) we have

$$\begin{aligned}
\delta(\delta(c))(I) &= \sum_{r=0}^{k+1} (-1)^r \delta(c)(I_r) \\
&= \sum_{r=0}^{k+1} (-1)^r \left(\sum_{s<r} (-1)^s c(I_{r,s}) + \sum_{s>r} (-1)^{s-1} c(I_{r,s}) \right).
\end{aligned}$$

Thus the term $c(I_{r,s})$ occurs twice in the sum, but occurs once with opposite signs, and hence $\delta(\delta(c))(I) = 0$.

The cochain complex

$$0 \longrightarrow C^0(\mathcal{U}; \mathbf{R}) \xrightarrow{\ \delta\ } C^1(\mathcal{U}; \mathbf{R}) \xrightarrow{\ \delta\ } \cdots \xrightarrow{\ \delta\ } C^k(\mathcal{U}; \mathbf{R}) \xrightarrow{\ \delta\ } \cdots \tag{5.8.2}$$

is called the ***Čech cochain complex*** of the cover $\mathcal{U}$. The ***Čech cohomology groups*** of the cover $\mathcal{U}$ are the cohomology groups of the Čech cochain complex:

$$H^k(\mathcal{U};\mathbf{R}) := \frac{\ker(\delta\colon C^k(\mathcal{U};\mathbf{R}) \to C^{k+1}(\mathcal{U};\mathbf{R}))}{\operatorname{im}(\delta\colon C^{k-1}(\mathcal{U};\mathbf{R}) \to C^k(\mathcal{U};\mathbf{R}))}.$$

The rest of the section is devoted to proving the following theorem.

Theorem 5.8.3. *Let X be a manifold and $\mathcal{U}$ a finite good cover of X. Then for all $k \geq 0$ we have isomorphisms*

$$H^k(\mathcal{U};\mathbf{R}) \cong H^k(X).$$

Remarks 5.8.4.

(1) The definition of $H^k(\mathcal{U};\mathbf{R})$ only involves the nerve $N(\mathcal{U})$ of this cover so Theorem 5.8.3 is in effect a proof of the claim we made above: *the cohomology of X is in principle determined by the intersection properties of the U_i's.*
(2) Theorem 5.8.3 gives us another proof of an assertion we proved earlier: if X admits a finite good cover the cohomology groups of X are finite dimensional.

The proof of Theorem 5.8.3 will involve an interesting de Rham theoretic generalization of the Čech cochain complex (5.8.2). Namely, for k and ℓ non-negative integers we define a ***Čech cochain of degree k with values in Ω^ℓ*** to be a map c which assigns each $I \in N^k(\mathcal{U})$ an ℓ-form $c(I) \in \Omega^\ell(U_I)$.

The set of these cochains forms vector space $C^k(\mathcal{U};\Omega^\ell)$. We will show how to define a coboundary operator

$$\delta\colon C^{k-1}(\mathcal{U};\Omega^\ell) \to C^k(\mathcal{U};\Omega^\ell)$$

similar to the Čech coboundary operator (5.8.1). To define this operator, let $I \in N^k(\mathcal{U})$ and for $0 \leq r \leq k$ let $\gamma_r\colon \Omega^\ell(U_{I_r}) \to \Omega^\ell(U_I)$ be the restriction map defined by

$$\omega \mapsto \omega|_{U_I}.$$

(Note that the restriction $\omega|_{U_I}$ is well-defined since $U_I \subset U_{I_r}$.) Thus given a Čech cochain $c \in C^{k-1}(\mathcal{U};\Omega^\ell)$ we can, mimicking (5.8.1), define a Čech cochain $\delta(c) \in C^k(\mathcal{U};\Omega^\ell)$ by setting

$$\delta(c)(I) := \sum_{r=0}^{k} \gamma_r(c(I_r)) = \sum_{r=0}^{k} c(I_r)|_{U_I}. \tag{5.8.5}$$

In other words, except for the γ_r's the definition of this cochain is formally identical with the definition (5.8.1). We leave it as an exercise that the operators

$$\delta\colon C^{k-1}(\mathcal{U};\Omega^\ell) \to C^k(\mathcal{U};\Omega^\ell)$$

satisfy $\delta \circ \delta = 0$.[1]

Thus, to summarize, for every non-negative integer ℓ we have constructed a Čech cochain complex with values in Ω^ℓ

$$(5.8.6)\qquad 0 \longrightarrow C^0(\mathcal{U};\Omega^\ell) \xrightarrow{\delta} C^1(\mathcal{U};\Omega^\ell) \xrightarrow{\delta} \cdots \xrightarrow{\delta} C^k(\mathcal{U};\Omega^\ell) \xrightarrow{\delta} \cdots.$$

Our next task in this section will be to compute the cohomology of the complex (5.8.6). We'll begin with the zeroth-cohomology group of (5.8.6), i.e., the kernel of the coboundary operator

$$\delta\colon C^0(\mathcal{U};\Omega^\ell) \to C^1(\mathcal{U};\Omega^\ell).$$

An element $c \in C^0(\mathcal{U};\Omega^\ell)$ is, by definition, a map which assigns to each $i \in \{1,\dots,d\}$ an element $c(i) =: \omega_i$ of $\Omega^\ell(U_i)$.

Now let $I = (i_0, i_1)$ be an element of $N^1(\mathcal{U})$. Then, by definition, $U_{i_0} \cap U_{i_1}$ is non-empty and

$$\delta(c)(I) = \gamma_{i_0}\omega_{i_1} - \gamma_{i_1}\omega_{i_0}.$$

Thus $\delta(c) = 0$ if and only if

$$\omega_{i_0}|_{U_{i_0}\cap U_{i_1}} = \omega_{i_1}|_{U_{i_0}\cap U_{i_1}}$$

for all $I \in N^1(\mathcal{U})$. This is true if and only if each ℓ-form $\omega_i \in \Omega^\ell(U_i)$ is the restriction to U_i of a globally defined ℓ-form ω on $U_1 \cup \cdots \cup U_d = X$. Thus

$$\ker(\delta\colon C^0(\mathcal{U};\Omega^\ell) \to C^1(\mathcal{U};\Omega^\ell)) = \Omega^\ell(X).$$

Inserting $\Omega^\ell(X)$ into the sequence (5.8.6) we get a new sequence

$$(5.8.7)\qquad 0 \longrightarrow \Omega^\ell(X) \hookrightarrow C^0(\mathcal{U};\Omega^\ell) \xrightarrow{\delta} \cdots \xrightarrow{\delta} C^k(\mathcal{U};\Omega^\ell) \xrightarrow{\delta} \cdots,$$

and we prove the following theorem.

Theorem 5.8.8. *Let X be a manifold and $\mathcal{U}$ a finite good cover of X. Then for each non-negative integer ℓ, the sequence (5.8.7) is exact.*

We've just proved that the sequence (5.8.7) is exact in the first spot. To prove that (5.8.7) is exact in its kth spot, we will construct a chain homotopy operator

$$Q\colon C^k(\mathcal{U};\Omega^\ell) \to C^{k+1}(\mathcal{U};\Omega^\ell)$$

[1] *Hint:* Except for keeping track of the γ_r's the proof is identical to the proof of the analogous result for the coboundary operator (5.8.1).

with the property that

(5.8.9) $$\delta Q(c) + Q\delta(c) = c$$

for all $c \in C^k(\mathcal{U};\Omega^\ell)$. To define this operator we'll need a slight generalization of Theorem 5.2.22.

Theorem 5.8.10. *Let X be a manifold with a finite good cover $\{U_1,\dots,U_N\}$. Then there exist functions $\phi_1,\dots,\phi_N \in C^\infty(X)$ such that $\operatorname{supp}(\phi_i) \subset U_i$ for $i = 1,\dots,N$ and $\sum_{i=1}^N \phi_i = 1$.*

Proof. Let $(\rho_i)_{i\geq 1}$, where $\rho_i \in C_0^\infty(X)$, be a partition of unity subordinate to the cover $\mathcal{U}$ and let ϕ_1 be the sum of the ρ_j's with support in U_1:

$$\phi_1 := \sum_{\substack{j\geq 1 \\ \operatorname{supp}(\rho_j)\subset U_1}} \rho_j.$$

Deleting those ρ_j such that $\operatorname{supp}(\rho_j) \subset U_1$ from the sequence $(\rho_i)_{i\geq 1}$, we get a new sequence $(\rho_i')_{i\geq 1}$. Let ϕ_2 be the sum of the ρ_j''s with support in U_1:

$$\phi_2 := \sum_{\substack{j\geq 1 \\ \operatorname{supp}(\rho_j')\subset U_2}} \rho_j'.$$

Now we delete those ρ_j' such that $\operatorname{supp}(\rho_j') \subset U_2$ from the sequence $(\rho_i')_{i\geq 1}$ and construct ϕ_3 by the same method we used to construct ϕ_1 and ϕ_2, and so on. □

Now we define the operator Q and prove Theorem 5.8.8.

Proof of Theorem 5.8.8. To define the operator (5.8.9) let $c \in C^k(\mathcal{U};\Omega^\ell)$. Then we define $Qc \in C^{k-1}(\mathcal{U};\Omega^\ell)$ by defining its value at $I \in N^{k-1}(I)$ to be

$$Qc(I) := \sum_{i=0}^{d} \phi_i c(i, i_0, \dots, i_{k-1}),$$

where $I = (i_0,\dots,i_{k-1})$ and the ith summand on the right is defined to be equal to 0 when $U_i \cap U_I$ is empty and defined to be the product of the function ϕ_i and the ℓ-form $c(i, i_0,\dots,i_{k-1})$ when $U_i \cap U_I$ is non-empty. (Note that in this case, the multi-index $(i, i_0,\dots,i_{k-1})$ is in $N^k(\mathcal{U})$ and so by definition $c(i, i_0,\dots,i_{k-1})$ is an element of $\Omega^\ell(U_i \cap U_I)$. However, since ϕ_i is supported on U_i, we can extend the ℓ-form $\phi_i c(i, i_0,\dots,i_{k-1})$ to U_I by setting it equal to zero outside of $U_i \cap U_I$.)

To prove (5.8.9) we note that for $c \in C^k(\mathcal{U};\Omega^\ell)$ we have

$$Q\delta c(i_0,\dots,i_k) = \sum_{i=1}^{d} \phi_i \delta c(i, i_0,\dots,i_k),$$

so by (5.8.5) the right-hand side is equal to

(5.8.11) $$\sum_{i=1}^{d} \phi_i \gamma_i c(i_0,\dots,i_k) + \sum_{i=1}^{d}\sum_{s=0}^{k} (-1)^{s+1} \phi_i \gamma_{i_s} c(i, i_0,\dots,\hat{i}_s,\dots,i_k),$$

where $(i, i_0, \ldots, \hat{\imath}_s, \ldots, i_k)$ is the multi-index $(i, i_0, \ldots, i_k)$ with i_s deleted. Similarly,

$$\begin{aligned}\delta Qc(i_0, \ldots, i_k) &= \sum_{s=0}^{k}(-1)^s \gamma_{i_s} Qc(i_0, \ldots, \hat{\imath}_s, \ldots, i_k)\\ &= \sum_{i=1}^{d}\sum_{s=0}^{k}(-1)^s \phi_i \gamma_{i_s} c(i, i_0, \ldots, \hat{\imath}_s, \ldots, i_k).\end{aligned}$$

However, this is the negative of the second summand of (5.8.11); so, by adding these two summands we get

$$(Q\delta c + \delta Qc)(i_0, \ldots, i_k) = \sum_{i=1}^{d} \phi_i \gamma_i c(i_0, \ldots, i_k)$$

and since $\sum_{i=1}^{d} \phi_i = 1$, this identity to

$$(Q\delta c + \delta Qc)(I) = c(I),$$

or, since $I \in N^k(\mathcal{U})$ is any element, $Q\delta c + \delta Qc = c$.

Finally, note that Theorem 5.8.8 is an immediate consequence of the existence of this chain homotopy operator. Namely, if $c \in C^k(\mathcal{U}; \Omega^\ell)$ is in the kernel of δ, then

$$c = Q\delta(c) + \delta Q(c) = \delta Q(c),$$

so c is in the image of $\delta : C^{k-1}(\mathcal{U}; \Omega^\ell) \to C^k(\mathcal{U}; \Omega^\ell)$.

To prove Theorem 5.8.3, we note that in addition to the exact sequence

$$0 \longrightarrow \Omega^\ell(X) \hookrightarrow C^0(\mathcal{U}; \Omega^\ell) \xrightarrow{\delta} \cdots \xrightarrow{\delta} C^k(\mathcal{U}; \Omega^\ell) \xrightarrow{\delta} \cdots,$$

we have another exact sequence

(5.8.12)

$$0 \to C^k(\mathcal{U}; \mathbf{R}) \xrightarrow{d} C^k(\mathcal{U}; \Omega^0) \xrightarrow{d} C^k(\mathcal{U}; \Omega^1) \xrightarrow{d} \cdots \xrightarrow{d} C^k(\mathcal{U}; \Omega^\ell) \xrightarrow{d} \cdots,$$

where the d's are defined as follows: given $c \in C^k(\mathcal{U}; \Omega^\ell)$ and $I \in N^k(\mathcal{U})$, we have that $c(I) \in \Omega^\ell(U_I)$ and $d(c(I)) \in \Omega^{\ell+1}(U_I)$, and we define $d(c) \in C^k(\mathcal{U}; \Omega^{\ell+1})$ to be the map given by

$$I \mapsto d(c(I)).$$

It is clear from this definition that $d(d(c)) = 0$ so the sequence

$$C^k(\mathcal{U}; \Omega^0) \xrightarrow{d} C^k(\mathcal{U}; \Omega^1) \xrightarrow{d} \cdots \xrightarrow{d} C^k(\mathcal{U}; \Omega^\ell) \xrightarrow{d} \cdots \tag{5.8.13}$$

is a complex and the exactness of this sequence follows from the fact that, since $\mathcal{U}$ is a good cover, the sequence

$$\Omega^0(U_I) \xrightarrow{d} \Omega^1(U_I) \xrightarrow{d} \cdots \xrightarrow{d} \Omega^\ell(U_I) \xrightarrow{d} \cdots$$

is exact. Moreover, if $c \in C^k(\mathcal{U}; \Omega^0)$ and $dc = 0$, then for every $I \in N^k(\mathcal{U})$, we have $dc(I) = 0$, i.e., the 0-form $c(I)$ in $\Omega^0(U_I)$ is constant. Hence the map given by $I \mapsto c(I)$ is just a Čech cochain of the type we defined earlier, i.e., a map

$N^k(\mathcal{U}) \to \mathbf{R}$. Therefore, the kernel of this map $d\colon C^k(\mathcal{U};\Omega^0) \to C^k(\mathcal{U};\Omega^1)$ is $C^0(\mathcal{U};\mathbf{R})$; and, adjoining this term to the sequence (5.8.13) we get the exact sequence (5.8.12).

Lastly we note that the two operations we've defined above, the d operation

$$d\colon C^k(\mathcal{U};\Omega^\ell) \to C^k(\mathcal{U};\Omega^{\ell+1})$$

and the δ operation $\delta\colon C^k(\mathcal{U};\Omega^\ell) \to C^{k+1}(\mathcal{U};\Omega^\ell)$, commute. i.e., for $c \in C^k(\mathcal{U};\Omega^\ell)$ we have

$$\delta dc = d\delta c. \tag{5.8.14}$$

Proof of equation (5.8.14). For $I \in N^{k+1}(\mathcal{U})$

$$\delta c(I) = \sum_{r=0}^{k} (-1)^r \gamma_r c(I_r),$$

so if $c(I_r) =: \omega_{I_r} \in \Omega^\ell(U_{I_r})$, then

$$\delta c(I) = \sum_{r=0}^{k} (-1)^r \omega_{I_r}|_{U_I}$$

and

$$d\delta(I) = \sum_{r=0}^{k} d\omega_{I_r}|_{U_I} = \delta dc(I). \qquad \square$$

Putting these results together we get the following commutative diagram:

$$\begin{array}{ccccccccccc}
 & & \vdots & & \vdots & & \vdots & & \vdots & & \\
 & & \uparrow d & & \uparrow & & \uparrow & & \uparrow & & \\
0 & \to & \Omega^2(X) & \hookrightarrow & C^0(\mathcal{U};\Omega^2) & \xrightarrow{\delta} & C^1(\mathcal{U};\Omega^2) & \xrightarrow{\delta} & C^2(\mathcal{U};\Omega^2) & \to & \cdots \\
 & & \uparrow d & & \uparrow d & & \uparrow d & & \uparrow d & & \\
0 & \to & \Omega^1(X) & \hookrightarrow & C^0(\mathcal{U};\Omega^1) & \xrightarrow[\delta]{} & C^1(\mathcal{U};\Omega^1) & \xrightarrow[\delta]{} & C^2(\mathcal{U};\Omega^1) & \to & \cdots \\
 & & \uparrow d & & \uparrow d & & \uparrow d & & \uparrow d & & \\
0 & \to & \Omega^0(X) & \hookrightarrow & C^0(\mathcal{U};\Omega^0) & \xrightarrow[\delta]{} & C^1(\mathcal{U};\Omega^0) & \xrightarrow[\delta]{} & C^2(\mathcal{U};\Omega^0) & \to & \cdots \\
 & & \uparrow & & \uparrow & & \uparrow & & \uparrow & & \\
 & & 0 & \to & C^0(\mathcal{U};\mathbf{R}) & \xrightarrow[\delta]{} & C^1(\mathcal{U};\mathbf{R}) & \xrightarrow[\delta]{} & C^2(\mathcal{U};\mathbf{R}) & \to & \cdots \\
 & & & & \uparrow & & \uparrow & & \uparrow & & \\
 & & & & 0 & & 0 & & 0 & &
\end{array}$$

In this diagram all columns are exact except for the extreme left hand column, which is the usual de Rham complex, and all rows are exact except for the bottom row which is the usual Čech cochain complex.

Exercises for §5.8

Exercise 5.8.i. Deduce from this diagram that if a manifold X admits a finite good cover $\mathcal{U}$, then $H^k(\mathcal{U};\mathbf{R}) \cong H^k(X)$.

Hint: Let $c \in \Omega^{k+1}(X)$ be a form such that $dc = 0$. Show that one can, by a sequence of "chess moves" (of which the first few stages are illustrated in the (5.8.15) below), convert c into an element $\check{c}$ of $C^{k+1}(\mathcal{U};\mathbf{R})$ satisfying $\delta(\check{c}) = 0$

$$
\begin{array}{ccccccccc}
c & \longmapsto & c^{k+1,0} & & & & & & \\
 & & \uparrow d & & & & & & \\
 & & c^{k,0} & \overset{}{\underset{\delta}{\longmapsto}} & c^{k,1} & & & & \\
 & & & & \uparrow d & & & & \\
 & & & & c^{k-1,1} & \underset{\delta}{\longmapsto} & c^{k-1,2} & & \\
 & & & & & & \uparrow d & & \\
 & & & & & & c^{k-2,2} & \underset{\delta}{\longmapsto} & c^{k-2,3} \\
 & & & & & & & & \uparrow d \\
 & & & & & & & & c^{k-3,3} \underset{\delta}{\longmapsto} \cdots
\end{array}
\tag{5.8.15}
$$

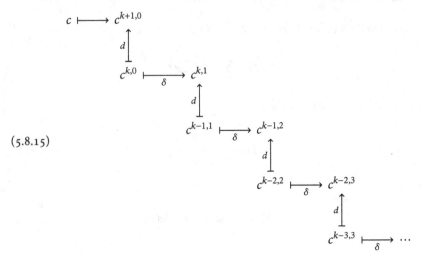

Appendix A

Bump Functions and Partitions of Unity

We will discuss in this appendix a number of "global-to-local" techniques in multi-variable calculus: techniques which enable one to reduce global problems on manifolds and large open subsets of $\mathbf{R}^n$ to local problems on small open subsets of these sets. Our starting point will be the function ρ on $\mathbf{R}$ defined by

$$\rho(x) := \begin{cases} e^{-\frac{1}{x}}, & x > 0, \\ 0, & x \le 0, \end{cases} \tag{A.1}$$

which is positive for x positive, zero for x negative and everywhere C^∞. (We will sketch a proof of this last assertion at the end of this appendix.) From ρ one can construct a number of other interesting C^∞ functions.

Example A.2. For $a > 0$ the function $\rho(x) + \rho(a - x)$ is positive for all x so the quotient $\rho_a(x)$ of $\rho(x)$ by this function is a well-defined C^∞ function with the properties:

$$\begin{cases} \rho_a(x) = 0 & \text{for } x \le 0, \\ 0 \le \rho_a(x) \le 1, & \\ \rho_a(x) = 1 & \text{for } x \ge a. \end{cases}$$

Example A.3. Let I be the open interval (a, b), and ρ_I the product of $\rho(x - a)$ and $\rho(b - x)$. Then $\rho_I(x)$ is positive for $x \in I$ and zero on the complement of I.

Example A.4. More generally let $I_1, \dots, I_n$ be open intervals and let $Q = I_1 \times \cdots \times I_n$ be the open rectangle in $\mathbf{R}^n$ having these intervals as sides. Then the function

$$\rho_Q(x) := \rho_{I_1}(x_1) \cdots \rho_{I_n}(x_n)$$

is a C^∞ function on $\mathbf{R}^n$ that's positive on Q and zero on the complement of Q.

Using these functions we will prove the following lemma.

Lemma A.5. *Let C be a compact subset of $\mathbf{R}^n$ and U an open set containing C. Then there exists function $\phi \in C_0^\infty(U)$ such that*

$$\begin{cases} \phi(x) \ge 0 & \textit{for all } x \in U, \\ \phi(x) > 0 & \textit{for } x \in C. \end{cases}$$

Proof. For each $p \in C$ let Q_p be an open rectangle with $p \in Q_p$ and $\overline{Q}_p \subset U$. The Q_p's cover C_i so, by Heine–Borel there exists a finite subcover $Q_{p_1}, \ldots, Q_{p_N}$. Now let $\phi := \sum_{i=1}^{N} \rho_{Q_{p_i}}$. □

This result can be slightly strengthened. Namely we claim the following theorem.

Theorem A.6. *There exists a function $\psi \in C_0^\infty(U)$ such that $0 \le \psi \le 1$ and $\psi = 1$ on C.*

Proof. Let ϕ be as in Lemma A.5 and let $a > 0$ be the greatest lower bound of the restriction of ϕ to C. Then if ρ_a is the function in Example A.2 the function $\rho_a \circ \phi$ has the properties indicated above. □

Remark A.7. The function ψ in this theorem is an example of a *bump function*. If one wants to study the behavior of a vector field $\boldsymbol{v}$ or a k-form ω on the set C, then by multiplying $\boldsymbol{v}$ (or ω) by ψ one can, without loss or generality, assume that $\boldsymbol{v}$ (or ω) is compactly supported on a small neighborhood of C.

Bump functions are one of the standard tools in calculus for converting global problems to local problems. Another such tool is *partitions of unity*.

Let U be an open subset of $\mathbf{R}^n$ and $\mathcal{U} = \{U_\alpha\}_{\alpha \in I}$ a covering of U by open subsets (indexed by the elements of the *index set* I). Then the partition of unity theorem asserts the following.

Theorem A.8. *There exists a sequence of functions $\rho_1, \rho_2, \ldots \in C_0^\infty(U)$ such that:*

(1) $\rho_i \ge 0$;
(2) *for every i, there is an $\alpha \in I$ with $\rho_i \in C_0^\infty(U_\alpha)$;*
(3) *for every $p \in U$, there exist a neighborhood U_p of p in U and an $N_p > 0$ such that $\rho_i|_{U_p} = 0$ for $u > N_p$;*
(4) $\sum \rho_i = 1$.

Remark A.9. Because of item (3) the sum in item (4) is well defined. We will derive this result from a somewhat simpler set-theoretical result.

Theorem A.10. *There exists a countable covering of U by open rectangles $(Q_i)_{i \ge 1}$ such that:*

(1) $\overline{Q}_i \subset U$;
(2) *for each i, there is an $\alpha \in I$ with $\overline{Q}_i \subset U_\alpha$;*
(3) *for every $p \in U$, there exist a neighborhood U_p of p in U and $N_p > 0$ such that $Q_i \cap U_p$ is empty for $i > N_p$.*

We first note that Theorem A.10 implies Theorem A.8. To see this note that the functions ρ_{Q_i} in Example A.4 above have all the properties indicated in Theorem A.8 except for property (4). Moreover since the Q_i's are a covering of U the sum $\sum_{i=1}^{\infty} \rho_{Q_i}$ is everywhere positive. Thus we get a sequence of ρ_i's satisfying (1)–(4) by taking ρ_i to be the quotient of ρ_{Q_i} by this sum.

Proof of Theorem A.10. Let $d(x, U^c)$ be the distance of a point $x \in U$ to the complement $U^c := \mathbf{R}^n \setminus U$ of U in $\mathbf{R}^n$, and let A_r be the compact subset of U consisting of points, $x \in U$, satisfying $d(x, U^c) \geq 1/r$ and $|x| \leq r$. By Heine–Borel we can find, for each r, a collection of open rectangles, $Q_{r,i}$, $i = 1, \ldots, N_r$, such that $\overline{Q}_{r,i}$ is contained in $\mathrm{int}(A_{r+1} \setminus A_{r-2})$ and in some U_α and such that the Q_i, r's are a covering of $A_r \setminus \mathrm{int}(A_{r-1})$. Thus the $Q_{r,i}$'s have the properties listed in Theorem A.10, and by relabeling, i.e., setting $Q_i = Q_{1,i}$ for $1 \leq i \leq N_i$, $Q_{N_1+i} := Q_{2,i}$, for $1 \leq i \leq N_2$, etc., we get a sequence, $Q_1, Q_2, \ldots$ with the desired properties. □

Applications

We will next describe a couple of applications of Theorem A.10.

Application A.11 (Improper integrals). Let $f : U \to \mathbf{R}$ be a continuous function. We will say that f is integrable over U if the infinite sum

$$\sum_{i=1}^{\infty} \int_U \rho_i(x) \, |f(x)| \, dx. \tag{A.12}$$

converges and if so we will define the ***improper integral*** of f over U to be the sum

$$\sum_{i=1}^{\infty} \int_U \rho_i(x) f(x) dx. \tag{A.13}$$

(Notice that each of the summands in this series is the integral of a compactly supported continuous function over $\mathbf{R}^n$ so the individual summands are well-defined. Also it's clear that if f itself is a compactly supported function on $\mathbf{R}^n$, (A.12) is bounded by the integral of $|f|$ over $\mathbf{R}^n$, so for every $f \in C_0(\mathbf{R}^n)$ the improper integral of f over U is well-defined.)

Application A.14 (An extension theorem for C^∞ maps). Let X be a subset of $\mathbf{R}^n$ and $f : X \to \mathbf{R}^m$ a continuous map. We will say that f is C^∞ if, for every $p \in X$, there exist an open neighborhood, U_p, of p in $\mathbf{R}^n$ and a C^∞ map, $g_p : U_p \to \mathbf{R}^m$ such that $g_p = f$ on $U_p \cap X$.

Theorem A.15 (Extension Theorem). *If $f : X \to \mathbf{R}^m$ is C^∞, there exist an open neighborhood U of X in $\mathbf{R}^n$ and a C^∞ map $g : U \to \mathbf{R}^m$, such that $g = f$ on X.*

Proof. Let $U = \bigcup_{p \in X} U_p$ and let $(\rho_i)_{i \geq 1}$ be a partition of unity subordinate to the covering $\mathcal{U} = \{U_p\}_{p \in X}$ of U. Then, for each i, there exists a p such that the support of ρ_i is contained in U_p. Let

$$g_i(x) = \begin{cases} \rho_i g_p(x), & x \in U_p, \\ 0, & x \in U_p^c. \end{cases}$$

Then $g = \sum_{i=1}^{\infty} g_i$ is well defined by item (3) of Theorem A.8 and the restriction of g to X is given by $\sum_{i=1}^{\infty} \rho_i f$, which is equal to f. □

Exercises for Appendix A

Exercise A.i. Show that the function (A.1) is C^∞.

Hints:

(1) From the Taylor series expansion

$$e^x = \sum_{k=0}^{\infty} \frac{x^k}{k!}$$

conclude that for $x > 0$

$$e^x \geq \frac{x^{k+n}}{(k+n)!}.$$

(2) Replacing x by $1/x$ conclude that for $x > 0$,

$$e^{1/x} \geq \frac{1}{(n+k)!} \frac{1}{x^{n+k}}.$$

(3) From this inequality conclude that for $x > 0$,

$$e^{-\frac{1}{x}} \leq (n+k)!\, x^{n+k}.$$

(4) Let $f_n(x)$ be the function

$$f_n(x) := \begin{cases} e^{-\frac{1}{x}} x^{-n}, & x > 0, \\ 0, & x \leq 0. \end{cases}$$

Conclude from (3) that for $x > 0$,

$$f_n(x) \leq (n+k)!\, x^k$$

for all k.

(5) Conclude that f_n is C^1 differentiable.

(6) Show that

$$\frac{d}{dx} f_n = f_{n+2} - n f_{n+1}.$$

(7) Deduce by induction that the f_n's are C^r differentiable for all r and n.

Exercise A.ii. Show that the improper integral (A.13) is well-defined independent of the choice of partition of unity.

Hint: Let $(\rho'_j)_{j\geq 1}$ be another partition of unity. Show that (A.13) is equal to

$$\sum_{i=1}^{\infty} \sum_{j=1}^{\infty} \int_U \rho_i(x) \rho'_j(x) f(x) dx.$$

Appendix B

The Implicit Function Theorem

Let U be an open neighborhood of the origin in $\mathbf{R}^n$ and let $f_1, \dots, f_k$ be C^∞ functions on U with the property $f_1(0) = \cdots = f_k(0) = 0$. Our goal in this appendix is to prove the following *implicit function theorem*.

Theorem B.1. *Suppose the matrix*

$$\left(\frac{\partial f_i}{\partial x_j}(0)\right) \tag{B.2}$$

is non-singular. Then there exist a neighborhood U_0 of 0 in $\mathbf{R}^n$ and a diffeomorphism $g\colon (U_0, 0) \hookrightarrow (U, 0)$ of U_0 onto an open neighborhood of 0 in U such that

$$g^\star f_i = x_i, \quad i = 1, \dots, k$$

and

$$g^\star x_j = x_j, \quad j = k+1, \dots, n.$$

Remarks B.3.

(1) Let $g(x) = (g_1(x), \dots, g_n(x))$. Then the second set of equations just say that

$$g_j(x_1, \dots, x_n) = x_j$$

for $j = k+1, \dots, n$, so the first set of equations can be written more concretely in the form

$$f_i(g_1(x_1, \dots, x_n), \dots, g_k(x_1, \dots, x_n), x_{k+1}, \dots, x_n) = x_i \tag{B.4}$$

for $i = 1, \dots, k$.

(2) Letting $y_i = g_i(x_1, \dots, x_n)$, equations (B.4) become

$$f_i(y_1, \dots, y_k, x_{k+1}, \dots, x_n) = x_i. \tag{B.5}$$

Hence what the implicit function theorem is saying is that, *modulo the assumption* (B.2) *the system of equations* (B.5) *can be solved for the y_i's in terms of the x_i's.*

(3) By a linear change of coordinates:

$$x_i \mapsto \sum_{r=1}^{k} a_{i,r} x_r, \quad i = 1, \dots, k,$$

we can arrange without loss of generality for the matrix (B.2) to be the identity matrix, i.e.,

$$\frac{\partial f_i}{\partial x_j}(0) = \delta_{i,j}, \quad 1 \le i, j \le k. \tag{B.6}$$

Our proof of this theorem will be by induction on k.

First, let's suppose that $k = 1$ and, for the moment, also suppose that $n = 1$. Then the theorem above is just the inverse function theorem of freshman calculus. Namely if $f(0) = 0$ and $df/dx(0) = 1$, there exists an interval (a, b) about the origin on which df/dx is greater than $1/2$, so f is strictly increasing on this interval and its graph looks like the curve in the Figure B.1 with $c = f(a) \le -\frac{1}{2}a$ and $d = f(b) \ge \frac{1}{2}b$.

The graph of the inverse function $g\colon [c, d] \to [a, b]$ is obtained from this graph by just rotating it through ninety degrees, i.e., making the y-axis the horizontal axis and the x-axis the vertical axis. (From the picture it's clear that $y = f(x) \iff x = g(y) \iff f(g(y)) = y$.)

Most elementary text books regard this intuitive argument as being an adequate proof of the inverse function theorem; however, a slightly beefed-up version of this proof (which is completely rigorous) can be found in [12, Chapter 12]. Moreover, as Spivak points out in [12, Chapter 12], if the slope of the curve in the Figure B.1 at the point (x, y) is equal to λ, the slope of the rotated curve at (y, x) is $1/\lambda$, so from this proof one concludes that if $y = f(x)$

$$\frac{dg}{dy}(y) = \left(\frac{df}{dx}(x)\right)^{-1} = \left(\frac{df}{dx}(g(y))\right)^{-1}. \tag{B.7}$$

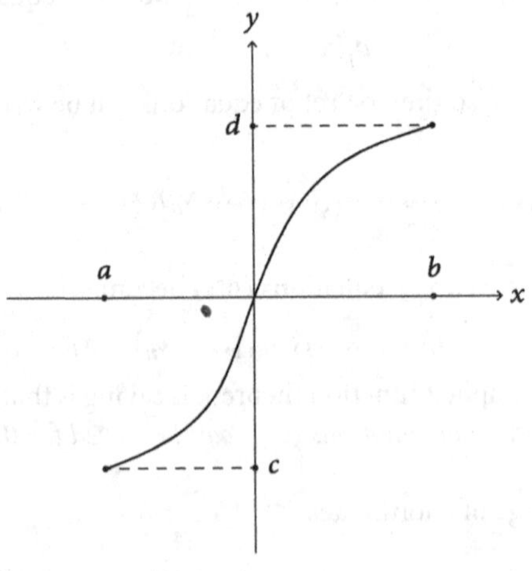

Figure B.1. The implicit function theorem.

Since f is a continuous function, its graph is a continuous curve and, therefore, since the graph of g is the same curve rotated by ninety degrees, g is also a continuous functions. Hence by (B.7), g is also a C^1-function and hence by (B.7), g is a C^2-function, etc. In other words g is in $C^\infty([c,d])$.

Let's now prove that the implicit function theorem with $k = 1$ and n arbitrary. This amounts to showing that if the function f in the discussion above depends on the parameters, $x_2, \dots x_n$ in a C^∞ fashion, then so does its inverse g. More explicitly let's suppose $f(x_1, \dots, x_n)$ is a C^∞ function on a subset of $\mathbf{R}^n$ of the form $[a,b] \times V$ where V is a compact, convex neighborhood of the origin in $\mathbf{R}^n$ and satisfies $\partial f/\partial x_1 \geq \frac{1}{2}$ on this set. Then by the argument above there exists a function, $g = g(y, x_2, \dots, x_n)$ defined on the set $[a/2, b/2] \times V$ with the property

(B.8) $$f(x_1, x_2, \dots, x_n) = y \iff g(y, x_2, \dots, x_n) = x_1.$$

Moreover by (B.7) g is a C^∞ function of y and

(B.9) $$\frac{\partial g}{\partial y}(y, x_2, \dots, x_n) = \frac{\partial f}{\partial x_1}(x_1, x_2, \dots, x_n)^{-1}$$

at $x_1 = g(y)$. In particular, since $\frac{\partial f}{\partial x_1} \geq \frac{1}{2}$

$$0 < \frac{\partial g}{\partial y} < 2$$

and hence

(B.10) $$|g(y', x_2, \dots, x_n) - g(y, x_2, \dots, x_n)| < 2|y' - y|$$

for points y and y' in the interval $[a/2, b/2]$.

The $k = 1$ case of Theorem B.1 is almost implied by (B.8) and (B.9) except that we must still show that g is a C^∞ function, not just of y, but of all the variables, $y, x_2, \dots, x_n$, and this we'll do by quoting another theorem from freshman calculus (this time a theorem from the second semester of freshman calculus).

Theorem B.11 (Mean Value Theorem in n variables). *Let U be a convex open set in $\mathbf{R}^n$ and $f : U \to \mathbf{R}$ a C^∞ function. Then for $a, b \in U$ there exists a point c on the line interval joining a to b such that*

$$f(b) - f(a) = \sum_{i=1}^{n} \frac{\partial f}{\partial x_i}(c)(b_i - a_i).$$

Proof. Apply the one-dimensional mean value theorem to the function

$$h(t) := f((1-t)a + tb). \qquad \square$$

Let's now show that the function g in (B.8) is a C^∞ function of the variables, y and x_2. To simplify our notation we'll suppress the dependence of f on $x_3, \dots, x_n$ and write f as $f(x_1, x_2)$ and g as $g(y, x_2)$. For $h \in (-\varepsilon, \varepsilon)$, ε small, we have

$$y = f(g(y, x_2 + h), x_2 + h) = f(g(y, x_2), x_2),$$

and, hence, setting $x_1' = g(y, x_2 + h)$, $x_1 = g(y, x_2)$ and $x_2' = x_2 + h$, we get from the mean value theorem

$$\begin{aligned} 0 &= f(x_1', x_2') - f(x_1, x_2) \\ &= \frac{\partial f}{\partial x_1}(c)(x_1' - x_1) + \frac{\partial f}{\partial x_2}(c)(x_2' - x_2) \end{aligned}$$

and therefore

$$g(y, x_2 + h) - g(y, x_2) = \left(-\frac{\partial f}{\partial x_1}(c)\right)^{-1} \frac{\partial f}{\partial x_2}(c)\, h \tag{B.12}$$

for some c on the line segment joining (x_1, x_2) to (x_1', x_2'). Letting h tend to zero we can draw from this identity a number of conclusions:

(1) Since f is a C^∞ function on the compact set $[a, b] \times V$ its derivatives are bounded on this set, so the right-hand side of (B.12) tends to zero as h tends to zero, i.e., for fixed y, g is continuous as a function of x_2.

(2) Now divide the right-hand side of (B.12) by h and let h tend to zero. Since g is continuous in x_2, the quotient of (B.12) by h tends to its value at (x_1, x_2). Hence for fixed y f is differentiable as a function of x_2 and

$$\frac{\partial g}{\partial x_2}(y, x_2) = -\left(\frac{\partial f}{\partial x_1}\right)^{-1}(x_1, x_2)\frac{\partial f}{\partial x_2}(x_1, x_2), \tag{B.13}$$

where $x_1 = g(y, x_2)$.

(3) Moreover, by the inequality (B.10) and the triangle inequality

$$\begin{aligned} |g(y', x_2') - g(y, x_2)| &\le |g(y', x_2') - g(y, x_2')| + |g(y, x_2') - g(y, x_2)| \\ &\le 2|y' - y| + |g(y, x_2') - g(y, x_2)|, \end{aligned}$$

hence g is continuous as a function of y *and* x_2.

(4) Hence by (B.9) and (B.13), g is a C^1-function of y and x_2. Repeating this argument inductively shows that g is a C_∞ function of y and x_2.

This argument works, more or less verbatim, for more than two x_i's and proves that g is a C^∞ function of $y, x_2, \ldots, x_n$. Thus with $f = f_1$ and $g = g_1$ Theorem B.1 is proved in the special case $k = 1$.

Proof of Theorem B.1 *for* $k > 1$. We'll now prove Theorem B.1 for arbitrary k by induction on k. By induction we can assume that there exist a neighborhood U_0' of the origin in $\mathbf{R}^n$ and a C^∞ diffeomorphism $\phi\colon (U_0', 0) \hookrightarrow (U, 0)$ such that

$$\phi^* f_i = x_i \tag{B.14}$$

for $2 \le i \le k$ and

$$\phi^* x_j = x_j \tag{B.15}$$

for $j = 1$ and $k + 1 \leq j \leq n$. Moreover, by (B.6)

$$\left(\frac{\partial}{\partial x_1}\phi^* f_1\right)(0) = \sum_{i=1}^{k} \frac{\partial f_i}{\partial x_1}(0)\frac{\partial}{\partial x_i}\phi_i(0) = \frac{\partial \phi_1}{\partial x_1} = 1$$

since $\phi_1 = \phi^* x_1 = x_1$. Therefore we can apply Theorem B.1, with $k = 1$, to the function, $\phi^* f_1$, to conclude that there exist a neighborhood U_0 of the origin in $\mathbf{R}^n$ and a diffeomorphism $\psi\colon (U_0, 0) \to (U_0', 0)$ such that $\psi^*\phi^* f_1 = x_1$ and $\psi^*\phi^* x_i = x_i$ for $1 \leq i \leq n$. Thus by (B.14) $\psi^*\phi^* f_i = \psi^* x_i = x_i$ for $2 \leq i \leq k$, and by (B.15) $\psi^*\phi^* x_j = \psi^* x_j = x_j$ for $k + 1 \leq j \leq n$. Hence if we let $g = \phi \circ \psi$ we see that:

$$g^* f_i = (\phi \circ \psi)^* f_i = \psi^* \phi^* f_i = x_i$$

for $i \leq i \leq k$ and

$$g^* x_j = (\phi \circ \psi)^* x_j = \psi^* \phi^* x_j = x_j$$

for $k + 1 \leq j \leq n$. □

We'll derive a number of subsidiary results from Theorem B.1. The first of these is the n-dimensional version of the inverse function theorem.

Theorem B.16. *Let U and V be open subsets of $\mathbf{R}^n$ and $\phi\colon (U, p) \to (V, q)$ a C^∞ map. Suppose that the derivative of ϕ at p*

$$D\phi(p)\colon \mathbf{R}^n \to \mathbf{R}^n$$

is bijective. Then ϕ maps a neighborhood of p in U diffeomorphically onto a neighborhood of q in V.

Proof. By pre-composing and post-composing ϕ by translations we can assume that $p = q = 0$. Let $\phi = (f_1, \ldots, f_n)$. Then the condition that $D\phi(0)$ be bijective is the condition that the matrix (B.2) be non-singular. Hence, by Theorem B.1, there exist a neighborhood V_0 of 0 in V, a neighborhood U_0 of 0 in U, and a diffeomorphism $g\colon V_0 \to U_0$ such that

$$g^* f_i = x_i$$

for $i = 1, \ldots, n$. However, these equations simply say that g is the inverse of ϕ, and hence that ϕ is a diffeomorphism. □

A second result which we'll extract from Theorem B.1 is the *canonical submersion theorem.*

Theorem B.17. *Let U be an open subset of $\mathbf{R}^n$ and $\phi\colon (U, p) \to (\mathbf{R}^k, 0)$ a C^∞ map. Suppose ϕ is a* ***submersion*** *at p, i.e., suppose its derivative*

$$D\phi(p)\colon \mathbf{R}^n \to \mathbf{R}^k$$

is onto. Then there exist a neighborhood V of p in U, a neighborhood U_0 of the origin in $\mathbf{R}^n$, and a diffeomorphism $g\colon (U_0, 0) \to (V, p)$ such that the map $\phi \circ g\colon (U_0, 0) \to (\mathbf{R}^n, 0)$ is the restriction to U_0 of the ***canonical submersion****:*

$$\pi\colon \mathbf{R}^n \to \mathbf{R}^k, \quad \pi(x_1, \dots, x_n) := (x_1, \dots, x_k).$$

Proof. Let $\phi = (f_1, \dots, f_k)$. Composing ϕ with a translation we can assume $p = 0$ and by a permutation of the variables $x_1, \dots, x_n$ we can assume that the matrix (B.2) is non-singular. By Theorem B.1 we conclude that there exists a diffeomorphism $g\colon (U_0, 0) \hookrightarrow (U, p)$ with the properties $g^* f_i = x_i$ for $i = 1, \dots, k$, and hence,

$$\phi \circ g(x_1, \dots, x_n) = (x_1, \dots, x_k). \qquad \square$$

As a third application of Theorem B.1 we'll prove a theorem which is similar in spirit to Theorem B.17, the *canonical immersion theorem.*

Theorem B.18. *Let U be an open neighborhood of the origin in $\mathbf{R}^k$ and $\phi\colon (U, 0) \to (\mathbf{R}^n, p)$ a C^∞ map. Suppose that the derivative of ϕ at 0*

$$D\phi(0)\colon \mathbf{R}^k \to \mathbf{R}^n$$

is injective. Then there exist a neighborhood U_0 of 0 in U, a neighborhood V of p in $\mathbf{R}^n$, and a diffeomorphism.

$$\psi\colon V \to U_0 \times \mathbf{R}^{n-k}$$

such that the map $\psi \circ \phi\colon U_0 \to U_0 \times \mathbf{R}^{n-k}$ is the restriction to U_0 of the ***canonical immersion***

$$\iota\colon \mathbf{R}^k \to \mathbf{R}^k \times \mathbf{R}^{n-k}, \quad \iota(x_1, \dots, x_k) = (x_1, \dots, x_k, 0, \dots, 0).$$

Proof. Let $\phi = (f_1, \dots, f_n)$. By a permutation of the f_i's we can arrange that the matrix

$$\left(\frac{\partial f_i}{\partial x_j}(0)\right), \quad 1 \le i, j \le k$$

is non-singular and by composing ϕ with a translation we can arrange that $p = 0$. Hence by Theorem B.16 the map

$$\chi\colon (U, 0) \to (\mathbf{R}^k, 0) \quad x \mapsto (f_1(x), \dots, f_k(x))$$

maps a neighborhood U_0 of 0 diffeomorphically onto a neighborhood V_0 of 0 in $\mathbf{R}^k$. Let $\psi\colon (V_0, 0) \to (U_0, 0)$ be its inverse and let

$$\gamma\colon V_0 \times \mathbf{R}^{n-k} \to U_0 \times \mathbf{R}^{n-k}$$

be the map

$$\gamma(x_1, \dots, x_n) := (\psi(x_1, \dots, x_k), x_{k+1}, \dots, x_n).$$

Then

$$\text{(B.19)} \qquad \begin{aligned}(\gamma \circ \phi)(x_1, \dots, x_k) &= \gamma(\chi(x_1, \dots, x_k), f_{k+1}(x), \dots, f_n(x)) \\ &= (x_1, \dots, x_k, f_{k+1}(x), \dots, f_n(x)).\end{aligned}$$

Now note that the map

$$h\colon U_0 \times \mathbf{R}^{n-k} \to U_0 \times \mathbf{R}^{n-k}$$

defined by

$$h(x_1, \ldots, x_n) := (x_1, \ldots, x_k, x_{k+1} - f_{k+1}(x), \ldots, x_n - f_n(x))$$

is a diffeomorphism. (Why? What is its inverse?) and, by (B.19),

$$(h \circ \gamma \circ \phi)(x_1, \ldots, x_k) = (x_1, \ldots, x_k, 0, \ldots, 0),$$

i.e., $(h \circ \gamma) \circ \phi = \iota$. Thus we can take the V in Theorem B.18 to be $V_0 \times \mathbf{R}^{n-k}$ and the ψ to be $h \circ \gamma$. □

Remark B.20. The canonical submersion and immersion theorems can be succinctly summarized as saying that every submersion "looks locally like the canonical submersion" and every immersion "looks locally like the canonical immersion".

Appendix C

Good Covers and Convexity Theorems

Let X be an n-dimensional submanifold of $\mathbf{R}^N$. Our goal in this appendix is to prove that X admits a good cover. To do so we'll need some basic facts about *convexity*.

Definition C.1. Let U be an open subset of $\mathbf{R}^n$ and ϕ a C^∞ function on U. We say that ϕ is ***convex*** if the matrix

$$\left[\frac{\partial^2 \phi}{\partial x_i \partial x_j}(p)\right] \tag{C.2}$$

is positive-definite for all $p \in P$.

Suppose now that U itself is convex and ϕ is a convex function on U which has as image the half open interval $[0, a)$ and is a proper map $U \to [0, a)$. (In other words, for $0 \leq c < a$, the set $\{ x \in U \mid \phi(x) \leq c \}$ is compact.)

Theorem C.3. *For* $0 < c < a$, *the set*

$$U_c := \{x \in U \mid \phi(x) < c\}$$

is convex.

Proof. For $p, q \in \partial U_c$ with $p \neq q$, let

$$f(t) := \phi((1-t)p + tq),$$

for $0 \leq t \leq 1$. We claim that

$$\frac{d^2 f}{dt^2}(t) > 0 \tag{C.4}$$

and

$$\frac{df}{dt}(0) < 0 < \frac{df}{dt}(1). \tag{C.5}$$

To prove (C.4) we note that

$$\frac{d^2 f}{dt^2}(t) = \sum_{1\le i,j\le n} \frac{\partial^2 \phi}{\partial x_i \partial x_j}((1-t)p + tq)(p_i - q_i)(p_j - q_j)$$

and that the right-hand side is positive because of the positive-definiteness of (C.2).

To prove (C.5) we note that $\frac{df}{dt}$ is strictly increasing by (C.4). Hence $\frac{df}{dt}(0) > 0$ would imply that $\frac{df}{dt}(t) > 0$ for all $t \in [0,1]$ and hence would imply that f is strictly increasing. But

$$f(0) = f(1) = \phi(p) = \phi(q) = c,$$

so this would be a contradiction. We conclude that $\frac{df}{dt}(0) < 0$. A similar argument shows that $\frac{df}{dt}(1) > 0$.

Now we use equations (C.4) and (C.5) to show that U_c is convex. Since $\frac{df}{dt}$ is strictly increasing it follows from (C.5) that $\frac{df}{dt}(t_0) = 0$ for some $t_0 \in (0,1)$ and that $\frac{df}{dt}(t) < 0$ for $t \in [0, t_0)$ and $\frac{df}{dt}(t) > 0$ for $t \in (t_0, 1]$. Therefore, since $f(0) = f(1) = c$, we have $f(t) < c$ for all $t \in (0,1)$; so $\phi(x) < c$ on the line segment defined by $(1-t)p + tq$ for $t \in (0,1)$. Hence U_c is convex. □

Coming back to the manifold $X \subset \mathbf{R}^N$, let $p \in X$ and let $L_p := T_pX$. We denote by π_p the orthogonal projection

$$\pi_p \colon X \to L_p + p,$$

i.e., for $q \in X$ and $x = \pi_p(q)$, we have that $q - x$ is orthogonal to L_p (it's easy to see that π_p is defined by this condition). We will prove that this projection has the following convexity property.

Theorem C.6. *There exists a positive constant $c(p)$ such that if $q \in X$ satisfies $|p - q|^2 < c(p)$ and $c < c(p)$, then the set*

(C.7) $$U(q,c) := \{q' \in X \mid |q' - q| < c\}$$

is mapped diffeomorphically by π_p onto a convex open set in $L_p + p$.

Proof. We can, without loss of generality, assume that $p = 0$ and that

$$L = \mathbf{R}^n = \mathbf{R}^n \times \{0\}$$

in $\mathbf{R}^n \times \mathbf{R}^k = \mathbf{R}^N$, where $k := N - n$. Thus X can be described, locally over a convex neighborhood U of the origin in $\mathbf{R}^n$ as the graph of a C^∞ function $f \colon (U,0) \to (\mathbf{R}^k, 0)$ and since $\mathbf{R}^n \times \{0\}$ is tangent to X at 0, we have

(C.8) $$\frac{\partial f}{\partial x_i} = 0 \quad \text{for } i = 1, \dots, n.$$

Given $q = (x_q, f(x_q)) \in X$, let $\phi_q \colon U \to \mathbf{R}$ be the function

(C.9) $$\phi_q(x) := |x - x_q|^2 + |f(x) - f(x_q)|^2.$$

Then by (C.7) $\pi_p(U(q,c))$ is just the set

$$\{x \in U \mid \phi_q(x) < c\},$$

so to prove the theorem it suffices to prove that if $c(p)$ is sufficiently small and $c < c(p)$, this set is convex. However, at $q = 0$, by equations (C.8) and (C.9) we have

$$\frac{\partial^2 \phi_q}{\partial x_i \partial x_j} = \delta_{i,j},$$

and hence by continuity ϕ_q is convex on the set $\{ x \in U \mid |x|^2 < \delta\}$ provided that δ and $|q|$ are sufficiently small. Hence if $c(p)$ is sufficiently small, then $\pi_p(U(q,c))$ is convex for $c < c(p)$. □

We will now use Theorem C.6 to prove that X admits a good cover: for every $p \in X$, let

$$\varepsilon(p) := \frac{\sqrt{c(p)}}{3}$$

and let

$$U_p := \{q \in X \mid |p - q| < \varepsilon(p)\}.$$

Theorem C.10. *The U_p's are a good cover of X.*

Proof. Suppose that the intersection

$$U_{p_1} \cap \cdots \cap U_{p_k} \tag{C.11}$$

is non-empty and that

$$\varepsilon(p_1) \geq \varepsilon(p_2) \geq \cdots \geq \varepsilon(p_k).$$

Then if p is a point in the intersection $U_{p_1} \cap \cdots \cap U_{p_k}$, then for $i = 1, \ldots, k$ we have $|p - p_i| < \varepsilon(p_1)$, hence $|p_i - p_1| < 2\varepsilon(p_1)$. Moreover, if $q \in U_{p_i}$ and $|q - p_i| < \varepsilon(p_1)$, then $|q - p_1| < 3\varepsilon(p_1)$ and so $|q - p_1|^2 < c(p_1)$. Therefore the set U_{p_i} is contained in $U(p_i, c(p_1))$ and consequently by Theorem C.6 is mapped diffeomorphically onto a convex open subset of $L_{p_1} + p_1$ by the projection π_{p_1}. Consequently, the intersection (C.11) is mapped diffeomorphically onto a convex open subset of $L_{p_1} + p_1$ by π_{p_1}. □

Bibliography

1. V. I. Arnol'd, *Mathematical Methods of Classical Mechanics*, Graduate Texts in Mathematics, Vol. 60, Springer Science & Business Media, 2013.
2. G. Birkhoff and G.-C. Rota, *Ordinary Differential Equations*, 4th edn., John Wiley & Sons, 1989.
3. R. Bott and L. W. Tu, *Differential Forms in Algebraic Topology*, Graduate Texts is Mathematics, Vol. 82, Springer Science & Business Media, 2013.
4. V. Guillemin and A. Pollack, *Differential Topology*, AMS Chelsea Publishing Series, Vol. 370, American Mathematical Society, 2010.
5. V. Guillemin and S. Sternberg, *Symplectic Techniques in Physics*, Cambridge University Press, 1990.
6. N. V. Ivanov, A differential forms perspective on the Lax proof of the change of variables formula, *The American Mathematical Monthly* **112** (2005), no. 9, 799–806.
7. J. L. Kelley, *General Topology*, Graduate Texts in Mathematics, Vol. 27, Springer, 1975.
8. P. D. Lax, Change of variables in multiple integrals, *The American Mathematical Monthly* **106** (1999), no. 6, 497–501.
9. P. D. Lax, Change of variables in multiple integrals II, *The American Mathematical Monthly* **108** (2001), no. 2, 115–119.
10. J. R. Munkres, *Analysis on Manifolds*, Westview Press, 1997.
11. J. R. Munkres, *Topology*, 2nd edn., Prentice-Hall, 2000 (English).
12. M. Spivak, *Calculus on Manifolds: A Modern Approach to Classical Theorems of Advanced Calculus*, Mathematics Monographs, Westview Press, 1971.
13. F. W. Warner, *Foundations of Differentiable Manifolds and Lie Groups*, Graduate Texts in Mathematics, Vol. 94, Springer-Verlag New York, 1983.

Index of Notation

Glossary of Terminology

Printed in the United States
By Bookmasters